Feature Papers in *BioChem*

Feature Papers in *BioChem*

Guest Editors

Buyong Ma
Manuel Aureliano

MDPI

Basel • Beijing • Wuhan • Barcelona • Belgrade • Novi Sad • Cluj • Manchester

Guest Editors

Buyong Ma
School of Pharmacy
Shanghai Jiao Tong
University
Shanghai
China

Manuel Aureliano
Faculdade de Ciências e
Tecnologia
Universidade do Algarve
Faro
Portugal

Editorial Office
MDPI AG
Grosspeteranlage 5
4052 Basel, Switzerland

This is a reprint of the Special Issue, published open access by the journal *BioChem* (ISSN 2673-6411), freely accessible at: www.mdpi.com/journal/biochem/special_issues/764DYCV6IZ.

For citation purposes, cite each article independently as indicated on the article page online and using the guide below:

Lastname, A.A.; Lastname, B.B. Article Title. *Journal Name* **Year**, *Volume Number*, Page Range.

ISBN 978-3-7258-4464-7 (Hbk)
ISBN 978-3-7258-4463-0 (PDF)
https://doi.org/10.3390/books978-3-7258-4463-0

Contents

About the Editors

Buyong Ma

Prof. Dr. Buyong Ma is a Full Professor at Shanghai Jiaotong University, China. His research interests mainly focus on (1) molecular mechanisms of action and drug design; (2) computational biology of cancer and immunology; (3) protein dynamics and function; and (4) antibody/antigen design and protein engineering and design. He has more than 220 scientific publications which have received over 17000 citations (h-index 61 according to Scopus). He is a pioneer in the computational study of protein conformational dynamics and ligand interactions, protein interaction hot spots, and protein aggregation. He was also among the first to elucidate the structure and mechanism of protein amyloid formation.

Manuel Aureliano

Manuel Aureliano is a Full Professor of Biochemistry (Aggregation in Inorganic Biochemistry) with the Faculty of Sciences and Technology of the University of Algarve, Faro, Portugal, teaching biochemistry, inorganic biochemistry, muscle contraction, and advanced chemistry and biochemistry. He completed his undergraduate, master's, and PhD studies at the University of Coimbra, Coimbra, Portugal. He was the Director of the Biochemistry degree at the University of Algarve for about two decades, from 1998 to 2013 and from 2021 to 2025. At the Algarve Centre for Marine Sciences (CCMAR-Algarve), he investigates the applications of polyoxometalates, particularly decavanadate, in the environment and biomedicine. He was selected as an "Outstanding Reviewer" (top 10 reviewers) for the journal *Metallomics* for three consecutive years (2017, 2018, and 2019). To date, he has participated in the revision of more than 900 papers from about 140 journals. Moreover, he has served in various editorial roles, including Editor, Associate Editor, and/or Guest Editor, with many journals such as *BioChem*, *IJMS*, *Biology*, *Toxics*, *Molecules*, *Metals*, *Frontiers in Chemistry* (Medicinal and Pharmaceutical Chemistry), and *Frontiers in Molecular Biosciences* (Bioenergetics). In 2021, 2022, 2023, and 2024, he was included in the "Worlds Top 2% Scientists list" (impact-career and year). Recently (December 2024), he was awarded the third edition of the "UAlg Researcher Award" by the University of Algarve.

BioChem

MDPI

Editorial

Feature Papers in *BioChem*

Manuel Aureliano [1,2,*] and Buyong Ma [3,*]

1 Faculdade de Ciências e Tecnologia, Universidade do Algarve, Campus de Gambelas, 8005-139 Faro, Portugal
2 Centro de Ciências do Mar (CCMar), Universidade do Algarve, 8005-139 Faro, Portugal
3 School of Pharmacy, Shanghai Jiao Tong University, Shanghai 200240, China
* Correspondence: maalves@ualg.pt (M.A.); mabuyong@sjtu.edu.cn (B.M.)

1. Introduction and Scope

Biochemistry, or the chemistry of life, is an interdisciplinary science that uses strategies and methods from all exact and natural sciences. In that sense, Biochemistry is a discipline known to be challenging for students, typically due to the extent and complexity of the content [1]. However, in the last 10 years, at least fifteen Nobel Prizes in Chemistry, Physiology and Medicine have been awarded in the field and/or for applications of Biochemistry, which reflects the importance of this area of knowledge in contemporary societies [2–7]. Biochemistry connects essential metal ions such as Mg^{2+}, Ca^{2+}, Na^{+} and Fe^{2+}, and organic compounds, such as nucleic acids, sugars, lipids and proteins, to make biological systems possible, representing a unique platform for interdisciplinary collaborations in teaching and in research [8,9]. Moreover, in addition to biomolecules (proteins, nucleic acids, sugars and lipids), small molecules are necessary for cellular homeostasis, for example, carbonate ions and phosphate ions, responsible for stabilizing physiological pH (close to 7.0) or antioxidants that prevent the effects of cellular stress, as well as due to environmental pollutants, such as metals and/or microplastics [10–12].

Biochemistry, as it is an interdisciplinary science, is very dynamic, and its boundaries are very hard to define not only because they are constantly changing but also because they are dependent on the specificities of each time period and scientific environments. Nevertheless understanding life on earth and human health are everlasting topics in biochemistry. Thus, this Special Issue titled "Feature Papers in *BioChem*" is expected to showcase interdisciplinary research in the diverse and interconnected fields of molecular biology, cell biology, structural biology, nucleic acid biology, chemical biology, synthetic biology, disease biology, biophysics, metallomics (or inorganic biochemistry) and theoretical biochemistry.

This Special Issue included six articles and four reviews. Among the articles, a wide range of topics were addressed, including our understanding of otolith proteomics in Atlantic cod [13], juvenile idiopathic arthritis' impact on long-term fertility due to prolonged exposure to immunosuppressive therapies [14], the role of tricyclic isoquinoline derivatives as antibacterial agents [15], the production of aromatic compounds in *Pseudomonas putida* [16], hydrogels made with Tilapia fish skin to increase collagen production [17], and the use of N-Myristoyltransferase inhibitors (NMTi) as a novel antiviral strategy against mammarenaviruses [18]. Additionally, the four reviews papers described the role of oxidative stress in the pathogenesis of the most common gastrointestinal diseases [19], the biotechnological production of vanillin [20], non-steroidal anti-inflammatory drugs' (NSAIDs) effects in the elderly population [21] and the biochemical mechanisms and applications of multiple cysteine-based protein ligases [22]. In fact, these ten contributions represent very good examples of the wide range of biochemistry boundaries identified in the 21 century. So far (4 June 2025), these 10 contributions have gathered 5 citations and 16,576 views, indicating an average of 1658 views per publication. Note that all these contributions have been published after June 2024, with six published in 2025 alone.

2. Contributions

The first paper published in the present Special Issue (SI) is entitled "Proteomic Blueprint of Atlantic Cod (Gadus morhua) Otoliths Revealing Environmental Stress Insights through Label-Free Quantitative Shotgun Proteomics" [13]. The study states that most identified proteins deposited daily and influenced by the environment are not implicated in the biomineralization of otolith [13]. It highlights the potential for the otolith proteome to recreate details of fish life history at previously unrealized levels. This contribution from Professor Banoub and collaborators is a very good example of how the boundaries of biochemistry can be promoted and interact simultaneously through the work of very different departments/faculties and research institutes in Canada, namely, for biochemistry, the Fisheries Centre and, for medicine, the Institute of Research in Cancer and Chemistry. Professor Banoub's research group and its collaborators also carried out research about the environmental impact of bioplastic use [23]. The second paper published in the present Special Issue is entitled "Anti-Müllerian Hormone Serum Levels as Biomarker of Ovarian Reserve in Adult Women with Juvenile Idiopathic Arthritis Treated with csDMARDs and/or bDMARDs: A Pilot Study" [14], conducted by the research group of Professor Elisa Gremese from the Fondazione Policlinico Universitario A. Gemelli IRCCS, Rome, Italy. Professor Gremese was also involved in "The Gemelli Against COVID-19 Post-Acute Care Study Group" project, which described the importance of using an interdisciplinary approach coordinated by a geriatrician to manage the potential post-acute care needs of recovered COVID-19 patients [24]. Herein, it was suggested that ovarian reserve, as assessed by anti-Müllerian hormone serum levels, appears to be comparable between those with juvenile idiopathic arthritis and age-matched controls and does not appear to be influenced by disease characteristics or prior/concomitant exposure to immunosuppressive drugs.

The third review paper published in the present Special Issue is entitled "Synthesis and Investigation of Tricyclic Isoquinoline Derivatives as Antibacterial Agents" [15], conducted by the research group of Professor Andrew McDonagh from the University of Technology Sydney, Australia. In this paper, a series of six isoquinoline-based compounds were synthesized for developing broad-spectrum antibacterial compounds. The authors pointed out that although the compounds presented antibacterial properties, their cytotoxic properties against mammalian cell lines revealed some cytotoxic effects, suggesting limitations in their antibacterial applications without further development. Professor Andrew McDonagh's recent research interests have included studies of coumarin compounds for use as *Chlamydial* protease inhibitors and anti-chlamydial agents [25]. From Professor Caroline Ranquet of BGene Genetics, Grenoble, France, we received the publication entitled "Overproduction of Phenolic Compounds in *Pseudomonas putida* KT2440 Through Endogen Deregulation of the Shikimate Pathway" [16]. The introduction of specific engineered enzymes into a metabolically engineered *Pseudomonas putida* strain resulted in significantly increased production of p-coumaric acid. Recent collaborations involving Caroline Ranquet included research about the Fis family members that synergistically control the virulence of *Legionella pneumophila* [26]. Herein, through computational modeling and experimental validation, the authors identified specific amino-acid residues responsible for tyrosine-mediated feedback inhibition. By using targeted mutagenesis, the introduction of engineered 3-deoxy-D-arabino-heptulosonate 7-phosphate synthase (DAHP synthase) variants reduced the sensitivity to feedback inhibition [16].

Burns are the most common type of injury in everyday life, and wound healing presents several challenges and concerns in medicine. The studies from the research group of Professor Ozge Cevik, School of Medicine, Aydin Adnan Menderes University, Turkey, about hydrogels made with Tilapia fish (TL, *Oreochromis niloticus*) skin show an increase in collagen production in burn treatment [17]. TL or TL-alginate hydrogels (AGTL) were applied to a burn wound created in Sprague–Dawley rats for 7 and 14 days, and the levels of hydroxyproline, a critical element in tissue reorganization, along with the gene expression levels of COL1A1, COL3A1, MMP-2 and MMP-9 and the protein expression levels of the matrix metalloproteinases (MMPs) MMP-2 and MMP-9, were evaluated. It was suggested that biological substances in the TL structure, in conjunction with alginate, were effective in the healing and reorganization of the wound tissue [20]. From the same group, studies on the anticarcinogenic properties of green-synthesized zinc oxide nanoparticles (ZnONPs) synthesized employing fresh *Citrus aurantium* aqueous extract on human breast cancer cells (MDA-MB-231) were described [27].

The sixth paper contributed by Professor Juan de la Torre, The Scripps Research Institute, La Jolla, USA, proposed a model of the effect of the NMT inhibitor on mammarenavirus cell entry and budding. NMT isozymes facilitate the addition of myristic acid to the glycines of stable signal peptides (SSPs) and Z proteins, protecting them from proteasome-mediated degradation. The viral glycoprotein precursor (GPC) is co- and post-translationally processed to produce a stable signal peptide (SSP), and the mature GP1 and GP2 subunits, together with the SSP, form the spikes that decorate the virus surface and mediate cell entry via receptor-mediated endocytosis. Myristoylated SSP interacts with GP2 to facilitate the fusion event in the late endosome required to complete the virus cell entry process, whereas myristoylated Z directs the virus assembly and budding process. The inhibition of SSP and Z myristoylation by NMT inhibitors results in the proteasome-mediated degradation of SSP and Z, which results in the inhibition of virus multiplication [18]. From the same authors, studies about repurposing drugs for synergistic combination therapies to counteract monkeypox virus tecovirimat resistance have been submitted [28].

The eighth contribution was a review article titled "Oxidative Stress-Induced Gastrointestinal Diseases: Biology and Nanomedicines—A Review" [19] from Dra Maryam Rezvani, the University of Cagliari, Italy. In this review, it was revealed that although oxidative stress in each part of the digestive system manifests itself in a specific way, all these diseases arise from the imbalance between the generation of the reactive intermediates (especially reactive oxygen species) and the antioxidant defense system. It was further pointed out that the annual incidence and mortality statistics of gastrointestinal diseases worldwide emphasize an urgent need to find an effective and non-invasive treatment method to overcome these life-threatening problems. Another review paper entitled "Biotechnological Advances in Vanillin Production: From Natural Vanilla to Metabolic Engineering Platforms" [20] was conducted by the research group of Professor Isabel Desgagne-Penix, Université du Québec à Trois-Rivières, Canada. The review highlights the significance of vanillin in various markets, its diverse applications and the current state of bio-engineered production using both prokaryotic and eukaryotic biological systems. The authors aim was to provide a current and innovative overview of vanillin bioengineering across various host systems, with special consideration given to microalgae. Altogether, the use of these systems to support the biotechnological production of vanillin, while leveraging the photosynthetic capabilities of microalgae to capture CO_2 and convert it into biomass, can significantly reduce the overall carbon footprint [20]. A recent study from Professor Isabel Desgagne-Penix's research group discusses the potential of norbelladine derivatives for use in Alzheimer's disease research and for future pharmaceutical developments in the field [29].

Professor Hugo Ribeiro, the Faculty of Medicine, the University of Porto, Portugal, as well as several Health Units, submitted a systematic review entitled "Appropriate Prescription of Non-Steroidal Anti-Inflammatory Drugs in Geriatric Patients—A Systematic

Review" [21]. This study aimed to evaluate the cardiovascular, gastrointestinal and renal safety profiles of ibuprofen, naproxen, acemetacin, diclofenac, celecoxib and etoricoxib in elderly patients. From the 2086 articles identified, twenty studies analyzed cardiovascular (CV) safety, fourteen analyzed gastrointestinal (GI) safety, and four analyzed renal safety. When CV risk is the main concern, celecoxib or naproxen are a good first choice. In high-GI-risk groups, the combination of proton pump inhibitor (PPI) with naproxen or celecoxib use is recommended. When renal function is the focus, celecoxib remains the first line of therapy. Diclofenac use in elderly populations should be avoided. Celecoxib is a good choice for elderly patients for whom it is difficult to direct pain treatment based on a single known risk factor [21].

Last but not the least, from the Purdue University, USA, we received a review paper describing protein ligases as nature's gift to the field of protein synthesis and engineering. The generation of active proteins via recombinant expression or chemical total synthesis has limitations in terms of yield and functionality. However, nature has provided a solution to this problem through evolving protein ligases that catalyze the formation of amide bonds between peptides/proteins, which can be exploited by protein engineers to develop robust functional proteins [22]. Several biochemical mechanisms and applications of multiple cysteine-based protein ligases have been identified, such as sortase A, an transpeptidase enzyme; butelase-1, an asparaginyl endopeptidase; OaAEP1, similar to butelase-1; transglutaminase 2, a member of the transglutaminase protein family; E3 ligases; and inteins, believed to act as self-catalyzing enzymes able to undergo *cis*- or *trans*-splicing. When comparing the mechanism of intein splicing with that of native chemical ligation (NCL), the authors found many similarities, which suggest that nature inspired the design of this chemical ligation [22]. From the same research group, led by Professor Qingfei Zheng, a study about transglutaminase 2-mediated histone monoaminylation and its role in cancer was received [30].

3. Biochemistry Insights Across Diverse Scientific Fields

In this Special Issue, a total of 50 authors from 8 countries (France, Australia, Canada, Italy, Iran, Turkey, Portugal, and the USA) were involved, with the large majority being young researchers, thus pointing out the presence of a new generation of scientists in several biochemistry fields. Altogether, this Special Issue reflects, in the 21 century, the existence of distinct and emergent biochemical contributions crossing several biochemical boundaries and applications.

Overall, five interdisciplinary aspects of biochemistry insights can be highlighted in this Special Issue (Figure 1):

1. The Cod Otolith Study [13] and Anti-Müllerian Hormone (AMH) Research [14] both use proteins as biomarkers. The otolith proteome reflects environmental stress and life history, while AMH serves as a biomarker for ovarian reserve. These studies highlight biochemistry's role in translating molecular signatures (proteins/hormones) into diagnostic or ecological insights. The benefits of further making use of natural biomaterials are best illustrated by the fact that hydrogels made with Tilapia fish can be used to treat burn wounds [17].
2. *Pseudomonas putida* phenolic overproduction [16] and vanillin bioengineering [20] focus on manipulating metabolic pathways. Both use enzyme engineering to optimize yields of high-value compounds (p-coumaric acid, vanillin). These studies exemplify biochemistry's role in sustainable bioproduction.
3. The Isoquinoline Antibacterial [15] and Non-Steroidal Anti-Inflammatory Drugs Safety [21] Reviews address the biochemical challenge of balancing efficacy and safety.

In the study of burn wound treatment [17], natural biomaterials could safe sources of hydrogels.

4. Oxidative Stress in Gastrointestinal Diseases [19] and Phenolic Compound Production [16] intersect through reactive oxygen species (ROS) and antioxidants. Phenolics (like p-coumaric acid) are natural antioxidants, suggesting potential applications in mitigating oxidative stress-related gastrointestinal diseases. This finding aligns with the review's call for non-invasive treatments, possibly via nanomedicine or engineered antioxidants.
5. DAHP Synthase Engineering [16] conducts enzyme modulation by reducing feedback inhibition to boost phenolic synthesis. These studies highlight biochemistry's role in enzyme kinetics and inhibitor design for industrial/therapeutic applications (Figure 1). Related to the inhibitor design, the inhibitor of NMT isozymes, which is required for the myristoylation of SSP, can stop virus multiplication [18]. Biochemistry reveals nature's mechanism and helps us to learn from nature to improve protein engineering, such as by using cysteine-based protein ligases and intein splicing [22].

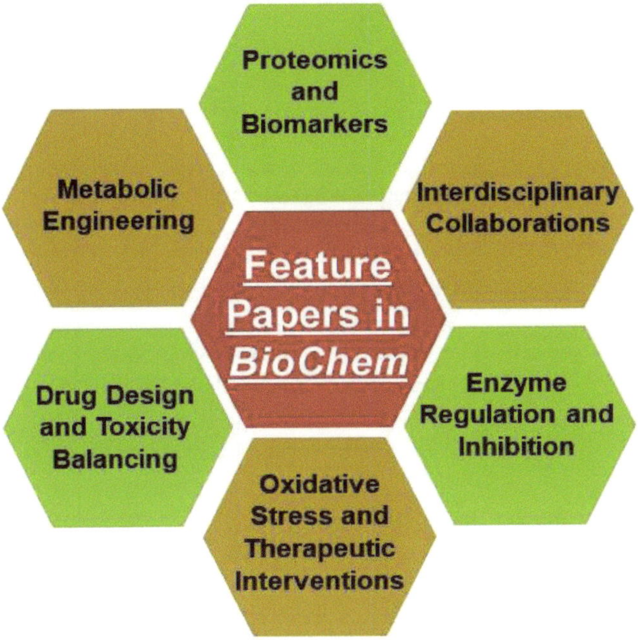

Figure 1. Feature papers in *BioChem*: Topics related to biochemistry boundaries.

This collection of Special Issue papers collectively illustrate biochemistry's central role in addressing global challenges—environmental monitoring, sustainable production, drug development and health diagnostics (Figure 1). The papers emphasize the interplay between molecular mechanisms, translational applications and interdisciplinary collaboration, showcasing biochemistry as a dynamic field at the boundary between life sciences, medicine and sustainability. Allegorically, biochemistry represents a key cog in the "clock of the knowledge", dynamically joining several scientific fields, such as chemistry, biology and medicine, among others, for perfect shared innovation and development (Figure 2).

Figure 2. Biochemistry represents a key cog in the "clock of the knowledge", dynamically joining several scientific wheels such as chemistry, biology and medicine, among others, for achieving perfect shared innovation and development.

Author Contributions: M.A. and B.M. contributed equally to all steps of this Editorial. All authors have read and agreed to the published version of the manuscript.

Funding: This study received Portuguese national funds from the Foundation for Science and Technology (FCT) through projects UIDB/04326/2020, UIDP/04326/2020 and LA/P/0101/2020 (M.A.).

Acknowledgments: The authors would like to acknowledge to all the contributing authors and reviewers. Thanks are also extended to João Mateus for providing the professional illustration of Figure 2.

Conflicts of Interest: The authors declare no conflicts of interest.

References

1. Costabile, M.; Simpson, B.; Turkanovic, J.; Hughes, B.P. Enhancing teaching effectiveness in biochemistry labs: Author reflections and improvement strategies. *Biochem. Mol. Biol. Educ.* **2024**, *52*, 559–568. [CrossRef]
2. Caterina, M.J.; Schumacher, M.A.; Tominaga, M.; Rosen, T.A.; Levine, J.D.; Julius, D. The capsaicin receptor: A heat-activated ion channel in the pain pathway. *Nature* **1997**, *389*, 816–824. [CrossRef]
3. Karikó, K.; Muramatsu, H.; Ludwig, J.; Weissman, D. Generating the optimal mRNA for therapy: HPLC purification eliminates immune activation and improves translation of nucleoside-modified, protein-encoding mRNA. *Nucleic Acids Res.* **2011**, *39*, e142. [CrossRef]
4. Lee, R.C.; Feinbaum, R.L.; Ambros, V. The *C. elegans* heterochronic gene lin-4 encodes small RNAs with antisense complementarity to lin-14. *Cell* **1993**, *75*, 843–854. [CrossRef]
5. Arnold, F.H. Directed evolution: Bringing new chemistry to life. *Angew. Chem. Int. Ed.* **2018**, *57*, 4143–4148. [CrossRef]
6. Lindahl, T. Instability and decay of the primary structure of DNA. *Nature* **1993**, *362*, 709–715. [CrossRef]
7. Jiang, L.; Althoff, E.A.; Clemente, F.R.; Doyle, L.; Röthlisberger, D.; Zanghellini, A.; Gallaher, J.L.; Betker, J.L.; Tanaka, F.; Barbas, C.F., III; et al. De novo computational design of retro-aldol enzymes. *Science* **2008**, *319*, 1387–1391. [CrossRef]
8. Maret, W. The quintessence of metallomics: A harbinger of a different life science based on the periodic table of the bioelements. *Metallomics* **2022**, *14*, mfac051. [CrossRef]

9. Poejo, J.; Gumerova, N.I.; Rompel, A.; Mata, A.M.; Aureliano, M.; Gutierrez-Merino, C. Unveiling the Agonistic Properties of Preyssler-Type Polyoxotungstates on Purinergic P2 Receptors. *J. Inorg. Biochem.* **2024**, *259*, 112640. [CrossRef]
10. Kadac-Czapska, K.; Ośko, J.; Knez, E.; Grembecka, M. Microplastics and Oxidative Stress-Current Problems and Prospects. *Antioxidants* **2024**, *13*, 579. [CrossRef]
11. Poljšak, B.; Fink, R. The protective role of antioxidants in the defence against ROS/RNS-mediated environmental pollution. *Oxid. Med. Cell Longev.* **2014**, *2014*, 671539. [CrossRef]
12. Koyama, H.; Kamogashira, T.; Yamasoba, T. Heavy Metal Exposure: Molecular Pathways, Clinical Implications, and Protective Strategies. *Antioxidants* **2024**, *13*, 76. [CrossRef]
13. Youssef, T.N.; Christian, S.L.; Rideout, R.; Adamack, A.; Thibault, P.; Bonneil, E.; Fridgen, T.D.; Banoub, J. Proteomic Blueprint of Atlantic Cod (*Gadus morhua*) Otoliths Revealing Environmental Stress Insights through Label-Free Quantitative Shotgun Proteomics. *BioChem* **2024**, *4*, 144–165. [CrossRef]
14. Di Mario, C.; Gigante, M.R.; Barini, A.; Petricca, L.; Barini, A.; Bianchi, A.; Alivernini, S.; Tolusso, B.; Gremese, E. Anti-Müllerian Hormone Serum Levels as Biomarker of Ovarian Reserve in Adult Women with Juvenile Idiopathic Arthritis Treated with csDMARDs and/or bDMARDs: A Pilot Study. *BioChem* **2024**, *4*, 313–322. [CrossRef]
15. Phillips, M.J.A.; Ung, A.T.; Harry, E.J.; Ashmore, J.; McDonagh, A.M. Synthesis and Investigation of Tricyclic Isoquinoline Derivatives as Antibacterial Agents. *BioChem* **2025**, *5*, 1. [CrossRef]
16. Merre, W.; Andrade, R.; Perot, C.; Chandor-Proust, A.; Ranquet, C. Overproduction of Phenolic Compounds in *Pseudomonas putida* KT2440 Through Endogen Deregulation of the Shikimate Pathway. *BioChem* **2025**, *5*, 4. [CrossRef]
17. Baydogan, B.; Kucuk, A.; Kozan, B.; Erdal, M.; Abas, B.I.; Cevik, O. Hydrogels Made with Tilapia Fish Skin Increase Collagen Production and Have an Effect on MMP-2/MMP-9 Enzymes in Burn Treatment. *BioChem* **2025**, *5*, 8. [CrossRef]
18. Witwit, H.; de la Torre, J.C. Mammarenavirus Z Protein Myristoylation and Oligomerization Are Not Required for Its Dose-Dependent Inhibitory Effect on vRNP Activity. *BioChem* **2025**, *5*, 10. [CrossRef]
19. Rezvani, M. Oxidative Stress-Induced Gastrointestinal Diseases: Biology and Nanomedicines—A Review. *BioChem* **2024**, *4*, 189–216. [CrossRef]
20. Tazon, A.W.; Awwad, F.; Meddeb-Mouelhi, F.; Desgagné-Penix, I. Biotechnological Advances in Vanillin Production: From Natural Vanilla to Metabolic Engineering Platforms. *BioChem* **2024**, *4*, 323–349. [CrossRef]
21. Costa, C.; Soares, D.; Borges, A.; Gonçalves, A.; Andrade, J.P.; Ribeiro, H. Appropriate Prescription of Non-Steroidal Anti-Inflammatory Drugs in Geriatric Patients—A Systematic Review. *BioChem* **2024**, *4*, 300–312. [CrossRef]
22. Ritsema, Y.; Li, H.; Zheng, Q. Protein Ligases: Nature's Gift for Protein/Peptide Synthesis. *BioChem* **2025**, *5*, 11. [CrossRef]
23. Atiwesh, G.; Mikhael, A.; Parrish, C.C.; Banoub, J.; Le, T.A.T. Environmental impact of bioplastic use: A review. *Heliyon* **2021**, *7*, e07918. [CrossRef]
24. Gemelli Against COVID-19 Post-Acute Care Study Group. Post-COVID-19 global health strategies: The need for an interdisciplinary approach. *Aging Clin. Exp. Res.* **2020**, *32*, 1613–1620. [CrossRef]
25. Phillips, M.J.A.; Huston, W.M.; McDonagh, A.M.; Rawling, T. 4-Chloroisocoumarins as Chlamydial Protease Inhibitors and Anti-Chlamydial Agents. *Molecules* **2024**, *29*, 1519. [CrossRef]
26. Andréa, C.; Bresson, J.; Ginévra, C.; Vianney, A.; Bailo, N.; Chapalain, A.; Attaiech, L.; Picq, K.; Ranquet, C.; Nasser, W.; et al. Fis family members synergistically control the virulence of *Legionella pneumophila. bioRxiv* **2024**, hal-04728706. [CrossRef]
27. Erdoğan, Ö.; Paşa, S.; Demirbolat, G.M.; Birtekocak, F.; Abbak, M.; Çevik, Ö. Synthesis, characterization, and anticarcinogenic potent of green-synthesized zinc oxide nanoparticles via *Citrus aurantium* aqueous peel extract. *Inorg. Nano-Met. Chem.* **2023**, *55*, 67–75. [CrossRef]
28. Witwit, H.; Cubitt, B.; Khafaji, R.; Castro, E.M.; Goicoechea, M.; Lorenzo, M.M.; Blasco, R.; Martinez-Sobrido, L.; de la Torre, J.C. Repurposing Drugs for Synergistic Combination Therapies to Counteract Monkeypox Virus Tecovirimat Resistance. *Viruses* **2025**, *17*, 92. [CrossRef]
29. Hashemian, S.M.; Merindol, N.; Paquin, A.; Singh, A.; Berthoux, L.; Daoust, B.; Desgagné-Penix, I. Synthesis, Characterization, and Biological Evaluation of *N*-Methyl Derivatives of Norbelladine. *Molecules* **2024**, *29*, 4442. [CrossRef]
30. Li, H.; Wu, J.; Zhang, N.; Zheng, Q. Transglutaminase 2-mediated histone monoaminylation and its role in cancer. *Biosci. Rep.* **2024**, *44*, BSR20240493. [CrossRef]

BioChem

MDPI

Article

Proteomic Blueprint of Atlantic Cod (*Gadus morhua*) Otoliths Revealing Environmental Stress Insights through Label-Free Quantitative Shotgun Proteomics

Trevena N. Youssef [1], Sherri L. Christian [1], Rick Rideout [2], Aaron Adamack [2], Pierre Thibault [3,4], Eric Bonneil [5], Travis D. Fridgen [6] and Joseph Banoub [1,2,6,*]

1 Biochemistry Department, Memorial University of Newfoundland, St. John's, NL A1B 3X9, Canada; tnyoussef@mun.ca (T.N.Y.); sherri@mun.ca (S.L.C.)
2 Fisheries and Oceans Canada, Northwest Atlantic Fisheries Centre, St. John's, NL A1C 5X1, Canada; rick.rideout@dfo-mpo.gc.ca (R.R.); aaron.adamack@dfo-mpo.gc.ca (A.A.)
3 Department of Medicine, University of Montreal, Montreal, QC H3C 3J7, Canada; pierre.thibault@umontreal.ca
4 Institute for Research in Immunology and Cancer (IRIC), University of Montreal, Montreal, QC H3T 1J4, Canada
5 Department of Chemistry, University of Montreal, Montreal, QC H3A 0B8, Canada; eric.bonneil@umontreal.ca
6 Chemistry Department, Memorial University of Newfoundland, St. John's, NL A1B 3X7, Canada; tfridgen@mun.ca
* Correspondence: joe.banoub@dfo-mpo.gc.ca

Abstract: Otoliths of the fish's inner ear serve as a natural chronological recorder because of their continuous formation marked by daily, monthly, and annual increments. Despite their importance, the comprehensive protein content of otoliths remains not fully identified. Using the label-free shotgun proteomics method with one-dimensional liquid chromatography coupled to electrospray ionization-orbitrap tandem mass spectrometry, we quantified a broad range of proteins, with individual otoliths containing between 1341 and 1839 proteins. The identified proteins could potentially serve as a blueprint for fish growth from embryo to adult. We quantified eleven heat-shock proteins (HSPs) in both sexes and several proteins impacted by endocrine disruptors, indicating the otolith's capacity to reflect environmental stress, potentially linked to climate change effects and altering of hormonal and neuroendocrine functions. Our bioinformatic ontology analysis confirmed the presence of proteins critical for various biological processes, including structural and enzymatic proteins. Protein–protein interaction (PPI) mapping also identified key interactions between the identified proteins. These findings significantly advance our understanding of otolith proteomics, offering a solid foundation for future work. Most of the identified proteins deposited daily and influenced by the environment were not implicated in the biomineralization of otolith, raising the potential for the otolith proteome to recreate details of fish life history at previously unrealized levels.

Keywords: otoliths; proteomics; biomineralization; quantitative shotgun analysis; *Gadus morhua*

1. Introduction

Atlantic cod (*Gadus morhua*) fishing is a significant economic resource in Canada, impacted by both overfishing and climate change [1–3]. The cod's inner ear plays a crucial role in its balance and auditory functions. It consists of three semicircular canals that detect head movements and are essential for orientation; these canals are filled with endolymph fluid and contain a crista, which is a sensor for rotational and angular motions. Additionally, the inner ear includes three otolithic organs, the saccule, lagena, and utricle, each containing a unique otolith or ear stone (sagitta, lapillus, and asteriscus) vital for auditory and equilibrium processes [4–6].

Otoliths are composed of calcium carbonate ($CaCO_3$) embedded with organic molecules such as complex polysaccharides and proteins, and are metabolically inert. They grow continuously through the daily accretion of concentric layers of the otolith matrix [7–9]. Otoliths are important tools in fisheries research, offering a window into the life history and ecological dynamics of fish species. Their multifaceted study includes macroscopic analyses, from shape and size, to microscopic examinations, such as growth increments and inorganic chemical composition, reflecting the fish's environment and behaviours [10]. The analysis of the otolith composition can reveal the existence of common environmental factors that influence fish growth, help uncover past climate conditions, and predict future environmental impacts on marine ecosystems [11,12]. Fish growth is a complex biological process influenced by a combination of intrinsic and extrinsic factors. The intrinsic, such as ontogeny and sex, and extrinsic, such as abiotic conditions of the environment or intra-specific interactions, may complicate inferences about climatic impact [11].

A notable study created a century-long growth biochronology (1908–2014) for Atlantic Cod growth in Icelandic waters from otoliths, and demonstrated that temperature variations significantly influenced growth, with younger fish benefiting from warmer temperatures while older fish experienced negative growth effects under similar conditions [11].

The sagitta, the largest otolith, has garnered significant attention in research due to its size and detailed information about the fish's age and growth patterns. Larger otoliths provide clearer and more distinct growth rings, offering accurate inter-specific morphological diversity [13,14].

Daily growth layers of otolith structures, formed by the gradual accumulation of new material, vary in thickness and composition depending on environmental conditions, some associated with proteins and some with the $CaCO_3$ component. This continuous addition from hatching to death provides a valuable record of various aspects of fish life history, and serves as a potential environmental record [15,16]. Proteins constitute a small portion of the otolith, estimated at around 2–3% [17,18]. As most of these proteins are unlikely to be directly involved in biomineralization, it suggests that a diversity of proteins present in the endolymph are trapped in the otolith during increment formation [17]. This raises the intriguing possibility that the otolith is not only archiving elemental markers of environmental history, but also protein markers of development and physiological change over an individual's lifetime [17,19]. Previous research has highlighted the significant roles of fifteen identified proteins in otolith formation and maintenance within vertebrates, focusing on three main groups: otolith matrix proteins, otolith anchoring proteins, and otolith regulatory proteins [15,20,21]. However, the comprehensive understanding of the otolith's whole protein composition has not yet been investigated.

The advancement in proteomics has been propelled by innovations in protein separation, mass spectrometry (MS) establishing the identification and analytical composition, and data analysis through bioinformatics. MS is central to extensive protein studies. The "bottom-up" approach dissects proteins into peptides via proteolysis for analysis, termed "shotgun proteomics", when applied to protein mixtures. This method assesses proteins indirectly by analyzing peptides from proteolytic digestion. In shotgun proteomics, peptides are fractionated, analyzed via LC-ESI-MS/MS, and identified by matching the characterized peptide sequences with the mass spectra against predicted spectra from a protein database [22–27].

We have recently demonstrated the feasibility of employing a shotgun proteomics approach to study the proteome of Atlantic Cod otolith key structures located in the fish's inner ears [17]. The data suggested that the otolith proteins could be used to discover the whole fish protein profile ranging from embryo to adult. We have shown that most of these proteins were not implicated in the biomineralization of otoliths, raising the potential for the otolith proteome to help recreate details of fish life history at previously unrealized levels [17]. This initial qualitative investigation of otolith proteins allowed the identification of two primary functional categories, significantly broadening our comprehension of their roles beyond mere biomineralization. The first category involved proteins integral to biochemical processes, primarily in synthesizing and degrading proteins. The second category included proteins that were instrumental in physiological processes. This data significantly broadened our understanding of the roles of otolith proteins beyond mere biomineralization. These recently identified proteins appear to have a significant influence on essential life processes, including but not limited to growth, development, metabolism, and the reproductive system within the otolith framework.

In this work, we present our current efforts to quantify the proteins of the cod otolith proteome from ten fish quantitatively, aiming to address the following objectives: (A) determine the quantity of either different, similar, or absent proteins present in the make-up of the otoliths obtained from both sexes, (B) establishment of the presence of indicators of heat stress present in the otoliths of both sexes, as a possible indicators of global warming of the Atlantic Ocean, and (C) establishment of the presence of endocrine disruptor protein bioindicators that can indicate the alternation of the hormonal and neuroendocrine fish functions.

2. Materials and Methods

2.1. Otoliths

The archived otolith collection at the Northwest Atlantic Fisheries Centre (St. John's, NL, Canada) contains thousands of Atlantic cod otoliths collected over several decades. The methods used to collect otoliths can vary slightly with the individual technician, but in general, otoliths were exposed via a dorsal incision in the fish's skull and were removed using forceps. Otoliths were blotted dry with paper towels and stored in small paper envelopes. For this study, we selected ten otoliths (five males and five females) collected in 2019 from the northeast coast of Newfoundland, Canada. Otoliths were washed several times with deionized water before protein extraction, as previously described [17]. Finally, each individual otolith was ground in a mortar using a pestle and then added to a fine powder.

2.2. Chemicals and Standards

All standards, samples, and buffers were prepared using ultra-pure Milli-Q H_2O (18.2 MΩ·cm, Merck Millipore, Darmstadt, Germany). All chemicals were purchased from Sigma Aldrich (Castle Hill, NSW, Australia) and were of the highest available purity. Mass spectrometric grade trypsin was obtained from Promega (Madison, WI, USA). MS-grade solvents for chromatography were obtained from Canadian Life Science (Peterborough, ON, Canada).

2.3. Otolith Protein Extraction

The sagitta otoliths were washed, cleaned, and dried as previously described [17]. Full powdered otoliths (2 mg) were suspended in 20% *w/v* trichloroacetic acid (TCA) (10 mL) and incubated overnight at room temperature. Samples were then centrifuged at 10,000× *g* for 10 min, and the supernatant was discarded. The crystals were washed with 100 μL of ice-cold acetone and recentrifuged. The supernatant and any remaining undissolved otolith were discarded, and the vacuum-dried pellet was processed for in-solution digestion. Briefly, 0.1 mg of the pellet was resuspended in denaturing buffer containing 8 M urea and 0.4 M ammonium bicarbonate (NH_4HCO_3). Then, 10 μL of 0.5 M dithiothreitol (DTT) was added and incubated for 30 min at 60 °C. After cooling for 5 min at room temperature, 20 μL of 0.7 M iodoacetamide ($IAcNH_2$) was added and incubated for 30 min. Next, the sample was diluted with 1.2 mL of H_2O followed by 10 μL of 0.1 M $CaCl_2$. For enzymatic digestion, 100 μL of 0.02 μg/μL trypsin (Promega Trypsin Gold, Mass Spectrometry Grade, Promega) prepared in 50 mM NH_4HCO_3 was added to each sample. The samples were incubated overnight at 37 °C in a shaker. The trypsin activity was inhibited by adding 1 μL of trifluoroacetic acid (TFA), and the samples were acidified to a pH below 3 with formic acid.

The samples were then desalted using Oasis HLB 3cc Extraction Cartridges (Waters, MA, USA) connected to a SuperlcoVisiprep DL manifold (Sigma-Aldrich, Darmstadt, Germany). The column was conditioned using 0.5 mL methanol followed by 1 mL of an elution buffer containing 50% ACN and 0.1% TFA, and then with 2 mL wash buffer prepared with 0.1% TFA. After loading the sample, the column was washed with 5 mL of wash buffer and eluted twice with 0.5 mL of 50% acetonitrile-0.1% TFA and once with 0.5 mL of 80% ACN-0.1% TFA. The eluant was dried using a vacuum concentrator (SpeedVac Concentrator, Thermo Electron Corp., Waltham, MA, USA). The dried peptide was reconstituted in 12 µL of resuspension buffer containing 5% acetonitrile and 0.1% formic acid.

2.4. Shotgun Proteomics by LC-ESI-MS/MS

The LC separation was carried out using the Ultimate 3000RSLCnano system (Dionex/Thermo Fisher Scientific, Waltham, MA, USA). For analysis, 2 µL (1 mg/mL) of the sample was injected onto an in-house packed capillary column (50 cm × 75 µm, pulled tip, ESI source solutions) packed with Jupiter C18 4 µm chromatographic media (Phenomenex, Torrance, CA, USA) at a flow rate of 300 nL/min. Chromatographic separation was performed by a 120 min method using solvent A (0.1% formic acid in MS-grade water) and solvent B (0.1% formic acid in MS-grade acetonitrile) from 5% to 30% for 90 min, then increasing to 55% for the next 12 min, and then to 95% for 8 min before being reduced to 5% B for the remainder of the 120 min run. The column oven temperature was set at 40 °C.

The ESI-MS and MS/MS spectra were obtained using an Orbitrap Fusion Lumos Tribrid Mass Spectrometer (ThermoScientific, Waltham, MA, USA) fitted with a Nanospray Flex Ion source and a FAIMSpro source. The detailed acquisition parameters for mass spectrometry analysis were as follows: For ionization, the spray voltage was 1.8 kV, and the ion transfer tube temperature was 300 °C. The data were acquired in data-dependent acquisition (DDA) mode with a full scan using Orbitrap at a resolution of 120,000 over a mass range of 400–2000 m/z. The FAIMS source was operated at three different compensation voltages (CV = 40, 60 and 80). The auto gain control (AGC) was set in standard mode, and maximum injection time at Auto mode for each MS/MS, acquisition of peaks with intensities above 5.0 e^3 were performed using normalized HCD collision energy of 35%. The cycle time was set at 1 s. The isolation window for MS/MS was set at 1.6 Da. The AGC target was set at standard mode with the maximum injection time mode as Auto. The precursor ions with positive charges 2 to 5 were selected for MS/MS analysis. After every single MS/MS acquisition, the dynamic exclusion time was set at 60 s with a mass tolerance of ±10 ppm.

The MS/MS raw files were acquired using Thermo Scientific Xcalibur 4.5 and Tune 3.5 (Waltham, MA, USA) and searched against the *Gadus morhua* protein database (Taxon ID 8049) downloaded from Uniport using Proteome Discoverer 2.5 (ThermoFisher, Waltham, MA, USA). The SequestHT search engine node was used for peptide and protein identification. Two missed cleavages for trypsin digestion were allowed with a 10 ppm precursor mass tolerance and fragment mass tolerance of 0.6 Da. Oxidization of methionine and N-terminal acetylation were set as dynamic modification, and carbamidomethyl cysteine was selected as a static modification.

2.5. Quantitative Analysis

This study was done using label-free proteomics quantification. The peptides are identified and searched against the *Gadus morhua* protein database (Taxon ID 8049) and searched for the MS^2 acquired during the DDA acquisition. The peptide quantification was performed by integrating the area under the curve of the MS^1 of the identified peptides using Proteome Discoverer 2.5 (ThermoFisher, Waltham, MA, USA). This was followed by populating the total area of all the identified peptides within a protein to derive protein abundances. The ratio of these abundances is then used to compare the different samples. This approach allows for relative quantitative analysis, comparing protein levels across samples (0.1 mg/mL), with careful standardization ensuring comparability despite not measuring absolute protein quantities.

To ensure precise comparisons in label-free proteomics quantification, this study implemented consistent protocols across all procedures, from sample prep to analysis.

Quantitative Expressed Protein Intensities Accuracy

The variance stabilization normalization (VSN) is a transformation technique employed in mass spectrometry data analysis to maintain consistent variance across the intensity spectrum [28]. VSN serves the dual purpose of background noise subtraction and data normalization, facilitating a clear relationship between mass spectrometry (MS) peak intensity and variance. Consequently, this enhances the accuracy of the analytical results by stabilizing variance. The missing value imputations were performed by replacing missing values with random values picked in the 5% lower end of the normal distribution of intensities. VSN was performed using Proteome Discoverer 2.5 (ThermoFisher, Waltham, MA, USA). Additionally, a p-value ≤ 0.05 was considered significant.

2.6. Statistical Analysis

The quantitative evaluation of the total number of protein expression differences between male and female groups was conducted using two-sample t-tests and then the calculation of 95% confidence intervals (CIs). The statistical analyses were carried out using various R software packages. These included tidyr [29], magrittr [30], plyr [31], dplyr [32], and ggplot2 [33], facilitating the creation of comparison tables to present the findings. Further normalization analysis, generation of heat maps, and creation of volcano plots (abundance ratio Adjusted p-Value: (Female)/(Male) by Benjamini-Hochberg procedure) were performed using Proteome Discoverer 2.5 (ThermoFsher, Waltham, MA, USA). A p-value ≤ 0.05 was considered significant.

2.7. Bioinformatics Analyses

Gene ontology (GO) functional analyses for the extracted otolith proteins from Atlantic cod (*Gadus morhua*) were performed using the ClueGO plugin (version 2.5.9) within Cytoscape software (version 3.9.1). Due to the absence of cod organisms in the ClueGO plugin program, Atlantic salmon (*Salmo Salar*) was used for analysis to obtain the conserved functional analysis with Atlantic cod (*Gadus morhua*). Biological functions, cellular components, immune system processes, molecular functions, and KEGG provide specific gene annotations with their corresponding functions within identified biological pathways. Finally, the STRING database was used to visualize the protein–protein interactions within Atlantic cod (*Gadus morhua*) (https://string-db.org/cgi/network?taskId=bybNmeXaeGFU&sessionId=boleulT8hMV6, accessed on 14 May 2024). To identify the top 20 proteins based on their degree of connectivity, the cytoHubba plugin was used.

3. Results and Discussion

The main purpose of this work was to provide a comprehensive analysis of the whole protein content of ten otoliths. As mentioned, we have formerly conducted a qualitative analysis of twelve otoliths [17], which helped us understand the composition of otolith proteins. We compared the total number of proteins identified in the previous study to those in the current study. We found differences in the total numbers; we speculate that this discrepancy is due to the ages of the samples. The previous study analyzed samples that were approximately 12 years old, while our current study includes samples ranging from 8 to 11 years old (Figure 1). However, the eleven HSPs were identified in both studies.

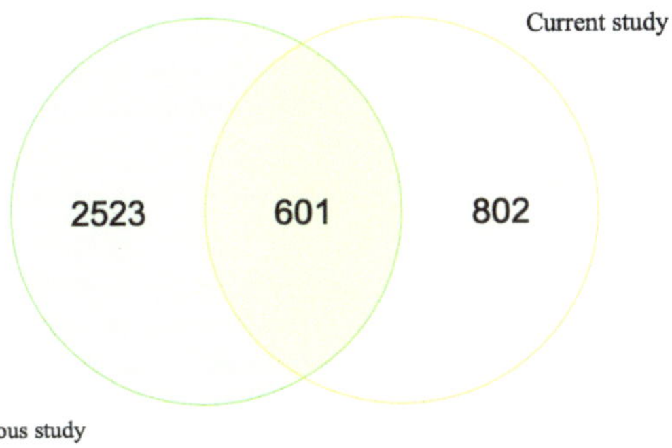

Figure 1. Comparison of the total number of otolith proteins between the current study and the previous one. This plot was generated by molbiotools online tools.

In the current research, we presented the quantitative results of the proteins identified in our previous qualitative study [17]. We focused on ten samples obtained from five males and five females collected in 2019 (Table S1, Supplementary Excel sheet), and we report interesting novel protein biomarkers obtained during this global quantitative analysis of either different, similar, or absent proteins in the make-up of the otoliths obtained from either sex. We highlight that the presence and variation of HSPs in marine organisms could potentially help predict how global climate change might impact species' metabolic costs [34]. Accordingly, in this manuscript, we suggest that the presence of HSPs in the otoliths of both sexes can be used as an indicator of global warming of the Atlantic Ocean. Furthermore, we quantified the proteins that could potentially be influenced by endocrine disruptors that can indicate alterations in the hormonal and neuroendocrine function of fish.

Analysis of our dataset revealed a range in the number of proteins per individual, varying from 1341 to 1839. Utilizing a two-sample t-test to compare the mean number of proteins between males and females, we found that there was no significant difference in the total number of proteins (Figure 2). This finding is supported by the 95% confidence intervals for females (1366.121, 1805.479) and for males (1327.221, 1700.779).

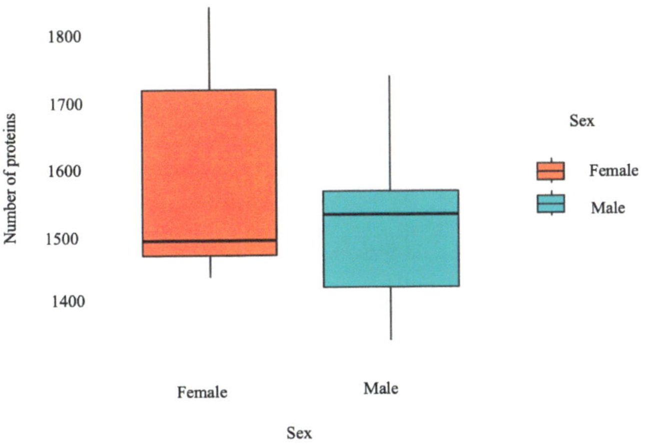

Figure 2. The total number of otolith proteins is similar between the sexes. This plot shows the distribution of protein counts for males and females. Despite apparent differences in medians and variability, a two-sample t-test determined that there is no statistically significant difference in the mean protein counts between sexes (p-value > 0.05). This plot was generated by R Studio using the package ggplot2 retrieved from [33].

In the following, we will describe the intensities of the protein blueprint of Atlantic cod otoliths that will reveal the different biological processes. It is out of the realm of possibility to discuss all of the quantified proteins in this rationale. For this reason, we shall deliberate about a few selected protein models that represent physiological, ontogenetic, evolutionary, and environmental processes.

3.1. Protein Expression between Sexes

In our study employing ESI-MS/MS with DDA, we conducted a comprehensive proteomic analysis across ten individuals (Table S2, Supplementary Excel sheet). We identified significantly upregulated and downregulated proteins in females compared to males with an abundance ratio (Female/Male) range of 0.000911165–401.7070581 through the volcano plot visualization (Figure 3 and Table S3, Supplementary Excel Sheet). Proteins with high abundance ratios (fold change > 2) and low adjusted p-values < 0.05 were considered significantly more abundant in females than males. A subset of these proteins exhibiting sex-specific expression differences (differentially expressed proteins between sexes), as visualized in the heatmap (Figure 4), which may have crucial roles in fish's physiological and reproductive processes.

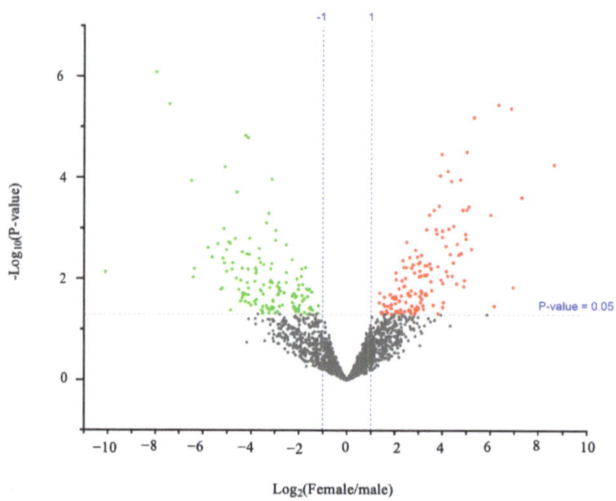

Figure 3. Protein expression is different between the sexes. Each dot expresses a specific protein. Proteins that appear in red are significantly upregulated in female samples, whereas those in green are significantly upregulated in male sexes at p-value < 0.05. This plot was generated by Proteome Discoverer 2.5.

3.1.1. Otolith Proteins with Substantial Female-Biased Expression

Among the proteins with substantial female-biased expression in otoliths, zona pellucida sperm-binding protein 4-like was notable for its high log2 intensity, as visualized in the heatmap (Figure 4) that showed $\geq$ 2-fold change in abundance between sexes with p-values $\leq$ 0.05. There is evidence that this protein, which is essential in forming the egg membrane and sperm–egg recognition, represents a key component of female fertilization [35]. In addition, the high presence of catenin (cadherin-associated protein), alpha 1, which is responsible for the conserved, calcium-dependent module crucial for cell–cell adhesion. Furthermore, it also plays a vital role in normal developmental processes and in maintaining tissue structure [36], which is likely related to its essential functions in reproductive processes.

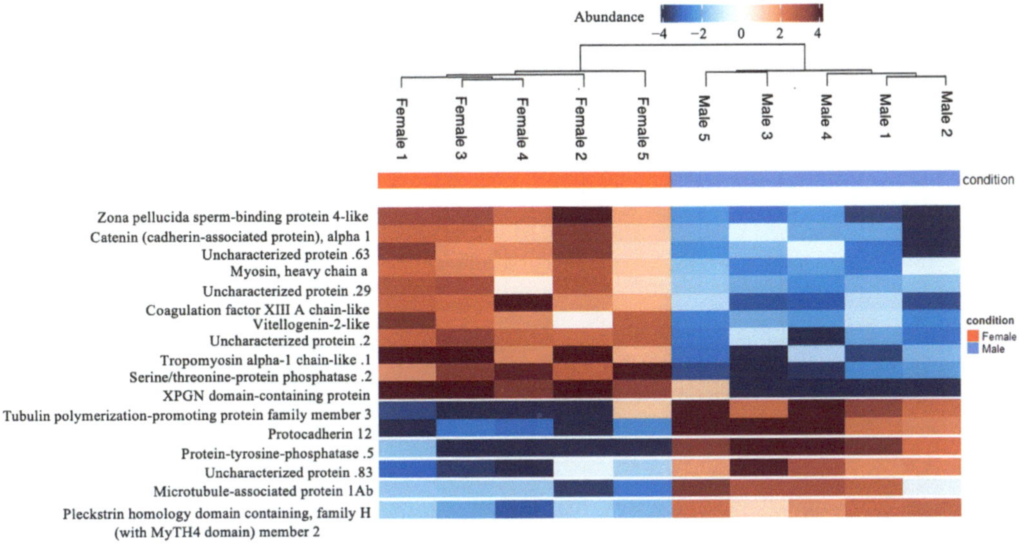

Figure 4. The differentially expressed proteins between sexes. The heatmap highlights proteins that showed the highest abundance between sexes with p-values ≤ 0.05. This visualization allows us to observe intensities of protein expression, with a colour gradient from blue lower expression to red representing higher expression levels. Each row corresponds to a protein, and each column to a sample, allowing us to compare the expression across our samples generated by Proteome Discoverer 2.5 software.

Myosin heavy chain a is a key contractile protein of the muscular system, and its expression is often a reliable indicator of muscle development and growth. In Atlantic salmon, as in many other species, the mRNA expression levels of myosin heavy chain correlate with the muscle's ability to grow and accrue protein rather than just increase in size [37]. It is expressed with a higher intensity in female otoliths, possibly indicating differences in muscle physiology or energy demands between the sexes (Figure 4). Tropomyosin alpha-1 chain-like is thought to be the master regulator of actin filament functions in the cytoskeleton. Coagulation factor XIII A chain-like protein (Figure 4) is important for blood coagulation and wound healing [38]. The vitellogenin-2-like proteins, a female marker, are phosphoglycolipoproteins synthesized in the livers of oviparous animals in response to circulating estrogens [17,39]. Serine/threonine-protein phosphatase is one of the key enzymes responsible for dephosphorylation in vertebrates involved in various cellular processes, including the cell cycle and signal transduction [40]. It also displayed varying expression levels, indicating potential differences in cell regulation. The expression of the XPG N-terminal domain-containing protein is a critical part of the XPG protein involved in DNA repair [41]. All these proteins display differential expression to support the unique physiological demands of females.

3.1.2. Otolith Proteins with Substantial Male-Biased Expression

Among the proteins with substantial male-biased expression, we identified protocadherin and protein-tyrosine-phosphatase (Figure 4). Protocadherins are cell adhesion molecules that belong to the cadherin superfamily and are expressed most prominently within the central nervous system, which suggests important neurobiological roles for these molecules [42]. A recent study suggested that genetic factors like protocadherins, which are expressed more in male cell lines, play a crucial role in the molecular basis of sex differences in the nervous system [43]. Protein-tyrosine-phosphatase is an enzyme that functions in a coordinated manner with protein tyrosine kinases to control signalling pathways that underlie a broad spectrum of fundamental physiological processes [44].

The plectstrin homology domain-containing protein, which is part of signal transduction in cells, exhibited a significant difference in expression, pointing to potential variances in cellular communication processes between genders [45]. Tubulin polymerization-promoting protein family member 3 is an intrinsically unstructured protein that induces tubulin polymerization [46]. It has a role in microtubule stabilization, cell division, and developmental processes [47]. Microtubule-associated protein 1Ab, comprising distantly related protein complexes with heavy and light chains, is believed to be involved in the regulation of the neuronal cytoskeleton [48]. Pleckstrin homology domain-containing family H (with MyTH4 Domain) member 2 is involved in crucial interactions with membranes and proteins, characterized by pleckstrin homology (PH) and MyTH4 domains crucial for cellular signalling. These domains aid in organizing the cytoskeleton, affecting cell shape, movement, and interactions between cell membranes and the cytoskeleton across various cellular activities [45]. These proteins' high expression reflects their importance in supporting the cellular processes specific to male Atlantic cod.

3.2. *Quantitative Analysis of the Total Protein Profile*

The comprehensive proteomic analysis across several individuals (Table S2, Supplementary Excel sheet) yielded 802 proteins consistently found across all individuals (Table S4, Supplementary Excel sheet), 202 highly abundant that were more than 2-fold increased in females (Table S5, Supplementary Excel sheet), and 90 proteins highly abundant that were more than 2-fold increased in males (Table S6, Supplementary Excel sheet). Furthermore, we also identified 81 proteins common only to males (Table S7, Supplementary Excel sheet) and 196 common only to females (Table S8, Supplementary Excel sheet), with 92 proteins exclusive to one individual, defined by sex (Table S9, Supplementary Excel sheet), which need further investigation, including genetic analysis, to understand the underlying reasons behind their expression. Therefore, we highlighted five proteins present within all males that were unique to males with high abundance (Table 1), and nine proteins unique to females with high abundance (Table 2). Lastly, our shotgun analysis also uncovered 214 uncharacterized proteins (Table S10, Supplementary Excel sheet).

Table 1. Five proteins that are unique in males.

Description	Accession	Female Count [a]	Male Count [b]	Abundances (Average): Male [c]
Charged multivesicular body protein 2Ba OS = Gadus morhua OX = 8049 GN = chmp2ba PE = 3 SV = 1	A0A8C4YUD6	0	5	261,448.36432
DNA damage-binding protein 1 OS = Gadus morhua OX = 8049 GN = ddb1 PE = 3 SV = 1	A0A8C5A9Z6	0	5	98,004.482948
Fibroblast growth factor OS = Gadus morhua OX = 8049 PE = 3 SV = 1	A0A8C4ZEA7	0	5	24,497.65508
Microtubule-associated protein 1Ab OS = Gadus morhua OX = 8049 GN = map1ab PE = 4 SV = 1	A0A8C5CU29	0	5	3,458,736.85
Uncharacterized protein OS = Gadus morhua OX = 8049 PE = 3 SV = 1	A0A8C5FA14	0	5	778,547.8001

[a] Female count: the count of females associated with each protein entry. [b] Male count: the count of males associated with each protein entry. [c] Abundances (average): quantitative protein abundance measurements were averaged for all-female groups.

Table 2. Nine proteins that are unique in females.

Description	Accession	Female Count [a]	Male Count [b]	Abundances (Average): Female [c]
Dihydrolipoamide acetyltransferase component of pyruvate dehydrogenase complex OS = Gadus morhua OX = 8049 GN = dbt PE = 3 SV = 1	A0A8C5C1K4	5	0	108,232.639474
Galectin OS = Gadus morhua OX = 8049 GN = LOC115545324 PE = 4 SV = 1	A0A8C5ACJ8	5	0	489,972.337868
Heterogeneous nuclear ribonucleoprotein A/Ba OS = Gadus morhua OX = 8049 GN = hnrnpaba PE = 4 SV = 1	A0A8C4YVY0	5	0	1,301,640.954998
Myosin light chain, phosphorylatable, fast skeletal muscle a OS = Gadus morhua OX = 8049 GN = mylpfa PE = 4 SV = 1	A0A8C5B8E2	5	0	495,851.125408
Nidogen 2a (osteonidogen) OS = Gadus morhua OX = 8049 PE = 4 SV = 1	A0A8C5D3Y0	5	0	1,349,773.1182679997
Polypeptide N-acetylgalactosaminyltransferase OS = Gadus morhua OX = 8049 GN = galnt6 PE = 3 SV = 1	A0A8C5BP60	5	0	42,686.3808994
SRSF protein kinase 2 OS = Gadus morhua OX = 8049 GN = srpk2 PE = 4 SV = 1	A0A8C5CES3	5	0	20,084.879869999997
Uncharacterized protein OS = Gadus morhua OX = 8049 PE = 4 SV = 1	A0A8C5CCX7	5	0	211,271.979502
Vinculin OS = Gadus morhua OX = 8049 PE = 3 SV = 1	A0A8C5A430	5	0	411,121.875006

[a] Female count: the count of females associated with each protein entry. [b] Male count: the count of males associated with each protein entry. [c] Abundances (average): quantitative protein abundance measurements were averaged for all-male groups.

3.2.1. Identified Otolith Protein Profile Common in Both Sexes

Some common abundant protein families were quantified in both sexes. In this study, among the proteins recognized, actins, were ubiquitous between sexes. α-actins are found in muscle tissues and are a significant constituent of the contractile apparatus. Developmental stage-specific muscle protein isoforms have also been reported for several fish species during the development [49–51]. Sixteen proteins related to the actin family were detected in this study (Table S11, Supplementary Excel sheet).

In addition, we quantified tropomyosins, which are actin-binding proteins that play a crucial role in regulating the actin cytoskeleton and muscle contraction [49,52]. Six proteins (tropomyosin 1 (alpha), tropomyosin alpha-3 chain-like, tropomyosin 4a, tropomyosin alpha-1 chain-like, tropomyosin 3, tropomyosin 2 (beta)) were detected in this study (Table S12, Supplementary Excel sheet).

We also detected that the tubulin proteins, which are the principal component of microtubules, are a heterodimer of two closely related proteins, α- and β-tubulin [49,53]. α- and β-tubulin were detected in this study (Table S13, Supplementary Excel sheet).

Similarly, we detected that the tyrosine 3-monooxygenase/tryptophan activation protein β-polypeptide belongs to the 14–3–3 family and plays a critical role in signal transduction by attaching to proteins containing phosphoserine. These are involved in different cellular processes, including cell cycle progression, survival pathways, and metabolic regulation [49,54,55]. Three proteins (tyrosine 3-monooxygenase/tryptophan 5-monooxygenase activation protein, beta polypeptide a, tyrosine 3-monooxygenase/tryptophan 5-monooxygenase activation protein, eta polypeptide, and Tyrosine 3-monooxygenase/tryptophan 5-monooxygenase activation protein, theta polypeptide b) were detected in this study (Table S14, Supplementary Excel sheet).

The keratin proteins were consistently present across all individuals. Keratin proteins are ubiquitous and varied in their types between different age groups. Specifically, type II keratins were predominantly observed in the younger group, whereas type I keratins were more prevalent in the older group [49,56–58]. Both types were detected in our data (Table S15, Supplementary Excel sheet).

Otolith Heat Shock Proteins

HSPs are expressed in response to various types of stress. This includes thermal, anoxia, acidosis, hypoxia, exposure to toxins, intense protein breakdown, and microbial infections [59,60]. They are classified into five families based on molecular weight as well as domain structures and functions: Hsp110, Hsp90, Hsp70, Hsp60, Hsp40, Hsp10, and small HSP families [60,61]. We identified and quantified eleven proteins belonging to the Hsp90, Hsp70, Hsp60, and Hsp40 families, as shown in the heatmap (Figure 5) and listed in Table S16, Supplementary Excel sheet.

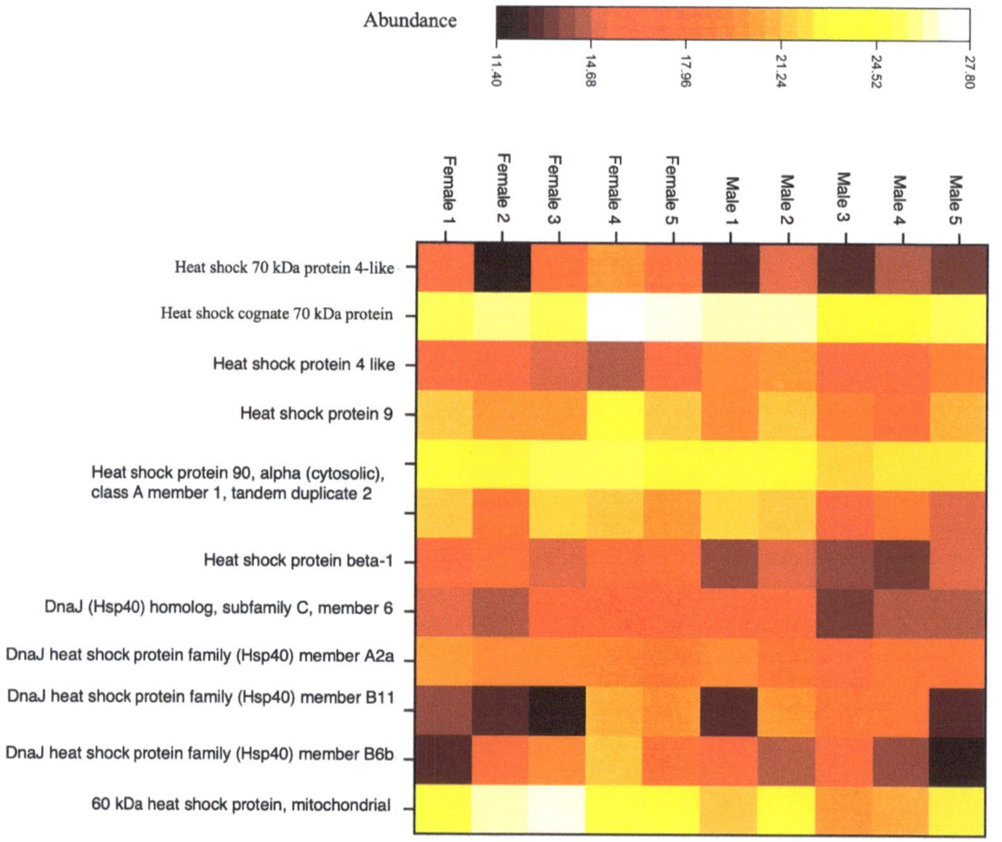

Figure 5. The expression of heat shock proteins within all individuals. Each column represents a sample, and each row corresponds to a specific protein, allowing us to compare the expressed protein absorbance across our samples. This figure was created by Proteome Discoverer 2.5 software.

Hsp90 can suppress thermal aggregation and facilitate protein folding by reducing misfolding via interactions with aggregation-prone unfolding intermediates [60,62–64]. Hsp70 assists in folding newly synthesized polypeptides, refolding of misfolded proteins, protein transport across organelles, and degradation of proteins. This protein is essential for maintaining cellular protein homeostasis, especially under stress conditions [65,66]. HSP60 proteins play a critical role in regulating mitochondrial protein homeostasis. HSP60 has been primarily considered to reside in the mitochondria, where HSP60 and HSP10 form a complex and facilitate mitochondrial protein folding. Studies on channel catfish have shown that HSP60 is implicated in the immune response to bacterial infections, suggesting it may be crucial for disease defense in fish [60,67]. Hsp40 proteins specifically interact with Hsp70, regulating its activity by stimulating ATP hydrolysis, which is essential for the chaperone's function in binding and releasing unfolded or misfolded protein substrates. This interaction is vital for maintaining cellular protein homeostasis, particularly under stress conditions that lead to protein damage [68].

As a result, HSPs are very crucial in maintaining cellular protein homeostasis through their chaperone functions. The presence of HSPs in both male and female fish reflects their fundamental role in cellular processes that are essential for the survival of all organisms, regardless of sex. These proteins are part of the cell's response to stress, assisting in protein folding, repair, and protection against damage caused by various stressors. Consequently, they play a crucial role in protecting the cod from the harmful effects of stress. [69–76]. Therefore, we predict that the presence of the eleven detected HSP chaperone proteins in our otolith samples could indicate climate change impacts.

Identification of Otolith Protein Bioindicators of Endocrine Disruptors

As mentioned in Section 3.1, a notable finding is the presence of vitellogenin-2-like protein, which directs egg development, showing a marked increase in abundance (193.2) in all female quantitative measurements [39]. However, its presence, albeit in lower abundance in males (38.1), indicates potential exposure to exogenous estrogens or estrogen mimics in aquatic environments [77–80]. This has been confirmed by previous studies of vitellogenin-2-like plasma levels in male vertebrates exposed to certain xenobiotic endocrine disruptors with estrogen-mimicking activity [81–84].

The zona pellucida sperm binding protein, which shows a focal role in the oocyte and gamete development, is the main predictor of fertilization capacity. Although the zona pellucida sperm binding protein-specific marker is present in higher abundance in females (192.4), it was also detected in males in low abundance (7.6) [79,80]. We speculate that the presence of vitellogenin-2-like protein and the zona pellucida sperm-binding protein in males is due to exposure to endocrine disruptors.

3.3. *Bioinformatics Analyses*

3.3.1. Gene Ontology (GO)

Next, we analyzed the proteins in both males and females based on annotated functions. Out of 1416 gene symbols, 328 genes had associated annotations from both sexes. We provide a conclusion chart (Figure 6), showing the visualization representing the number of genes associated with specific terms and functional groups.

A large portion of the biological function identified is mRNA transport (38.71%) (Figure 7A), [85]. The next most common biological process is the purine nucleoside monophosphate biosynthetic process (25.81%) [86].

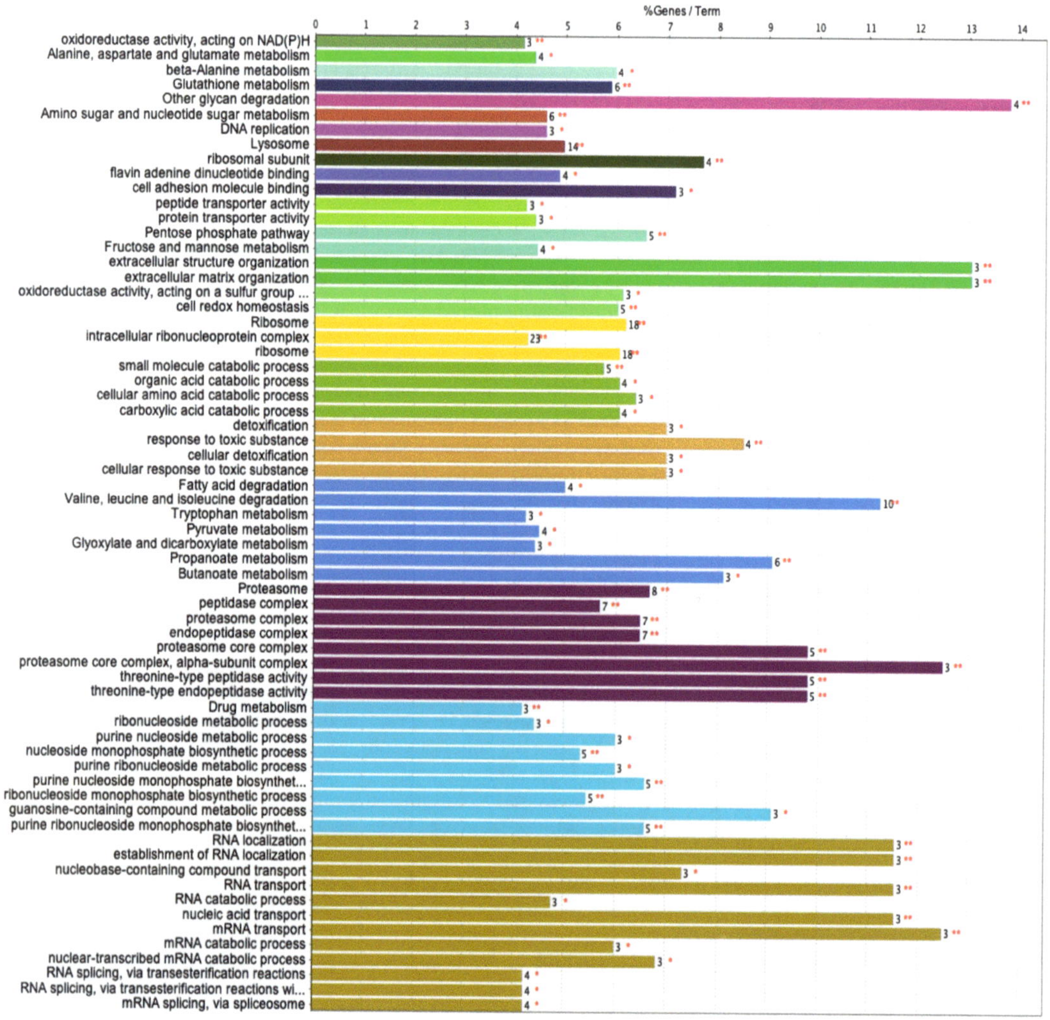

Figure 6. A conclusion chart provides a visualization representing the number of genes associated with specific terms and functional groups. Each bar represents a functional group, and the height of the bar corresponds to the number of genes related to the terms within that group. The label on each bar indicates the percentage of genes. * $p < 0.05$, ** $p < 0.01$.

A large portion of the cellular components (Figure 7B) is composed of genes that encode proteins associated with the proteasome complex (62.5%) [87]. Genes associated with the ribosomal subunits (12.5%) are critical for protein translation [88]. The smallest slice shown represents genes linked to intracellular ribonucleoprotein complexes (2.5%), which are involved in gene transcription [89].

A significant portion of the molecular functions of the proteins (Figure 7C) are associated with threonine-type endopeptidases (25.0%). Enzymes are vital in breaking down proteins by cutting internal peptide bonds in polypeptide chains. Another quarter of the genes is involved in the transport of peptides across cellular membranes. Oxidoreductase activity, acting on NAD(P)H (12.5%) and oxidoreductase activity, acting on a sulphur group of donors (12.5%); each category accounts for an equal proportion of the genes and indicates a significant role for oxidoreductase enzymes. Flavin adenine dinucleotide (FAD) binding (12.5%), this group of genes is associated with binding FAD, a cofactor involved in key metabolites for the maintenance of life and is involved in a wide range of physiological processes [90]. Cell adhesion molecule binding comprises 12.5%.

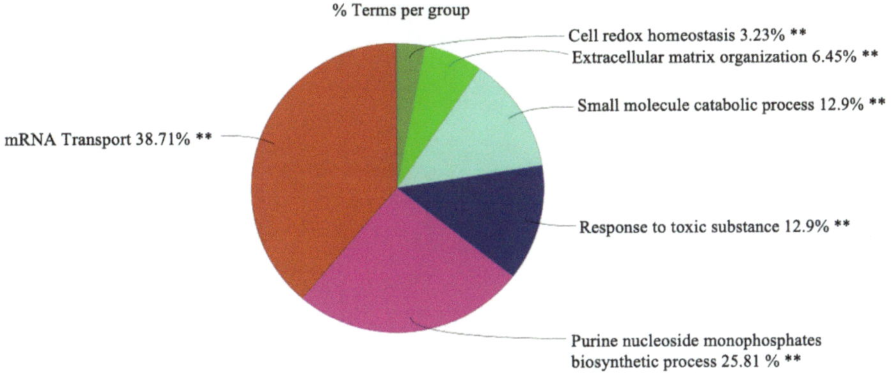

(A)

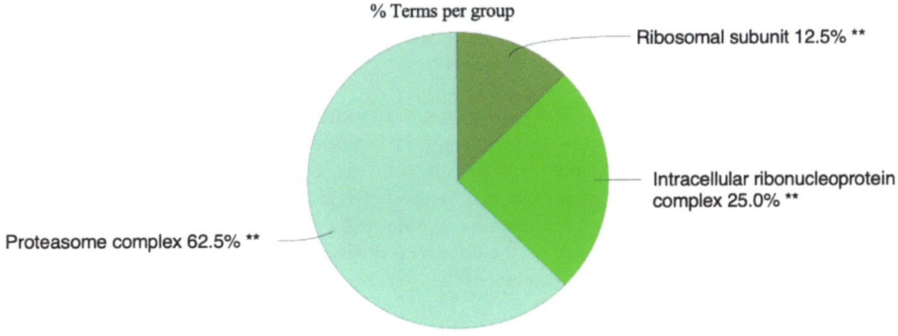

(B)

Figure 7. *Cont.*

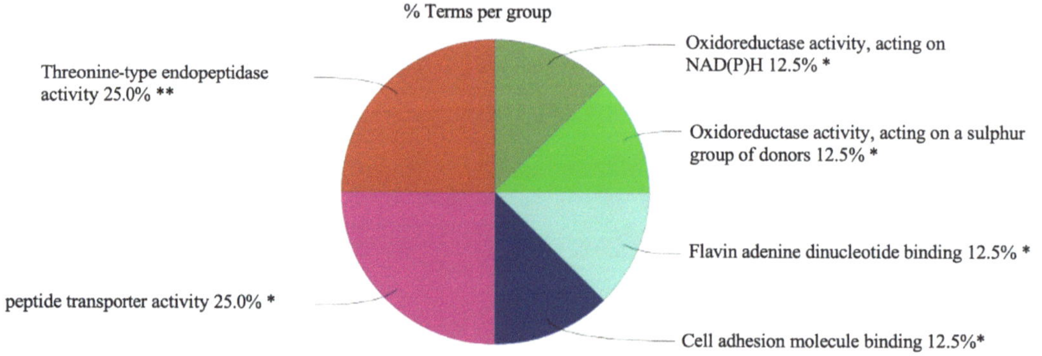

(C)

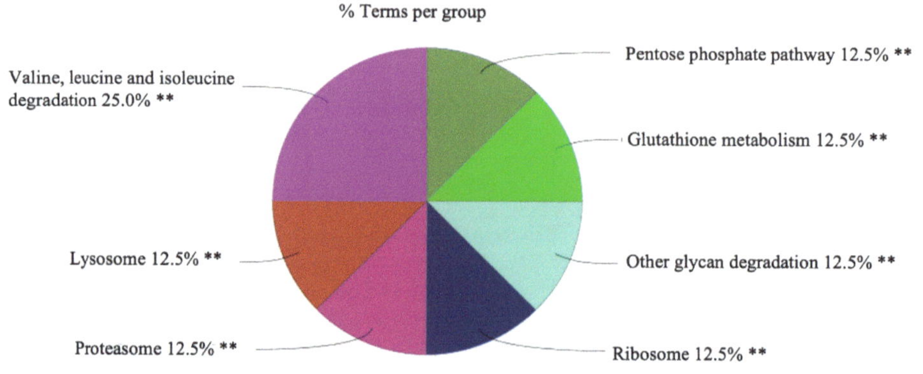

(D)

Figure 7. The distribution of identified genes across various functional groups. (**A**) This pie chart represents the distribution of genes across biological processes. (**B**) This pie chart represents the distribution of genes associated with specific cellular components. (**C**) This pie chart represents the distribution of identified genes across molecular functions. (**D**) This pie chart represents the distribution of identified genes categorized by their molecular functions, according to Kyoto Encyclopedia of Genes and Genomes (KEGG). The sizes of the slices are proportional to the number of genes involved in each process. Significant level of enrichment ** $p < 0.01$and * $p < 0.05$. This figure was created by Cytoscape software.

A significant proportion of proteins is involved in the catabolism of branched-chain amino acids (valine, leucine, and isoleucine) (25.0%) (Figure 7D), which not only act as building blocks for tissue protein (accounting for 35% of the essential amino acids in muscle), but also have other metabolic functions [91]. Lysosomes (12.5%), genes associated with lysosomal function, suggest involvement in degrading and recycling cellular waste, cellular signalling, and energy metabolism [92,93]. Proteasome (12.5%) indicates a notable representation of genes involved in the proteasome pathway, which is critical for the cell cycle, cell survival, and cellular homeostasis [94]. Pentose phosphate pathway (12.5%), a significant number of genes are involved in this pathway, which is crucial for nucleotide synthesis and the generation of NADPH for reductive biosynthesis [95]. Glutathione metabolism (12.5%) reflects genes involved in the synthesis and metabolism of glutathione, a major antioxidant that is critical in the regulation of the redox state of cells [96]. Glycan degradation (12.5%) genes are associated with the degradation of various glycans, complex carbohydrates that play major metabolic, structural, and physical roles in biological systems [97]. The ribosome comprises 2.5%, and a portion of the genes is related to the ribosome, indicating the importance of protein synthesis machinery in the cell [98].

3.3.2. Protein–Protein Interaction (PPI) Analysis

Our research used the STRING database to construct a comprehensive map of the PPI of the common otolith proteins within the *Gadus morhua*, listed in Table S17, Supplementary Excel sheet. This network elucidates potential key proteins in Atlantic cod biological processes and predicted PPIs, substantiated by evidence such as genetic co-occurrence and experimental validations [99]. Due to the complexity of our dataset, this study will focus on the top 20 proteins with the highest score, suggesting robust evidence from multiple sources, including experimental data and literature support [100] (Figure 8).

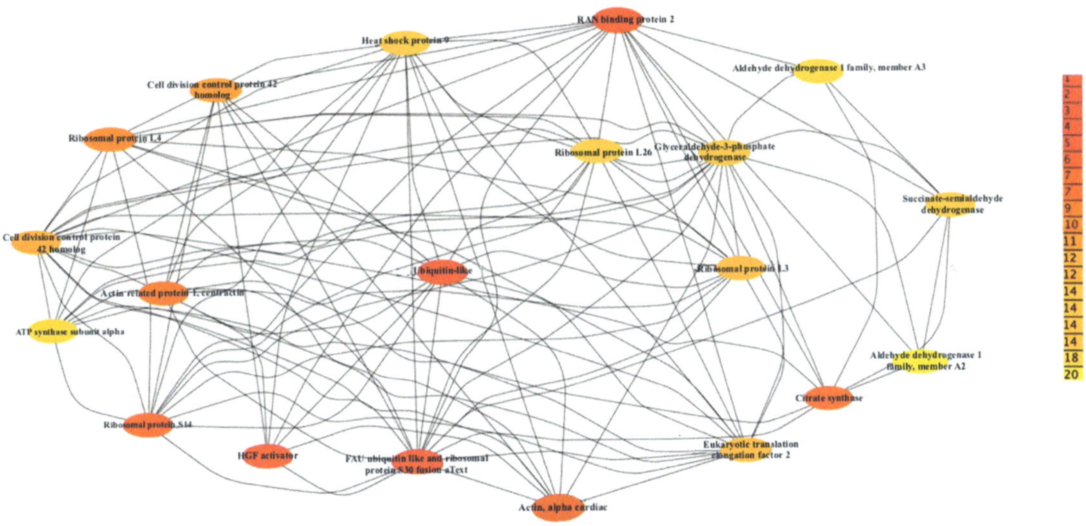

Figure 8. String network visualization of top 20 high-degree proteins in the otolith protein interaction map of Gadus morhua. This figure was created by Cytoscape software.

Key proteins identified include structural proteins such as Actin and Alpha Cardiac and enzymatic proteins like Citrate synthase and Glyceraldehyde-3-phosphate dehydrogenase (Table S18, Supplementary Excel Sheet). The category of ribosomal proteins includes Ribosomal protein S14 and Ribosomal protein L4. Additionally, signalling proteins like RAN binding protein 2, and heat shock proteins like Heat shock protein 9. Lastly, proteins such as the Eukaryotic translation elongation factor 2 are involved in the translation process.

This analysis enriches functional proteomic research in *Gadus morhua* and illustrates the potential to chronicle the extensive protein interactions that underpin vital physiological processes.

4. Conclusions and Future Directions

We have shown in this study the presence of five proteins within all males that represent uniqueness to males and nine proteins unique to females. In addition, 802 proteins were consistently found across all individuals.

In 2019, the bottom temperatures in NAFO Subarea 3K were significantly above normal, reflecting the broader impacts of global warming. The fall bottom temperatures in Subarea 3K were recorded to be between +0.5 °C and +2.5 °C above the long-term average of around 4 °C, marking a return to warmer anomalies not seen since 2011 [101–103]. This warming trend is consistent with the global patterns of ocean warming due to climate change [104]. Temperature variations in aquaculture environments can be substantial and can approach the upper critical thermal limits of 8–12 °C for Atlantic cod, which are lethal. These shifts can occur quickly, such as an ~8 °C increase in less than 12 h during thermocline inversions, particularly at depths where Atlantic cod tend to gather (≥5 m) [69,70]. Fish are unable to escape temperature fluctuations, inevitably facing stressful environments completely. This thermal stress profoundly affects gene transcription, targeting genes linked to oxidative stress response, apoptosis, protein folding, energy metabolism, synthesis, membrane fluidity, and immune functions [69–76]. The proteins encoded by these genes include some of the elements that regulate both the organismal and cellular stress responses. As a result, HSPs play a crucial role in protecting the fish against the deleterious effects of stress [69–76]. This is why we predict that the eleven HSP chaperone proteins detected in otolith could be an indicator of climate change (Table S19, Supplementary Excel sheet); however, this needs to be empirically determined.

Recent studies have brought to light significant evidence regarding the stress conditions in fish, focusing on the impact of endocrine-disrupting chemicals (EDCs), treated sewage, and other environmental factors [77,78]. This was discussed in the otolith's proteins of egg yolk precursors section. Although the zona pellucida sperm-binding protein-specific marker is present in higher abundance in females, it was also detected in males [79,80]. Likewise, environmental effluence and exposure to EDCs and other contaminants can be likely factors contributing to these observations [77–80].

Further comprehensive research is needed to elucidate the full range of influences that are central to the alteration of sex markers in fish, ensuring that we consider all possible sources of stress, whether they are pollutants or other environmental variables. This evidence collectively indicates a significant stress response in fish to environmental pollutants, warranting further investigation into the mechanisms of action and long-term consequences of these exposures on aquatic life. It is also vital to conduct further studies on the PPIs and their functional overlaps, to explore their evolutionary significance.

Finally, this study employs a quantitative shotgun proteomics approach and bioinformatic analysis to discover the unidentified proteins present in cod otoliths. Our results show that diverse proteins are stored in otolith that were accumulated, thereby enabling the discernment of both differences and similarities across protein profiles between sexes. We speculate that circulating proteins released by various cells in the body into the endolymph are captured by the otolith. In the future, we will explore the prospect of monitoring the growth of the fish by investigating the proteins present in each separate concentric layer of the otolith matrix.

In conclusion, this study suggests a broader applicative potential of otolith proteomics in marine research. By transcending traditional views of otoliths solely as chronometers, this study illuminates their capacity to provide invaluable insights into the proteomic profiles reflective of varying life stages and environmental interactions of cod. Despite these advances, our understanding of the otolith proteome remains nascent, with much terrain still to explore. Future endeavours will focus on comparing protein profiles of otoliths from different NAFO zones, trip, and depths. In the future, we plan to monitor the growth protein of the fish by investigating the presence of the protein concentric layers of the otolith matrix to follow the growth of the fish from embryo to adult. Such comparisons will be expected to substantially enrich our comprehension of the otolith proteome, opening new avenues for research and offering profound implications for marine and environmental science.

Supplementary Materials: The following supporting information can be downloaded at: https://www.mdpi.com/article/10.3390/biochem4020008/s1, Table S1: Overview of sample used in the current study.; Table S2 (1): 10 Otolith protein; Table S2 (2). 10 Otolith protein with protein sequences (Full data). Table S3: Volcano Plot; Table S4: 802 Proteins are always common between individuals.; Table S5: 202 Highly abundance proteins for female; Table S6: 90 Highly abundance proteins for males; Table S7: 81 Proteins Present in ≥X Males and zero females; Table S8: 196 Proteins Present in ≥X Males and zero females; Table S9: 92 Proteins identified in individual, either male or female; Table S10: 214 Uncharacterized proteins identified in our dataset.; Table S11: Sixteen proteins related to the actin family identified in this study; Table S12: Six proteins related to tropomyosins identified in this study; Table S13: α- and β-Tubulin identified in this study; Table S14: Three proteins related to the 14–3–3 family identified in this study; Table S15: Both keratin proteins types identified in our data; Table S16: List of heat-shocked proteins identified in our study; Table S17: The full Protein-Protein Interaction- STRING network in Figure S1.; Table S18: Top 20 in network ranked by Degree method; Figure S1: The full Protein-Protein Interaction- STRING network.

Author Contributions: Conceptualization, J.B., T.N.Y., R.R. and A.A.; methodology, J.B., S.L.C., P.T. and T.N.Y.; formal analysis, J.B., S.L.C., T.N.Y., P.T. and E.B.; writing—original draft preparation, J.B. and T.N.Y.; writing—review and editing, S.L.C., P.T., T.D.F., R.R., A.A. and T.D.F.; funding acquisition, J.B. and T.D.F. All authors have read and agreed to the published version of the manuscript.

Funding: This research received funding from DFO and MUN. Title—Protein Extractions of Atlantic Cod Otoliths. Contract No.—4500008253. Destination of Goods and Services: Fisheries and Oceans, Northwest Atlantic Fisheries Centre, East White Hills Road, A1C 5X1, St. John's, NL. Currency—CAD.

Institutional Review Board Statement: Not applicable.

Informed Consent Statement: Not applicable.

Data Availability Statement: The data presented in this study are available in the Supplementary Materials.

Acknowledgments: The authors would like to thank Atef Mansour, DFO NL. Science Branch, for his continuing encouragement and for providing financial support for this proteomics project. The particular otoliths used in this study were collected as part of the sentinel cod sampling program.

Conflicts of Interest: The authors declare no conflicts of interest.

References

1. Norin, T.; Canada, P.; Bailey, J.A.; Gamperl, A.K. Thermal biology and swimming performance of Atlantic cod (*Gadus morhua*) and haddock (*Melanogrammus aeglefinus*). *PeerJ* **2019**, *7*, e7784. [CrossRef]
2. Pinhorn, A.T. Fishery and biology of Atlantic cod (*Gadus morhua*) off the Southwest coast of Newfoundland. *J. Fish. Res. Board. Can.* **1969**, *26*, 3133–3164. [CrossRef]
3. Meidell, L.S.; Carvajal, A.K.; Rustad, T.; Falch, E. Upgrading marine oils from cod (*Gadus morhua*) on-board the deep-sea vessels-from waste to value. *Foods* **2023**, *12*, 1659. [CrossRef]
4. Hawkins, A.D.; Popper, A.N. Sound detection by Atlantic cod: An overview. *J. Acoust. Soc. Am.* **2020**, *148*, 3027. [CrossRef]
5. Dauphin, Y.; Dufour, E. Composition and properties of the soluble organic matrix of the otolith of a marine fish: *Gadus morhua Linne*, 1758 (*Teleostei, Gadidae*). *Comp. Biochem. Physiol. A Mol. Integr. Physiol.* **2003**, *134*, 551–561. [CrossRef]
6. Murayama, E.; Takagi, Y.; Ohira, T.; Davis, J.G.; Greene, M.I.; Nagasawa, H. Fish otolith contains a unique structural protein, otolin-1. *Eur. J. Biochem.* **2002**, *269*, 688–696. [CrossRef]
7. Lewis, L.S.; Huang, J.L.; Willmes, M.; Fichman, R.A.; Hung, T.-C.; Ellison, L.T.; Stevenson, T.A.; Teh, S.J.; Hammock, B.G.; Schultz, A.A.; et al. Visual, spectral, and microchemical quantification of crystalline anomalies in otoliths of wild and cultured delta smelt. *Sci. Rep.* **2022**, *12*, 1–12.
8. Degens, E.T.; Deuser, W.G.; Haedrich, R.L. Molecular structure and composition of fish otoliths. *Mar. Biol.* **1969**, *2*, 105–113. [CrossRef]
9. Payan, P.; De Pontual, H.; Bœuf, G.; Mayer-Gostan, N. Endolymph chemistry and otolith growth in fish. *Comptes Rendus. Palevol.* **2004**, *3*, 535–547. [CrossRef]
10. Ferri, J. Otoliths and Their Applications in Fishery Science. *Fishes* **2023**, *8*, 35. [CrossRef]
11. Smolinski, S.; Deplanque-Lasserre, J.; Hjorleifsson, E.; Geffen, A.J.; Godiksen, J.A.; Campana, S.E. Century-long cod otolith biochronology reveals individual growth plasticity in response to temperature. *Sci. Rep.* **2020**, *10*, 16708. [CrossRef]
12. Smoliński, S.; Mirny, Z. Otolith biochronology as an indicator of marine fish responses to hydroclimatic conditions and ecosystem regime shifts. *Ecol. Indic.* **2017**, *79*, 286–294. [CrossRef]

13. D'Iglio, C.; Albano, M.; Famulari, S.; Savoca, S.; Panarello, G.; Di Paola, D.; Perdichizzi, A.; Rinelli, P.; Lanteri, G.; Spano, N.; et al. Intra- and interspecific variability among congeneric Pagellus otoliths. *Sci. Rep.* **2021**, *11*, 16315. [CrossRef]
14. Santos, L.; Vaz-dos-Santos, A.M. Insights of otoliths morphology to reveal patterns of Teleostean fishes in the southern Atlantic. *Fishes* **2022**, *8*, 21. [CrossRef]
15. Lundberg, Y.W.; Xu, Y.; Thiessen, K.D.; Kramer, K.L. Mechanisms of otoconia and otolith development. *Dev. Dyn.* **2015**, *244*, 239–253. [CrossRef]
16. Serre, S.H.; Nielsen, K.E.; Thomsen, T.B.; Hüssy, K. Analysis of cod otolith microchemistry by continuous line transects using LA-ICP-MS. *Geol. Surv. Den. Greenl. Bull.* **2018**, *41*, 91–94. [CrossRef]
17. Rideout, R.M.; Youssef, T.N.; Adamack, A.T.; John, R.; Cohen, A.M.; Fridgen, T.D.; Banoub, J.H. Qualitative shotgun proteomics strategy for protein expression profiling of fish otoliths. *BioChem* **2023**, *3*, 102–117. [CrossRef]
18. Campana, S.E. Chemistry and composition of fish otoliths:pathways, mechanisms and applications. *Mar. Ecol. Prog. Ser.* **1999**, *188*, 263–297. [CrossRef]
19. Thomas, O.R.B.B.; Swearer, S.E.; Kapp, E.A.; Peng, P.; Tonkin-Hill, G.Q.; Papenfuss, A.; Roberts, A.; Bernard, P.; Roberts, B.R.; Tonkin-Hill, G.Q.; et al. The inner ear proteome of fish. *FEBS J.* **2019**, *286*, 66–81. [CrossRef]
20. Huang, S.; Qian, S. Advances in otolith-related protein research. *Front. Neurosci.* **2022**, *16*, 1–9.
21. Bielak, K.; Benkowska-Biernacka, D.; Ptak, M.; Stolarski, J.; Kalka, M.; Ozyhar, A.; Dobryszycki, P. Otolin-1, an otolith- and otoconia-related protein, controls calcium carbonate bioinspired mineralization. *Biochim. Biophys. Acta Gen. Subj.* **2023**, *1867*, 130327. [CrossRef]
22. Zhang, Y.; Fonslow, B.R.; Shan, B.; Baek, M.C.; Yates, J.R., 3rd. Protein analysis by shotgun/bottom-up proteomics. *Chem. Rev.* **2013**, *113*, 2343–2394. [CrossRef]
23. Wolters, D.A.; Washburn, M.P.; Yates, J.R. An automated multidimensional protein identification technology for shotgun proteomics. *Anal. Chem.* **2001**, *73*, 5683–5690. [CrossRef]
24. Yates, J.R., 3rd. Mass spectral analysis in proteomics. *Annu. Rev. Biophys. Biomol. Struct.* **2004**, *33*, 297–316. [CrossRef]
25. Link, A.J.; Eng, J.; Schieltz, D.M.; Carmack, E.; Mize, G.J.; Morris, D.R.; Garvik, B.M.; Yates, J.R., 3rd. Direct analysis of protein complexes using mass spectrometry. *Nat. Biotechnol.* **1999**, *17*, 676–682. [CrossRef]
26. Yates, J.R., 3rd. Mass spectrometry and the age of the proteome. *J. Mass. Spectrom.* **1998**, *33*, 1–19. [CrossRef]
27. Banoub, J.H.; Youssef, T.; Mikhael, A.; Mikhae, A. Proteomic technology applications for fisheries. *Tech. Report. Fish. Aquat. Sci.* **2022**, *3465*, xi + 668 p.
28. Valikangas, T.; Suomi, T.; Elo, L.L. A systematic evaluation of normalization methods in quantitative label-free proteomics. *Brief. Bioinform.* **2018**, *19*, 1–11. [CrossRef]
29. Wickham, H.; Henry, L. tidyr: Tidy Messy Data. R Package Version 1.1.3. 2021. Available online: https://CRAN.R-project.org/package=tidyr (accessed on 14 May 2024).
30. Bache, S.M.; Wickham, H. magrittr: A Forward-Pipe Operator for R. R Package Version 2.0.1. 2021. Available online: https://CRAN.R-project.org/package=magrittr (accessed on 14 May 2024).
31. Wickham, H. The Split-Apply-Combine Strategy for Data Analysis. *J. Stat. Softw.* **2011**, *40*, 1–29. [CrossRef]
32. Wickham, H.; François, R.; Henry, L.; Müller, K. dplyr: A Grammar of Data Manipulation. R Package Version 1.1.4. 2023. Available online: https://github.com/tidyverse/dplyr (accessed on 14 May 2024).
33. Hadley, W. *ggplot2: Elegant Graphics for Data Analysis*; Springer: New York, NY, USA, 2016.
34. Tomanek, L. Variation in the heat shock response and its implication for predicting the effect of global climate change on species' biogeographical distribution ranges and metabolic costs. *J. Exp. Biol.* **2010**, *213*, 971–979. [CrossRef]
35. Litscher, E.S.; Wassarman, P.M. The fish egg's zona pellucida. *Curr. Top. Dev. Biol.* **2018**, *130*, 275–305.
36. Nathaniel Clarke, D.; Lowe, C.J.; James Nelson, W. The cadherin-catenin complex is necessary for cell adhesion and embryogenesis in Nematostella vectensis. *Dev. Biol.* **2019**, *447*, 170–181. [CrossRef]
37. Hevrøy, E.M.; Jordal, A.E.O.; Hordvik, I.; Espe, M.; Hemre, G.I.; Olsvik, P.A. Myosin heavy chain mRNA expression correlates higher with muscle protein accretion than growth in Atlantic salmon, Salmo salar. *Aquaculture* **2006**, *252*, 453–461. [CrossRef]
38. Casas, L.; Saborido-Rey, F.; Ryu, T.; Michell, C.; Ravasi, T.; Irigoien, X. Sex change in Clownfish: Molecular insights from transcriptome analysis. *Sci. Rep.* **2016**, *6*, 35461. [CrossRef]
39. Cohen, A.M.; Mansour, A.A.H.; Banoub, J.H. 'De novo' sequencing of Atlantic cod vitellogenin tryptic peptides by matrix-assisted laser desorption/ionization quadrupole time-of-flight tandem mass spectrometry: Similarities with haddock vitellogenin. *Rapid Commun. Mass. Spectrom.* **2005**, *19*, 2454–2460. [CrossRef]
40. Liu, W.B.; Yan, Q.; Liu, F.Y.; Tang, X.C.; Chen, H.G.; Liu, J.; Nie, L.; Zhang, X.W.; Ji, W.K.; Hu, X.H.; et al. Protein serine/threonine phosphotase-1 is essential in governing normal development of vertebrate eye. *Curr. Mol. Med.* **2012**, *12*, 1361–1371. [CrossRef]
41. Schärer, O.D. XPG: Its Products and Biological Roles. In *Molecular Mechanisms of Xeroderma Pigmentosum*; Advances in Experimental Medicine and Biology; Springer: New York, NY, USA, 2009; pp. 83–92.
42. Pancho, A.; Aerts, T.; Mitsogiannis, M.D.; Seuntjens, E. Protocadherins at the crossroad of signaling pathways. *Front. Mol. Neurosci.* **2020**, *13*, 117. [CrossRef]
43. Pottmeier, P.; Nikolantonaki, D.; Lanner, F.; Peuckert, C.; Jazin, E. Sex-biased gene expression during neural differentiation of human embryonic stem cells. *Front. Cell Dev. Biol.* **2024**, *12*, 1341373. [CrossRef]

44. Tonks, N.K. Protein tyrosine phosphatases: From genes, to function, to disease. *Nat. Rev. Mol. Cell Biol.* **2006**, *7*, 833–846. [CrossRef]
45. Lenoir, M.; Kufareva, I.; Abagyan, R.; Overduin, M. Membrane and protein interactions of the Pleckstrin homology domain superfamily. *Membr.* **2015**, *5*, 646–663. [CrossRef]
46. Staverosky, J.A.; Pryce, B.A.; Watson, S.S.; Schweitzer, R. Tubulin polymerization-promoting protein family member 3, Tppp3, is a specific marker of the differentiating tendon sheath and synovial joints. *Dev. Dyn.* **2009**, *238*, 685–692. [CrossRef]
47. Orosz, F. On the tubulin polymerization promoting proteins of zebrafish. *Biochem. Biophys. Res. Commun.* **2015**, *457*, 267–272. [CrossRef]
48. Noiges, R.; Eichinger, R.; Kutschera, W.; Fischer, I.; Nemeth, Z.; Wiche, G.; Propst, F. Microtubule-associated protein 1A (MAP1A) and MAP1B: Light chains determine distinct functional properties. *J. Neurosci.* **2002**, *22*, 2106–2114. [CrossRef]
49. Sveinsdóttir, H.; Vilhelmsson, O.; Gudmundsdóttir, Á. Proteome analysis of abundant proteins in two age groups of early Atlantic cod (*Gadus morhua*) larvae. *Comp. Biochem. Physiol. Part. D Genom. Proteom.* **2008**, *3*, 243–250. [CrossRef]
50. Focant, B.; Huriaux, F.; Baras, E.; Vandewalle, P. Expression of myofibrillar proteins and parvalbumin isoforms in white muscle of the developing turbot Scophthalmus maximus (*Pisces, Pleuronectiformes*). *Basic. Appl. Myol.* **2000**, *10*, 269–278.
51. Huriaux, F.; Baras, E.; Vandewalle, P.; Focant, B. Expression of myofibrillar proteins and parvalbumin isoforms in white muscle of dorada during development. *J. Fish. Biol.* **2003**, *62*, 774–792. [CrossRef]
52. Perry, S.V. Vertebrate tropomyosin: Distribution, properties and function. *J. Muscle Res. Cell Motil.* **2001**, *22*, 5–49. [CrossRef]
53. Oehlmann, V.D.; Berger, S.; Sterner, C.; Korsching, S.I. Zebrafish beta tubulin 1 expression is limited to the nervous system throughout development, and in the adult brain is restricted to a subset of proliferative regions. *Gene Expr. Patterns* **2004**, *4*, 191–198.
54. Meek, S.E.; Lane, W.S.; Piwnica-Worms, H. Comprehensive proteomic analysis of interphase and mitotic 14-3-3-binding proteins. *J. Biol. Chem.* **2004**, *279*, 32046–32054. [CrossRef]
55. Van Hemert, M.J.; Steensma, H.Y.; van Heusden, G.P. 14-3-3 proteins: Key regulators of cell division, signalling and apoptosis. *Bioessays* **2001**, *23*, 936–946. [CrossRef]
56. Suzuki, K.; Sato, K.; Katsu, K.; Hayashita, H.; Kristensen, D.B.; Yoshizato, K. Novel Rana keratin genes and their expression during larval to adult epidermal conversion in bullfrog tadpoles. *Differentiation* **2001**, *68*, 44–54. [CrossRef]
57. Suzuki, K.; Utoh, R.; Kotani, K.; Obara, M.; Yoshizato, K. Lineage of anuran epidermal basal cells and their differentiation potential in relation to metamorphic skin remodeling. *Dev. Growth Differ.* **2002**, *44*, 225–238. [CrossRef]
58. Ishida, Y.; Suzuki, K.; Utoh, R.; Obara, M.; Yoshizato, K. Molecular identification of the skin transformation center of anuran larval skin using genes of Rana adult keratin (RAK) and SPARC as probes. *Dev. Growth Differ.* **2003**, *45*, 515–526. [CrossRef]
59. Jeyachandran, S.; Chellapandian, H.; Park, K.; Kwak, I.-S. A Review on the involvement of heat shock proteins (Extrinsic Chaperones) in response to stress conditions in aquatic organisms. *Antioxidants* **2023**, *12*, 1444. [CrossRef]
60. Xie, Y.; Song, L.; Weng, Z.; Liu, S.; Liu, Z. Hsp90, Hsp60 and sHsp families of heat shock protein genes in channel catfish and their expression after bacterial infections. *Fish. Shellfish. Immunol.* **2015**, *44*, 642–651. [CrossRef]
61. Song, L.; Li, C.; Xie, Y.; Liu, S.; Zhang, J.; Yao, J.; Jiang, C.; Li, Y.; Liu, Z. Genome-wide identification of Hsp70 genes in channel catfish and their regulated expression after bacterial infection. *Fish. Shellfish. Immunol.* **2016**, *49*, 154–162. [CrossRef]
62. Chiosis, G.; Digwal, C.S.; Trepel, J.B.; Neckers, L. Structural and functional complexity of HSP90 in cellular homeostasis and disease. *Nat. Rev. Mol. Cell Biol.* **2023**, *24*, 797–815. [CrossRef]
63. Didenko, T.; Duarte, A.M.; Karagoz, G.E.; Rudiger, S.G. Hsp90 structure and function studied by NMR spectroscopy. *Biochim. Biophys. Acta* **2012**, *1823*, 636–647. [CrossRef]
64. Street, T.O.; Lavery, L.A.; Agard, D.A. Substrate binding drives large-scale conformational changes in the Hsp90 molecular chaperone. *Mol. Cell* **2011**, *42*, 96–105. [CrossRef]
65. Rosenzweig, R.; Nillegoda, N.B.; Mayer, M.P.; Bukau, B. The Hsp70 chaperone network. *Nat. Rev. Mol. Cell Biol.* **2019**, *20*, 665–680. [CrossRef]
66. Roberts, R.J.; Agius, C.; Saliba, C.; Bossier, P.; Sung, Y.Y. Heat shock proteins (chaperones) in fish and shellfish and their potential role in relation to fish health: A review. *J. Fish. Dis.* **2010**, *33*, 789–801. [CrossRef]
67. Duan, Y.; Tang, H.; Mitchell-Silbaugh, K.; Fang, X.; Han, Z.; Ouyang, K. Heat shock protein 60 in cardiovascular physiology and diseases. *Front. Mol. Biosci.* **2020**, *7*, 73. [CrossRef]
68. Song, L.; Zhang, J.; Li, C.; Yao, J.; Jiang, C.; Li, Y.; Liu, S.; Liu, Z. Genome-wide identification of hsp40 genes in channel catfish and their regulated expression after bacterial infection. *PLoS ONE* **2014**, *9*, e115752. [CrossRef]
69. Hori, T.S.; Gamperl, A.K.; Afonso, L.O.; Johnson, S.C.; Hubert, S.; Kimball, J.; Bowman, S.; Rise, M.L. Heat-shock responsive genes identified and validated in Atlantic cod (*Gadus morhua*) liver, head kidney and skeletal muscle using genomic techniques. *BMC Genom.* **2010**, *11*, 72. [CrossRef]
70. Gollock, M.J.; Currie, S.; Petersen, L.H.; Gamperl, A.K. Cardiovascular and haematological responses of Atlantic cod (*Gadus morhua*) to acute temperature increase. *J. Exp. Biol.* **2006**, *209*, 2961–2970. [CrossRef]
71. Perez-Casanova, J.C.; Rise, M.L.; Dixon, B.; Afonso, L.O.; Hall, J.R.; Johnson, S.C.; Gamperl, A.K. The immune and stress responses of Atlantic cod to long-term increases in water temperature. *Fish. Shellfish. Immunol.* **2008**, *24*, 600–609. [CrossRef]
72. Kassahn, K.S.; Caley, M.J.; Ward, A.C.; Connolly, A.R.; Stone, G.; Crozier, R.H. Heterologous microarray experiments used to identify the early gene response to heat stress in a coral reef fish. *Mol. Ecol.* **2007**, *16*, 1749–1763. [CrossRef]

73. Buckley, B.A.; Gracey, A.Y.; Somero, G.N. The cellular response to heat stress in the gobyGillichthys mirabilis: A cDNA microarray and protein-level analysis. *J. Exp. Biol.* **2006**, *209*, 2660–2677. [CrossRef]
74. Podrabsky, J.E.; Somero, G.N. Changes in gene expression associated with acclimation to constant temperatures and fluctuating daily temperatures in an annual killifish Austrofundulus limnaeus. *J. Exp. Biol.* **2004**, *207*, 2237–2254. [CrossRef]
75. Basu, N.; Todgham, A.E.; Ackerman, P.A.; Bibeau, M.R.; Nakano, K.; Schulte, P.M.; Iwama, G.K. Heat shock protein genes and their functional significance in fish. *Gene* **2002**, *295*, 173–183. [CrossRef]
76. Yiangou, M.; Paraskeva, E.; Hsieh, C.C.; Markou, E.; Victoratos, P.; Scouras, Z.; Papaconstantinou, J. Induction of a subgroup of acute phase protein genes in mouse liver by hyperthermia. *Biochim. Biophys. Acta* **1998**, *1396*, 191–206. [CrossRef]
77. Gross-Sorokin, M.Y.; Roast, S.D.; Brighty, G.C. Assessment of feminization of male fish in English rivers by the Environment Agency of England and Wales. *Environ. Health Perspect.* **2006**, *114* (Suppl. S1), 147–151. [CrossRef]
78. Mills, L.J.; Gutjahr-Gobell, R.E.; Horowitz, D.B.; Denslow, N.D.; Chow, M.C.; Zaroogian, G.E. Relationship between reproductive success and male plasma vitellogenin concentrations in cunner, Tautogolabrus adspersus. *Env. Health Perspect.* **2003**, *111*, 93–100. [CrossRef]
79. Soffker, M.; Tyler, C.R. Endocrine disrupting chemicals and sexual behaviors in fish-a critical review on effects and possible consequences. *Crit. Rev. Toxicol.* **2012**, *42*, 653–668. [CrossRef]
80. Montjean, D.; Neyroud, A.S.; Yefimova, M.G.; Benkhalifa, M.; Cabry, R.; Ravel, C. Impact of endocrine disruptors upon non-genetic inheritance. *Int. J. Mol. Sci.* **2022**, *23*, 3350. [CrossRef]
81. Cohen, A.M.; Mansour, A.A.; Banoub, J.H. Absolute quantification of Atlantic salmon and rainbow trout vitellogenin by the 'signature peptide' approach using electrospray ionization QqToF tandem mass spectrometry. *J. Mass. Spectrom.* **2006**, *41*, 646–658. [CrossRef]
82. Cohen, A.M.; Jahouh, F.; Sioud, S.; Rideout, R.M.; Morgan, M.J.; Banoub, J.H. Quantification of Greenland halibut serum vitellogenin: A trip from the deep sea to the mass spectrometer. *Rapid Commun. Mass. Spectrom.* **2009**, *23*, 1049–1060. [CrossRef]
83. Banoub, J.; Thibault, P.; Mansour, A.; Cohen, A.; Heeley, D.H.; Jackman, D. Characterisation of the intact rainbow trout vitellogenin protein and analysis of its derived tryptic and cyanogen bromide peptides by matrix-assisted laser desorption/ionisation time-of-flight-mass spectrometry and electrospray ionisation quadrupole/time. *Eur. J. Mass. Spectrom.* **2003**, *9*, 509–524. [CrossRef]
84. Cohen, A.M.; Banoub, J.H. Application of Mass Spectrometry for the Analysis of Vitellogenin, a Unique Biomarker for Xenobiotic Compounds. *NATO Sci. Peace Secur. Ser. A: Chem. Biol.* **2010**, 301–318. [CrossRef]
85. Katahira, J. Nuclear export of messenger RNA. *Genes* **2015**, *6*, 163–184. [CrossRef]
86. Zhao, H.; Chiaro, C.R.; Zhang, L.; Smith, P.B.; Chan, C.Y.; Pedley, A.M.; Pugh, R.J.; French, J.B.; Patterson, A.D.; Benkovic, S.J. Quantitative analysis of purine nucleotides indicates that purinosomes increase de novo purine biosynthesis. *J. Biol. Chem.* **2015**, *290*, 6705–6713. [CrossRef]
87. Tanaka, K. The proteasome: Overview of structure and functions. *Proc. Jpn. Acad. Ser. B Phys. Biol. Sci.* **2009**, *85*, 12–36. [CrossRef]
88. Gregory, B.; Rahman, N.; Bommakanti, A.; Shamsuzzaman, M.; Thapa, M.; Lescure, A.; Zengel, J.M.; Lindahl, L. The small and large ribosomal subunits depend on each other for stability and accumulation. *Life Sci. Alliance* **2019**, *2*, e201800150. [CrossRef]
89. Dreyfuss, G.; Philipson, L.; Mattaj, I.W. Ribonucleoprotein particles in cellular processes. *J. Cell Biol.* **1988**, *106*, 1419–1425. [CrossRef]
90. Moreno, A.; Taleb, V.; Sebastian, M.; Anoz-Carbonell, E.; Martinez-Julvez, M.; Medina, M. Cofactors and pathogens: Flavin mononucleotide and flavin adenine dinucleotide (FAD) biosynthesis by the FAD synthase from Brucella ovis. *IUBMB Life* **2022**, *74*, 655–671. [CrossRef]
91. Zhang, S.; Zeng, X.; Ren, M.; Mao, X.; Qiao, S. Novel metabolic and physiological functions of branched chain amino acids: A review. *J. Anim. Sci. Biotechnol.* **2017**, *8*, 10. [CrossRef]
92. Moore, M.N.; Koehler, A.; Lowe, D.; Viarengo, A. Chapter thirty-three lysosomes and autophagy in aquatic animals. In *Methods in Enzymology*; Academic Press: Cambridge, MA, USA, 2008; pp. 581–620.
93. Bonam, S.R.; Wang, F.; Muller, S. Lysosomes as a therapeutic target. *Nat. Rev. Drug Discov.* **2019**, *18*, 923–948. [CrossRef]
94. Hu, G.; Shu, Y.; Luan, P.; Zhang, T.; Chen, F.; Zheng, X. Genomic analysis of the proteasome subunit gene family and their response to high density and saline-alkali stresses in grass carp. *Fishes* **2022**, *7*, 350. [CrossRef]
95. Sharkey, T.D. Pentose phosphate pathway reactions in photosynthesizing cells. *Cells* **2021**, *10*, 1547. [CrossRef]
96. Li, S.; Liu, Y.; Li, B.; Ding, L.; Wei, X.; Wang, P.; Chen, Z.; Han, S.; Huang, T.; Wang, B.; et al. Physiological responses to heat stress in the liver of rainbow trout (*Oncorhynchus mykiss*) revealed by UPLC-QTOF-MS metabolomics and biochemical assays. *Ecotoxicol. Environ. Saf.* **2022**, *242*, 113949. [CrossRef]
97. Varki, A. Biological roles of glycans. *Glycobiology* **2017**, *27*, 3–49. [CrossRef]
98. Wang, W.; Nag, S.; Zhang, X.; Wang, M.H.; Wang, H.; Zhou, J.; Zhang, R. Ribosomal proteins and human diseases: Pathogenesis, molecular mechanisms, and therapeutic implications. *Med. Res. Rev.* **2015**, *35*, 225–285. [CrossRef]
99. Pattin, K.A.; Moore, J.H. Role for protein-protein interaction databases in human genetics. *Expert. Rev. Proteom.* **2009**, *6*, 647–659. [CrossRef]
100. Wimalagunasekara, S.S.; Weeraman, J.; Tirimanne, S.; Fernando, P.C. Protein-protein interaction (PPI) network analysis reveals important hub proteins and sub-network modules for root development in rice (Oryza sativa). *J. Genet. Eng. Biotechnol.* **2023**, *21*, 69. [CrossRef]

101. Cyr, F.; Snook, S.; Bishop, C.; Galbraith, P.S.; Pye, B.; Chen, N.; Han, G. Physical oceanographic conditions on the Newfoundland and Labrador shelf during 2019. *DFO Can. Sci. Advis. Sec. Res. Doc.* **2021**, iv + 54. Available online: https://waves-vagues.dfo-mpo.gc.ca/library-bibliotheque/40960754.pdf (accessed on 14 May 2024).
102. Cyr, F.; Colbourne, E.; Holden, J.; Snook, S.; Han, G.; Chen, N.; Bailey, W.; Higdon, J.; Lewis, S.; Pye, B.; et al. Physical oceanographic conditions on the Newfoundland and Labrador shelf during 2017. *DFO Can. Sci. Advis. Sec. Res.* **2019**, iv + 58. Available online: https://waves-vagues.dfo-mpo.gc.ca/library-bibliotheque/40888113.pdf (accessed on 14 May 2024).
103. Cyr, F.; Galbraith, P.S. A climate index for the Newfoundland and Labrador shelf. *Earth Syst. Sci. Data* **2021**, *13*, 1807–1828. [CrossRef]
104. Alexander, M.A.; Shin, S.I.; Scott, J.D.; Curchitser, E.; Stock, C. The response of the Northwest Atlantic ocean to climate change. *J. Clim.* **2020**, *33*, 405–428. [CrossRef]

BioChem

Article

Anti-Müllerian Hormone Serum Levels as Biomarker of Ovarian Reserve in Adult Women with Juvenile Idiopathic Arthritis Treated with csDMARDs and/or bDMARDs: A Pilot Study

Clara Di Mario [1,†], Maria Rita Gigante [2,†], Angelina Barini [3], Luca Petricca [2], Antonella Barini [3], Antonio Bianchi [4,5], Stefano Alivernini [1,2,6], Barbara Tolusso [1] and Elisa Gremese [1,6,*]

1 Immunology Research Core Facility, Gemelli Science and Technology Park, Fondazione Policlinico Universitario A. Gemelli IRCCS, 00168 Rome, Italy
2 Division of Rheumathology, Fondazione Policlinico Universitario A. Gemelli IRCCS, 00168 Rome, Italy
3 Chemistry, Biochemistry and Molecular Biology Clinic, Fondazione Policlinico Universitario A. Gemelli IRCCS, 00168 Rome, Italy
4 Endocrinology and Diabetes Department, Fondazione Policlinico Universitario A. Gemelli, IRCCS, 00168 Rome, Italy
5 Department of Translational Medicine and Surgery, Catholic University of the Sacred Heart, 00168 Rome, Italy
6 Clinical Immunology Unit, Fondazione Policlinico Universitario A. Gemelli-IRCCS, 00168 Rome, Italy
* Correspondence: elisa.gremese@unicatt.it
† These authors contributed equally to this work.

Abstract: Background/Objectives: Juvenile idiopathic arthritis (JIA) is a chronic childhood disease that often persists into the reproductive years. JIA may impact long-term fertility due to the prolonged exposure to immunosuppressive therapies. **Methods:** A total of 35 adult JIA female patients of childbearing age and 20 age-matched healthy controls were studied to test their anti-Müllerian hormone (AMH) serum levels as a biomarker of ovarian reserve. Demographic characteristics, disease duration, previous and current treatments, disease activity (DAS44), and a health assessment questionnaire (HAQ) were recorded. **Results:** JIA patients had a mean age of 22.3 ± 2.9 years, a disease duration of 12.3 ± 6.1 years, and a DAS44 of 1.24 ± 0.61. No differences were found in AMH serum levels between JIA and controls (5.78 ± 2.37 ng/mL vs. 6.60 ± 2.68 ng/mL, respectively; $p = 0.17$). Among the patients, 22 (62.9%) were receiving a stable dose of methotrexate (MTX) and 19 (54.3%) a dose of TNFα inhibitors. No difference in AMH serum levels was observed between JIA patients who were or were not exposed to MTX ($p = 0.29$) or to TNFα inhibitors ($p = 0.50$). **Conclusions:** Ovarian reserve as assessed by AMH serum levels appears to be comparable between those with JIA and age-matched controls and does not appear to be influenced by disease characteristics or prior/concomitant exposure to immunosuppressive drugs.

Keywords: juvenile idiopathic arthritis; anti-Müllerian hormone; immunosuppressive drugs

1. Introduction

Juvenile Idiopathic Arthritis (JIA) refers to all forms of chronic arthritis of unknown origin lasting for at least 6 weeks in children and adolescents [1,2], being the most common arthropathy of childhood. It affects hundreds of thousands of children worldwide, representing a leading cause of short- and long-term disability [2–4]. Disease onset is usually before 16 years of age, and the disease often persists into reproductive age [2]. Therefore, in this category of patients, fertility preservation should be considered among the priority outcomes, as several studies have shown that patients with JIA or childhood-onset Systemic Lupus Erythematosus (SLE) had an increased risk of decreased reproductive potential as a consequence of their chronic inflammatory condition or drug exposure. However, an estimate of the actual risk of infertility in such patients is still unknown [5]. Fertility has been found to be unaffected in women with JIA [5], but it may be potentially compromised by physical, hormonal, or immunological mechanisms, as well as pharmacological treatments such as prolonged exposure to immunosuppressive therapy in young female patients [6,7]. Conversely, several studies have investigated fertility in female rheumatoid arthritis (RA) patients, particularly in a case-control study with age-matched controls, reporting no relationship between RA and nulliparity, infertility, oral contraceptive use, and adverse pregnancy outcomes [8].

Multiple autoimmune mechanisms are involved in the decrease of the ovarian follicle pool, resulting in premature ovarian failure (POF) (i.e., failure of ovarian function before the age of 40) [9,10]; in fact, the presence of antiovarian antibodies has been reported in premature ovarian aging [11,12]. As previously observed in SLE patients, ovarian function can be impaired by the presence of antibodies directed against the corpus luteum (anti-CoL) [13] and can be decreased by autoimmune oophoritis, leading to a reduction in hormones involved in the ovarian reserve [14]. In JIA, anti-CoL antibodies have rarely been detected, but a decrease in hormones related to ovarian function and reserve has been observed [15].

The quantity and quality of the ovarian follicle pool represent the ovarian reserve; they decay with age, resulting in a decrease in the woman's reproductive functions [16]. There is considerable individual variability in the age of menopause and subfertility [17], so chronological age is not a good predictor of ovarian reserve. To date, there is no proven marker directly related to ovarian follicle number, although serum levels of follicle-stimulating hormone (FSH), inhibin B, estradiol, and anti-Müllerian hormone (AMH) and the assessment of the ovarian antral follicle count (AFC) by transvaginal ultrasound evaluation have emerged as interesting biomarkers for the prediction of ovarian reserve [18].

These tests have some restrictions, which include lack of sensitivity and, in some cases, their dependence on the phase of menstrual cycle [19–21]. In a recent set of recommendations, the Practice Committee of the American Society for Reproductive Medicine stated that combined ovarian reserve test models do not consistently improve predictive ability more than a single ovarian reserve test, suggesting measurement of AMH as a reliable screening test for poor ovarian reserve [19]. In this context, several studies have shown a good predictive value in the evaluation of the ovarian follicle pool by serum levels of AFC and AMH, better than FSH [20,21]. AMH is a growth and differentiation factor of the transforming growth factor β family, produced in the granulosa cells of preantral and small antral follicles [22]. In women, AMH serum levels are undetectable at birth, reaching a maximum level after puberty and then progressively decreasing during reproductive life, until AMH becomes undetectable in menopause [23]. Moreover, it has been observed that the phase of the menstrual cycle does not influence AMH serum levels [24,25]. All these aspects make AMH a good predictive biomarker that reflects well the ovarian reserve over time.

To date, multiple pieces of evidence from the cancer field have demonstrated that AMH serum levels decrease during and after chemotherapy treatment in women, indicating AMH as an early and sensitive biomarker of ovarian reserve damage [26,27].

In recent years, several studies have been published on AMH as a biomarker of ovarian reserve in autoimmune rheumatic diseases [28–39] and it has been reported that chronic inflammatory joint diseases, such as RA and spondyloarthritis (SpA), negatively impact the ovarian reserve [34,37]. In JIA adult patients, a significant decrease in AMH levels appears to be unaffected by immunosuppressive therapies like methotrexate (MTX) or biological agents [15].

The aims of this study were (i) to evaluate AMH serum levels in a cohort of JIA adult women of childbearing age compared with age-matched healthy subjects, and (ii) to assess the influence of previous exposure to conventional disease-modifying anti-rheumatic drugs (cDMARDs) and/or biological DMARDs (bDMARDs), as well as of disease parameters on the ovarian reserve in adult women of childbearing age with JIA.

2. Materials and Methods

2.1. Subjects

Thirty-five consecutive female adult patients with a diagnosis of JIA according to ILAR criteria [40] were enrolled in the study at the Fondazione Policlinico Universitario A. Gemelli-IRCCS in Rome. Each enrolled patient had to meet the following inclusion criteria: age between 18 and 30 years, regular menses with no diagnosis of ovarian failure, no current or previous exposure to cytotoxic therapy or radiation for cancer, no previous diagnosis of polycystic ovarian syndrome (PCOS) [41], and no use of oral contraceptives or current pregnancy. At study entry, each enrolled JIA patient underwent peripheral blood sampling for serum AMH determination (see below). Complete ovarian function was assessed during the early follicular phase of the menstrual cycle (between the first and fifth day of menses) with estradiol and follicle-stimulating hormone (FSH). Moreover, clinical and demographic characteristics, disease duration, prior and current therapies, 44-joint disease activity score (DAS), and a health assessment questionnaire (HAQ) were collected at the time of sampling. All participants were asked to answer questions about their lifestyle and their gynecological and obstetric history, including age at menarche, menstrual cycle duration, gynecological disorders, number of full-term pregnancies, and number of miscarriages. Twenty age-matched healthy controls were included as a comparison group. The subjects who participated in this study were all Caucasian women. The study protocol was approved by the Ethics Committee of the Università Cattolica del Sacro Cuore (Ethics Committee protocol number: 0047470/17—31 October 2017) and all subjects provided signed informed consent.

2.2. Determination of Ovarian Reserve: Anti-Müllerian Hormone (AMH) Assay

At study entry, each enrolled subject underwent peripheral blood sampling. The blood was collected in a serum separator tube (BD Vacutainer™ SST™ II Advance) containing spray-coated silica and a polymer gel for serum separation. After peripheral blood drawing, each sample was centrifuged at 300 g within 15 min of collection and stored at −80 °C until assayed [36]. Serum AMH levels were determined with a 2-stage enzyme-linked immunosorbent assay using a commercially bought kit (AMH Gen II ELISA, Beckman Coulter, Inc., Brea, CA, USA) according to the manufacturer's revised protocol [42,43]. Each AMH assessment was run in duplicate for each sample [37,38]. The intra-assay CV (coefficient of variation) was 3.7% (at 3.8 ng/mL) and 3.4% (at 16.4 ng/mL). The interassay CV was 4.4% (at 3.8 ng/mL) and 3.4% (at 16.4 ng/mL). The assay range was 0.4–21 ng/mL and the sensitivity was 0.08 ng/mL [36]. All experiments were performed in accordance with relevant guidelines and regulations.

2.3. Statistical Analysis

Data were analyzed using SPSS Statistics 22.0 (IBM, Armonk, NY, USA) and Prism software 8.0 (Graph-Pad, San Diego, CA, USA). Categorical and quantitative variables were registered as frequencies, percentage, mean ± standard deviation (SD) as appropriate. The nonparametric Mann–Whitney U test was used to compare the continuous variables. The nonparametric χ^2 test was used to assess the differential distribution of the categorical variables. The Spearman's rank correlation was used to analyze the relationship between AMH serum levels and inflammatory and clinical parameters. p-values < 0.05 were considered statistically significant.

Regarding sample size calculation, this prospective interventional cohort study aimed to evaluate the potential differences in serum AMH levels between young adult women with JIA compared to age-matched healthy controls as its primary endpoint. To date, knowledge about ovarian reserve in women with JIA is scarce, and only one study by Ferreira et al. [15] addressed this topic on a small sample of JIA patients, albeit without significant variation between patients and age-matched controls. The Practice Committee of the American Society for Reproductive Medicine (ASRM) and the Endocrine Society in 2020 [19,44–46] suggested measuring AMH as a reliable screening for poor ovarian reserve, providing 1.5–3.0 ng/mL as normal values up to 25 years, with mean values around 2.0 ± 0.4 ng/mL. Thus, assuming a 20% reduction in AMH serum levels in JIA patients, with a power of 80% (type II beta error = 20%) and type I alpha error of 5%, a 95% two-sided confidence interval and a case-to-control ratio of 2:1, a group sample size of 27 and 14 would achieve 81.2% power to detect the hypothesized difference. Hypothesizing a potential dropout of 30% due to laboratory testing issues, we planned to enroll 35 women with JIA (cases) and 20 age-matched controls, respectively. The sample size calculation was performed using PASS2022 statistical software 22.0.3 with a two-sided Mann–Whitney U test [47].

3. Results

3.1. Demographic and Clinical Characteristic of the Cohort of Adult Female JIA Patients of Childbearing Age

A total of 35 consecutive JIA patients and 20 age-matched controls were included in the analysis. Table 1 summarizes the study population characteristics. The mean age of patients at baseline was 22.3 ± 2.9 years, comparable to the control group (23.1 ± 1.9 years, $p = 0.07$). The mean age at disease onset was 9.5 ± 5.0 years and the mean disease duration at the time of peripheral blood drawing was 12.3 ± 6.1 years (Figure 1). Considering the disease phenotype, systemic onset was observed in four (11.4%) patients, oligoarticular onset in nine (25.7%), polyarticular onset in 20 (57.1%), enthesis-related arthritis in one (2.8%) and psoriatic arthritis in one patient (2.8%), respectively. At enrolment, the mean DAS was 1.24 ± 0.61 and the HAQ was 0.30 ± 0.67 (Table 1).

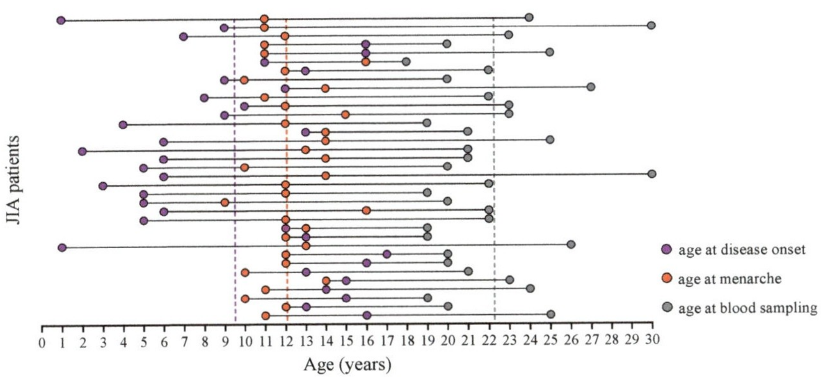

Figure 1. Scheme showing disease course and menarche timeline of individual adult JIA patients belonging to the enrolled study cohort.

Considering the previous and concomitant pharmacological treatments in the JIA cohort, 11 (31.4%) patients were taking low-dose prednisone (less than 7.5 mg/daily), 22 (62.9%) had previous exposure to methotrexate (MTX) for a mean period of 2.4 ± 3.4 years, and 19 (54.3%) were treated with anti-TNFα drugs for 2.1 ± 2.7 years during their disease course. Eight JIA (22.9%) patients were exposed to combination therapy (MTX + anti-TNFα) and ten (29.4%) to one or more other csDMARD (i.e., azathioprine, leflunomide, sulphasalazine or cyclosporine, respectively) (Table 1). Regarding ovarian function, no difference was found between JIA patients and healthy controls in FSH and Estradiol serum levels. As for their reproductive background, there were no differences between JIA adult patients compared to controls in mean age at menarche [12.1 ± 1.7 years for patients and 12.9 ± 1.6 years for the control group ($p = 0.21$)]; none of the patients and controls had had full-term pregnancy or miscarriage.

Table 1. Demographic and clinical characteristics of the adult female JIA cohort.

Demographic and Clinical Characteristics	JIA (n = 35)	Controls (n = 20)	p
Age, years	22.3 ± 2.9	23.1 ± 1.9	0.07
Age at disease onset, years	9.5 ± 5.0	-	-
Disease duration, years	12.3 ± 6.1	-	-
BMI (kg/m^2)	22.7 ± 5.2	22.1 ± 4.3	0.91
Smoking habit, n (%)	10 (28.6)	3 (13.6)	0.2
HAQ	0.30 ± 0.67	-	-
DAS44	1.24 ± 0.61	-	-
ESR, mm/1 h	17.4 ± 13.8	-	-
CRP, mg/L	3.9 ± 7.2	-	-
Systemic onset, n (%)	4 (11.4)	-	-
Oligoarticular onset, n (%)	9 (25.7)	-	-
Polyarticular onset, n (%)	20 (57.1)	-	-
Enthesitis-related arthritis, n (%)	1 (2.8)	-	-
Psoriatic arthritis, n (%)	1 (2.8)	-	-
Treatment (previous or current)			
MTX, n (%)	22 (62.9)	-	-
MTX treatment duration, years	2.4 ± 3.4	-	-
Anti-TNFα, n (%)	19 (54.3)	-	-
Anti-TNFα treatment duration, years	2.09 ± 2.71	-	-
MTX + anti-TNFα, n (%)	8 (22.9)	-	-
Other csDMARDs, n (%)	10/34 (29.4)	-	-
Prednisone (less than 7.5 mg/day)	11 (31.4%)	-	-
Gynecological and obstetric data			
Age at menarche, years	12.1 ± 1.7	12.9 ± 1.6	0.31
Subjects with full-term pregnancy, n (%)	0 (0)	0 (0)	-
Subjects with miscarriage, n (%)	0 (0)	0 (0)	-
FSH, mUI/mL	4.3 ± 2.4	4.9 ± 3.7	0.84
Estradiol pg/mL	97.9 ± 49.6	111.5 ± 46.8	0.44
AMH, ng/mL	5.8 ± 2.4	6.6 ± 2.8	0.17

Values are shown as mean ± SD, unless otherwise indicated. BMI: body mass index; DAS: disease activity score; ESR: erythrocyte sedimentation rate; CRP: C-reactive protein; MTX: methotrexate; csDMARDs: conventional synthetic disease-modifying antirheumatic drugs; FSH: follicle-stimulating hormone AMH: anti-Müllerian hormone.

3.2. *Association between AMH Serum Levels with Clinical Characteristic and Pharmacological Treatments in Adult Female JIA Patients Cohort*

Evaluation of AMH serum levels revealed that JIA patients did not differ from the control group (5.78 ± 2.37 ng/mL vs. 6.60 ± 2.68 ng/mL, respectively; $p = 0.17$) (Figure 2A). At disease onset, 12 (35.3%) JIA patients had already experienced menarche, with AMH serum levels similar between patients who experienced menarche before or after disease onset (5.28 ± 2.02 ng/mL and 6.17 ± 2.51 ng/mL; $p = 0.50$). Moreover, there was no correlation between baseline AMH serum levels and age (R = 0.09; $p = 0.61$), as well as age at disease onset (R = −0.37; $p = 0.83$) and disease duration (R = 0.03; $p = 0.88$) in the JIA cohort. Based on pharmacological treatments, JIA patients receiving MTX (cumulative MTX dose >5 g) showed similar AMH serum levels compared to JIA patients not receiving MTX (4.99 ± 2.30 vs. 6.24 ± 2.34 ng/mL in MTX users and MTX nonusers, respectively, $p = 0.29$) (Figure 2B). Similarly, JIA patients on anti-TNFα therapy did not show different AMH serum levels from those not using anti-TNFα therapy (6.14 ± 2.68 vs. 5.34 ± 1.93 ng/mL in anti-TNFα-exposed and anti-TNFα-unexposed, respectively, $p = 0.50$) (Figure 2C).

Looking at treatment exposure, no correlation was found between AMH serum levels and duration of MTX therapy (R = 0.24; $p = 0.16$) or anti-TNFα therapy (R = 0.33; $p = 0.06$).

Finally, no difference was found in AMH serum levels among JIA patients treated with MTX alone (n = 8; 6.08 ± 1.30 ng/mL, $p = 0.59$), anti-TNFα alone (n = 5; 5.60 ± 2.51 ng/mL; $p = 0.37$), and MTX + anti-TNFα combination (n = 8; 6.74 ± 2.60 ng/mL; $p = 0.91$) compared to controls (Figure 2D). Despite the limited number of patients, the results are in line with data already published in the literature. Furthermore, no correlations were found between

AMH serum levels and duration of exposure to MTX alone ($p = 0.10$) or anti-TNFα alone ($p = 0.08$), respectively.

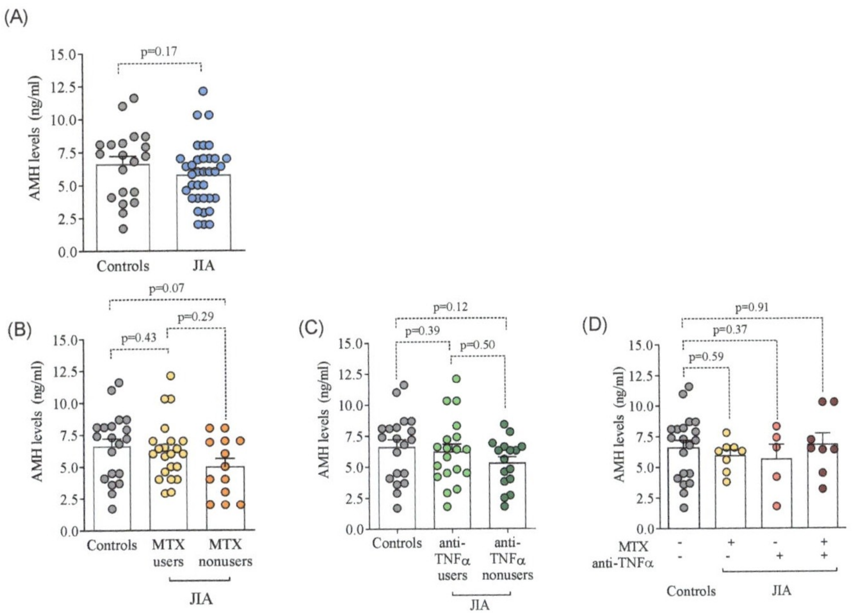

Figure 2. (**A–D**). Anti-Müllerian hormone serum levels in adult women JIA patients and healthy controls. (**A**) AMH serum levels in adult women JIA patients and age-matched controls enrolled in the study. (**B**,**C**) AMH serum levels in adult women JIA patients stratified based on MTX (**B**) and anti-TNFα exposure (**C**) compared to those that did not receive these drugs and to controls. (**D**) AMH serum levels in JIA patients treated with MTX, anti-TNFα and MTX/anti-TNFα combination, compared to controls. AMH: Anti-Müllerian hormone; TNF: tumor necrosis factor; JIA: juvenile idiopathic arthritis; MTX: methotrexate.

4. Discussion

In this pilot study, we evaluated the serum levels of AMH, a biomarker of ovarian reserve, in a cohort of adult women with JIA, during reproductive age, compared with age-matched healthy controls, to determine whether concomitant chronic inflammatory disease and/or exposure to conventional and/or biological DMARDs could have a detrimental effect on their ovarian reserve. We found that AMH serum levels in the JIA adult cohort did not differ from the aged-matched controls, regardless of prior or concomitant medications.

The AMH serum level has emerged as a good surrogate for estimating ovarian reserve, although recent data on decreased ovarian reserve as assessed by AMH in patients with rheumatic diseases have yielded some conflicting results [29–33]. In particular, in a study conducted on patients with early RA, Brouwer et al. [34] found a comparable concentration in AMH serum levels between the study cohort of RA patients and healthy controls at the time of diagnosis and at a six-month follow-up, as reported in long-standing RA patients [48], while Henes et al. found a significant reduction in ovarian reserve as measured by AMH level in RA patients compared with controls, as well as in spondyloarthritis and Behçet's disease patients [49].

To date, there are few data on the assessment of fertility in patients with JIA, in their postpubertal age, with AMH used as a biomarker of ovarian reserve [50]. Ferreira et al. documented a significantly lower ovarian reserve in JIA patients, not associated with the hypothalamic–pituitary–gonadal axis, compared to controls, deducing that the disease itself could have a negative influence on the patients' ovarian reserve [15]. In our study, we found no association between AMH serum levels and age in both JIA patients and healthy controls, and, in contrast to Ferreira et al., no significant difference in AMH serum levels was found between JIA patients and age-matched controls, suggesting a limited effect of this concomitant rheumatic disease on patients' ovarian reserve. This result could be explained by the narrow range of age distribution of the patients in our study and by the strict selection criteria that excluded patients undergoing cytotoxic drugs, which is relevant as it can influence the ovarian reserve parameters. Furthermore, focusing on the therapeutic profile, we found no significant differences in AMH serum levels among JIA patients exposed to MTX or bDMARDs alone or to MTX + anti-TNFα combination therapies compared to age-matched controls. These results are in agreement with a recent study in which no effect of biological drugs used to treat JIA was observed on AMH, FSH, LH, and estradiol levels even though the cohort of JIA patients was prepubertal and pubertal [50], thus suggesting the hypothesis that exposure to MTX or/and bDMARDs does not affect the ovarian reserve at each different woman's reproductive phase.

There are few data on the effects of csDMARDs and bDMARDs therapy on ovarian reserve. Scrivo et al. did not observe a change in AMH, FSH, and LH serum levels in SpA after 8 months of treatment with anti-TNFα, assuming that these drugs do not affect fertility [51]. It was observed that Balb/c mice exposed to MTX (5 g/m^2) could undergo destruction of primordial follicles [52], and Araujo and colleagues [53] observed that in adult patients with SLE diagnosed in childhood, a high cumulative dose of MTX could have a detrimental consequence on ovarian reserve, while no effect of MTX on ovarian reserve has been documented in JIA patients. However, as shown by Tomioka et al., in a cohort of 23 JIA patients, the use of nonsteroidal anti-inflammatory drugs (NSAIDs) may induce luteinized unruptured follicle (LUF) syndrome without compromising ovarian reserve [54].

The present study should be interpreted in the light of some considerations. The term JIA refers to a clinically heterogeneous group of arthritides with different clinical and patho-physiological manifestations, and the small sample size did not allow us to stratify AMH values according to the different clinical phenotypes of JIA. Another consideration is the stringent selection criteria of our cohort of JIA patients aged 18–30 years without a diagnosis of ovarian failure and the exclusion of cytotoxic therapies. In addition, this is a cross-sectional study with no data on AMH serum levels available prior to exposure to drugs administered during the JIA course.

In conclusion, in young adult women with JIA, the disease itself and the exposure to immunosuppressive therapies do not seem to compromise ovarian reserve as assessed by the AMH serum levels, and this may be relevant information also in light of the transition of JIA patients from pediatric age to adulthood [55]. Future prospective studies in larger cohorts of JIA patients are needed to determine whether long-term use of immunosuppressive therapy could influence ovarian reserve and to evaluate possible changes in AMH serum levels before and after exposure to conventional and biologic therapies, also in light of new treatments and strategies in disease management. Furthermore, observation of patients in a long-term follow-up, including data on pregnancies, is necessary to evaluate the impact on their future fertility.

Author Contributions: Conceptualization, C.D.M., M.R.G. and E.G.; Methodology, C.D.M., A.B. (Angelina Barini), A.B. (Antonella Barini) and B.T.; Formal analysis, C.D.M., M.R.G., L.P., S.A., B.T. and E.G.; Data curation, M.R.G., L.P. and S.A.; Writing—original draft, C.D.M., M.R.G., A.B. (Antonio Bianchi) and S.A.; Writing—review & editing, E.G.; Visualization, A.B. (Antonio Bianchi); Supervision, E.G. All authors have read and agreed to the published version of the manuscript.

Funding: This research received no external funding.

Institutional Review Board Statement: The study was conducted in accordance with the Declaration of Helsinki and approved by the Ethics Committee of the Fondazione Policlinico Universitario A. Gemelli (protocol code 0047470/17; date of approval: 31 October 2017).

Informed Consent Statement: Informed consent was obtained from all subjects involved in the study.

Data Availability Statement: Data is contained within the article.

Conflicts of Interest: The authors declare no conflicts of interest.

References

1. Giancane, G.; Consolaro, A.; Lanni, S.; Davì, S.; Schiappapietra, B.; Ravelli, A. Juvenile Idiopathic Arthritis: Diagnosis and Treatment. *Rheumatol. Ther.* **2016**, *3*, 187–207. [CrossRef] [PubMed]
2. Ravelli, A.; Marini, A. Juvenile Idiopathic Arthritis. *Lancet* **2007**, *369*, 767–778. [CrossRef] [PubMed]
3. Gremese, E.; Fedele, A.L.; Alivernini, S.; Ferraccioli, G. Ultrasound assessment as predictor of disease relapse in children and adults with arthritis in clinical stable remission: New findings but still unmet needs. *Ann. Rheum. Dis.* **2018**, *77*, 1391–1393. [CrossRef] [PubMed]
4. Gualano, B.; Bonfa, E.; Pereira, R.M.; Silva, C.A. Physical activity for paediatric rheumatic diseases: Standing up against old paradigms. *Nat. Rev. Rheumatol.* **2017**, *13*, 368–379. [CrossRef]
5. Hersh, A.; von Scheven, E.; Yelin, E. Adult outcomes of childhood-onet rheumatic disease. *Nat. Rev. Rheumatol.* **2011**, *75*, 290–295. [CrossRef]
6. Ostensen, M.; Almberg, K.; Koksvik, H.S. Sex, reproduction, and gynecological disease in young adults with a history of juvenile chronic arthritis. *J. Rheumatol.* **2000**, *27*, 1783–1787.
7. Ostensen, M. New insights into sexual functioning and fertility in rheumatic diseases. *Best. Pract. Res. Clin. Rheumatol.* **2004**, *18*, 219–232. [CrossRef]
8. Leroy, C. Immunosuppressive drugs and fertility. *Orph. J. Rare Dis.* **2015**, *10*, 136. [CrossRef]
9. Pope, J.E.; Bellamy, N.; Stevens, A. The lack of associations between rheumatoid arthritis and both nulliparity and infertility. *Semin. Arthritis. Rheum.* **1999**, *28*, 342–350. [CrossRef]
10. Hoek, A.; Schoemaker, J.; Drexhage, H.A. Premature ovarian failureand ovarian autoimmuity. *Endocr. Rev.* **1997**, *18*, 107–134.
11. Aysha Vandana, J.; Deepti, G. Premature Ovarian Failure: An Association with Autoimmune Diseases. *J. Clin. Diag. Res.* **2016**, *10*, QC10–QC12.
12. Nacak, G.B.; Ozkaya, E.; Abide, C.Y.; Bilgic, B.E.; Devranoglu, B.; Iscan, R.C. The impact of autoimmunity-related early ovarian aging on ICSI cycle outcome. *Gynecol. Endocrinol.* **2018**, *4*, 940–943. [CrossRef] [PubMed]
13. Pasoto, S.G.; Viana, V.S.; Mendonca, B.B.; Yoshinari, N.H.; Bonfa, E. Anti-corpus luteum antibody: A novel serological marker for ovarian dysfunction in systemic lupus erythematosus? *J. Rheumatol.* **1999**, *26*, 1087–1093. [PubMed]
14. Carp, H.J.A.; Selmi, C.; Shoenfeld, Y. The autoimmune bases of infertility and pregnancy loss. *J. Autoimm.* **2012**, *38*, 266–274. [CrossRef]
15. Ferreira, G.R.V.; Tomioka, R.B.; Aikawa, N.E.; Leon, E.P.; Maciel, G.A.R.; Serafini, P.C.; Baracat, E.C.; Goldenstein-Schainberg, C.; Pereira, R.M.R.; Bonfá, E.; et al. Ovarian Reserve in Young Juvenile Idiopathic Arthritis Patients. *Mod. Rheum.* **2019**, *29*, 447–451. [CrossRef]
16. te Velde, E.R.; Scheffer, G.J.; Dorland, M.; Broekmans, F.J.; Fauser, B.C. Developmental and endocrine aspects of a normal ovarian aging. *Mol. Cell. Endocrinol.* **1998**, *145*, 67–73. [CrossRef]
17. te Velde, E.R.; Dorland, M.; Broekmans, F.J. Age at menopause as a marker of reproductive ageing. *Maturitas* **1998**, *30*, 119–125. [CrossRef]
18. Fleming, R.; Seifer, D.B.; Frattarelli, J.L.; Ruman, J. Assessing ovarian response: Antral follicle count versus anti-Mullerian hormone. *Reprod. Biomed. Online* **2015**, *31*, 486–496. [CrossRef]
19. Practice Committee of the American Society for Reproductive Medicine. Testing and interpreting measures of ovarian reserve: A committee opinion. *Fertil. Steril.* **2015**, *103*, e9–e17. [CrossRef]
20. Visser, J.A.; de Jong, F.H.; Laven, J.S.E.; Themmen, A.P. Anti-Mullerian hormone: A new marker for ovarian function. *Soc. Reprod. Fertil.* **2006**, *13*, 1–9.
21. Wiweko, B.; Prawesti, D.M.P.; Hestiantoro, A.; Sumapraja, K.; Natadisastra, M.; Baziad, A. Chronological age vs biological age: An age related normogram for antral follicl count; FSH and anti-Mullerian hormone. *J. Assist. Reprod. Genet.* **2013**, *30*, 1563–1567. [CrossRef] [PubMed]
22. Fiçicioglu, C.; Kutlu, T.; Baglam, E.; Bakacak, Z. Early follicular antimüllerian hormone as an indicator of ovarian reserve. *Fertil. Steril.* **2006**, *85*, 592–596. [CrossRef] [PubMed]
23. Vigier, B.; Picard, J.Y.; Tran, D.; Legeai, L.; Josso, N. Production of anti-Müllerian hormone: Another homology between Sertoli and granulosa cells. *Endocrinology* **1984**, *114*, 1315–1320. [CrossRef] [PubMed]
24. La Marca, A.; Volpe, A. Anti-Mullerian hormone (AMH) in female reproduction: Is measurement of circulating AMH a useful tool? *Clinical. Endocrin.* **2006**, *64*, 603–610. [CrossRef]
25. Cook, C.L.; Siow, Y.; Taylor, S.; Fallat, M.E. Serum müllerian-inhibiting substance levels during normal mestrual cycles. *Fertil. Steril.* **2000**, *73*, 859–861. [CrossRef]
26. Dewailly, D.; Andersen, C.Y.; Balen, A.; Broekmans, F.; Dilaver, N.; Fanchin, R.; Griesinger, G.; Kelsey, T.W.; La Marca, A.; Lambalk, C.; et al. The physiology and clinical utility of anti-Mullerian hormone in women. *Hum. Reprod. Update* **2014**, *20*, 370–385. [CrossRef]

27. Brougham, M.F.; Crofton, P.M.; Johnson, E.J.; Evans, N.; Anderson, R.A.; Wallace, W.H.B. Anti Müllerian Hormone is a marker of gonadotoxicity in pre- and postpuberal girls treated for cancer: A prospective study. *J. Clin. Endocrinol. Metab.* **2012**, *97*, 2059–2067. [CrossRef]
28. Bala, J.; Seth, S.; Dhankhar, R.; Ghalaut, V.S. Chemotherapy: Impact on Anti-Müllerian Hormone Levels in Breast Carcinoma. *J. Clin. Diagn. Res.* **2016**, *10*, BC19–BC21. [CrossRef]
29. Lawrenz, B.; Henes, J.C.; Henes, M.; Neunhoeffer, E.; Schmalzing, M.; Fehm, T.; Kïtter, I. Impact of systemic lupus erythematosus on ovarian reserve in premenopausal women: Evaluation by using anti-Mullerian hormone. *Lupus* **2011**, *20*, 1193–1197. [CrossRef]
30. Mok, C.C.; Chan, P.T.; To, C.H. Anti-müllerian hormone and ovarian reserve in systemic lupus erythematosus. *Arthritis Rheum.* **2013**, *65*, 206–210. [CrossRef]
31. Gasparin, A.A.; Souza, L.; Siebert, M.; Xavier, R.M.; Chakr, R.M.S.; Palominos, P.E.; Brenol, J.C.; Monticielo, O.A. Assessment of anti-Müllerian hormone levels in premenopausal patients with systemic lupus erythematosus. *Lupus* **2016**, *25*, 227–232. [CrossRef] [PubMed]
32. Mont'Alverne, A.R.; Pereira, R.M.R.; Yamakami, L.Y.S.; Viana, V.S.T.; Baracat, E.C.; Bonfà, E.; Silva, C.A. Reduced ovarian reserve in patients with Takayasu arteritis. *J. Rheumatol.* **2014**, *41*, 2055–2059. [CrossRef] [PubMed]
33. Yamakami, L.Y.S.; Serafini, P.C.; de Araujo, D.B.; Bonfà, E.; Leon, E.P.; Baract, E.C.; Silva, C.A. Ovarian reserve in women with primary antiphospholipid syndrome. *Lupus* **2014**, *23*, 862–867. [CrossRef] [PubMed]
34. Brouwer, J.; Laven, J.S.; Hazes, J.M.; Schipper, I.; Dolhain, R.J. Levels of serum anti-Mullerian hormone, a marker for ovarian reserve, in women with rheumatoid arthritis. *Arthritis Care Res.* **2013**, *65*, 1534–1538. [CrossRef]
35. Clark, C.A.; Laskin, C.A.; Cadesky, K. Anti-Mullerian hormone: Reality check. *Hum. Reprod.* **2014**, *29*, 1845. [CrossRef]
36. Di Mario, C.; Petricca, L.; Gigante, M.R.; Barini, A.; Barini, A.; Varriano, V.; Paglionico, A.; Cattani, P.; Ferraccioli, G.; Tolusso, B.; et al. Anti-Müllerian hormone serum levels in systemic lupus erythematosus patients: Influence of the disease severity and therapy on the ovarian reserve. *Endocrine* **2019**, *63*, 369–375. [CrossRef]
37. Alexander, V.M.; Ashley-Martin, J.; Riley, J.K.; Cooper, A.R.; Ratts, V.S.; Jungheim, E.S. Association between arthritis treatments and ovarian reserve: A prospective study. *Reprod. Biomed. Online* **2021**, *42*, 1203–1210. [CrossRef]
38. Han, Y.F.; Yan, Y.; Wang, H.Y.; Chu, M.Y.; Sun, K.; Feng, Z.W.; Feng, H. Effect of systemic lupus erythematosus on the ovarian reserve: A systematic review and meta-analysis. *Jt. Bone Spine* **2024**, *91*, 105728. [CrossRef]
39. Notaro, A.L.G.; Lira Neto, F.T.; Bedoschi, G.M.; Silva, M.J.D.; Nunes, M.C.; Monteiro, C.C.P.; Figueiroa, J.N.; Souza, A.S.R. Evaluation of ovarian reserve in women with thyroid autoimmunity. *JBRA Assist. Reprod.* **2024**, *28*, 442–449.
40. Petty, R.E.; Southwood, T.R.; Baum, J.; Betthay, E.; Glass, D.N.; Manners, P.; Maldonado-Cocco, J.; Suarez-Almazor, M.; Orozco-Alcala, J.; Prieur, A.M. Revision of the proposed classification criteria for juvenile idiopathic arthritis: Durban, 1997. *J. Rheumatol.* **1998**, *25*, 1991–1994.
41. Rotterdam ESHRE/ASRM-Sponsored PCOS Consensus Workshop Group. Revised 2003 consensus on diagnostic criteria and long-term health risks related to polycystic ovary syndrome. *Fertil. Steril.* **2004**, *81*, 19–25. [CrossRef]
42. Craciunas, L.; Roberts, S.A.; Yates, A.P.; Smith, A.; Fitzgerald, C.; Pemberton, P.W. Modification of the Beckman-Coulter second-generation enzyme-linked immunosorbent assay protocol improves the reliability of serum antimüllerian hormone measurement. *Fertil. Steril.* **2015**, *103*, 554–559. [CrossRef] [PubMed]
43. MHRA. Urgent Field Safety Notice—FSN 20434-3 AMH Gen II ELISA (REF A79765) in 2013. Available online: http://www.hpra.ie/img/uploaded/documents/fsn/FSNJun2013/V16335_FSN.pdf (accessed on 15 August 2024).
44. Welsh, P.; Smith, K.; Nelson, S.M. A single-centre evaluation of two new anti-Mullerian hormone assays and comparison with the current clinical standard assay. *Hum. Reprod.* **2014**, *29*, 1035–1041. [CrossRef] [PubMed]
45. Dewailly, D.; Laven, J. AMH as the primary marker for fertility. *Eur. J. Endocrinol.* **2019**, *181*, D45–D51. [CrossRef] [PubMed]
46. Moolhuijsen, L.M.E.; Visser, J.A. Anti-Müllerian Hormone and Ovarian Reserve: Update on Assessing Ovarian Function. *J. Clin. Endocrinol. Metab.* **2020**, *105*, 3361–3373. [CrossRef] [PubMed]
47. *PASS 2022 Power Analysis and Sample Size Software 22.0.3*, NCSS, LLC.: Kaysville, UT, USA, 2022.
48. Lopez-Corbeto, M.; Martinez-Mateu, S.; Pluma, A.; Ferrer, R.; Lopez-Lasanta, M.; De Agustin, J.J.; Barceló, M.; Julià, A.; Marsal, S. The ovarian reserve as measured by the anti-Mullerian hormone is not diminished in patients with rheumatoid arthritis compared to the healthy population. *Clin. Exp. Rheumatol.* **2021**, *39*, 337–343. [CrossRef]
49. Henes, M.; Froeschlin, J.; Taran, F.A.; Brucker, S.; Rall, K.K.; Xenitidis, T.; Igney-Oertel, A.; Lawrenz, B.; Henes, J.C. Ovarian reserve alterations in premenopausal women with chronic inflammatory rheumatic diseases: Impact of rheumatoid arthritis, Behçet's disease and spondyloarthritis on anti-Müllerian hormone levels. *Rheumatology* **2015**, *54*, 1709–1712. [CrossRef]
50. Ozer, Y.; Yildiz, M.; Turan, H.; Tarcin, G.; Bingol Aydin, D.; Gunalp, A.; Haslak, F.; Kilic Konte, E.; Aslan, E.; Koker, O.; et al. Ovarian reserve in children with juvenile idiopathis arthritis using biologic disease-modifying anti-rheumatic drugs. *Clin. Rheumatol.* **2024**, *43*, 399–406. [CrossRef]
51. Scrivo, R.; Anastasi, E.; Castellani, C.; Conti, F.; Angeloni, A.; Granato, T. Ovarian reserve in patients with spondyloarthritis: Impact of biological disease modifying anti-rheumatic drugs on fertility status. *Clin. Exp. Rheum.* **2022**, *40*, 1738–1743. [CrossRef]
52. Gol, M.; Saygili, U.; Koyuncuoglu, M.; Uslu, T. Influence of hightdose methotrexate therapy on the primordial follicle of the mouse ovary. *J. Obstet. Gynaecol. Res.* **2009**, *35*, 429–433. [CrossRef]

53. de Araujo, D.B.; Yamakami, L.Y.M.; Aikawa, N.E.; Bonfà, E.; Viana, V.S.T.; Pasoto, S.G.; Pereira, R.M.; Serafin, P.C.; Borba, E.F.; Silva, C.A. Ovarian reserve in adult patients with childhood-onset lupus: A possible deleterious effect of methotrexate? *Scand. J. Rheumatol.* **2014**, *43*, 503–511. [CrossRef]
54. Tomioka, R.B.; Ferreira, G.R.V.; Aikawa, N.E.; Maciel, G.A.R.; Serafini, P.C.; Sallum, A.M.; Campos, L.M.A.; Goldestein-Schainberg, C.; Bonfá, E.; Silva, C.A. Non-steroidal anti-inflammatory drug induces luteinized unruptured follicle syndrome in young female juvenile idiopathic arthritis patients. *Clin. Rheumatol.* **2018**, *37*, 2869–2873. [CrossRef]
55. Ravelli, A.; Sinigaglia, L.; Cimaz, R.; Alessio, M.; Breda, L.; Cattalini, M.; Consolaro, A.; Conti, F.; Cortis, E.; D'Angelo, S.; et al. Tansitional care of young people with juvenile idiopathic arthritis in Italy: Results of a Delphi consensus survey. *Clin. Exp. Rheumatol.* **2019**, *37*, 1084–1091.

BioChem

MDPI

Article

Synthesis and Investigation of Tricyclic Isoquinoline Derivatives as Antibacterial Agents

Matthew J. A. Phillips [1,*], Alison T. Ung [1], Elizabeth J. Harry [2], Jason Ashmore [1] and Andrew M. McDonagh [1]

1 School of Mathematical and Physical Sciences, Faculty of Science, University of Technology Sydney, Sydney, NSW 2007, Australia; alison.ung@uts.edu.au (A.T.U.); jason.ashmore-1@uts.edu.au (J.A.); andrew.mcdonagh@uts.edu.au (A.M.M.)

2 Australian Institute for Microbiology & Infection, University of Technology Sydney, Sydney, NSW 2007, Australia; elizabeth.harry@uts.edu.au

* Correspondence: matthew.phillips-1@uts.edu.au

Abstract: Isoquinoline derivatives exhibit a range of biological properties, including antibacterial activity, and are thus attractive as a scaffold for developing broad-spectrum antibacterial compounds. A series of six isoquinoline-based compounds were synthesized using the reaction of 6,7-dimethoxy-1-methyl-3,4-dihydroisoquinoline with dimethyl acetylenedicarboxylate (DMAD) to provide the tricyclic (2Z)-[2-oxo-5,6-dihydropyrrolo[2,1,a]isoquinolin-3-ylidene]-2-ethanoate. The [2 + 3] cycloaddition of DMAD with C-6 and C-7 substituted 1-methyl-3,4-dihydroisoquinolines proceeded using aryl ethers or unsubstituted compounds, but not with amine, amide or nitro moieties at the C-7 position. Compounds **8d** and **8f** were found to have antibacterial properties against some Gram-positive pathogens (*Staphylococcus aureus*—**8d** = 16 μg/mL, **8f** = 32 μg/mL; *Streptococcus pneumoniae*—**8f** = 32 μg/mL; and *Enterococcus faecium*—**8d** = 128 μg/mL, **8f** = 64 μg/mL). Evaluation of their cytotoxic properties against mammalian cell lines revealed some cytotoxic effects (**8b** and **8d**, 125 μM, 24 h, HEp-2 cells) and (**8a**, **8b**, **8d** = 125 μM, **8f** = 62.5 μM, 24 h, McCoy B cells), suggesting limitations in their antibacterial applications without further development.

Keywords: antibacterial; isoquinoline; 3,4-dihydroisoquinoline; cytotoxicity

1. Introduction

Many bacterial pathogens are drug-resistant. A select few, known as the ESKAPE pathogens (*Enterococcus faecium*, *Staphylococcus aureus*, *Klebsiella pneumoniae*, *Acinetobacter baumanni*, *Pseudomonas aeruginosa*, and *Enterobacter species*), present a particular threat due to their worldwide prevalence [1]. The World Health Organization (WHO) has listed *A. baumanni*, *P. aeruginosa*, and *Enterobacter* species as priority pathogens requiring new therapeutic treatments due to their resistance to carbapenem and extended spectrum β-lactamase resistance. *Enterococcus faecium* and *S. aureus* remain a high priority, while other common infections, such as *Streptococcus pneumoniae*, are of increasing concern [2].

Very few antibacterial drugs have been approved in recent years. This has exacerbated the need for new drugs with novel structures. The WHO has published three main criteria that new antibiotic compounds should meet to reduce the risk of pathogens developing antibacterial resistance. New antibacterial agents should (i) have no cross-resistance (NCR), (ii) be a new chemical class, and (iii) target a new mode of action [3].

As of 2021, 27 antibacterial agents were in Phase I–III clinical trials that target priority pathogens, where 13 have noted activity against Gram-negative pathogens that are often harder to treat. However, many of these 27 drugs are either derivatives of a parent molecular class, such as macrolides, tetracyclines, and β-lactamase inhibitors, or combinatorial therapeutics [4]. Molecular derivatization can be useful in the rapid development of new antibiotics, but because the same mechanism is targeted, this typically leads to resistance [5]. Combinatorial therapies frequently reduce the rate of resistance and, in some cases, effectively eliminate drug resistance. However, this increases the therapeutic complexity due to the potential of drug–drug interactions and drug antagonism [6,7].

Recent advancements have shown that antibacterial agents can incorporate isoquinolines into their structures with good efficacy [8] and, in some cases, without susceptibility to resistance in *S. aureus* after 30 days [9]. These include berberine, a natural product that may interfere with FtsZ polymerization [10], an antibacterial target, and key enzyme in bacterial replication. Isoquinoline-based *N*-ethyl ureas can inhibit Gram-positive and Gram-negative pathogens and interfere with DNA gyrase and topoisomerase IV activity [11].

No approved antibacterial agents to date have utilized isoquinoline compounds, although US Food and Drug Authority (FDA) approved drugs such as papaverine (an antispasmodic), which has been used since the 1800s. Other examples include praziquantel (an anthelmintic), solifenacin (an anticholinergic), and lifitegrast for dry-eye management. The successful use of these drugs as therapeutic agents indicates that druglike molecules can be constructed around an isoquinoline core structure. An attractive property of isoquinoline compounds is their straightforward synthesis, which has been investigated since the late 19th century. Reported reactions for the preparation of isoquinoline compounds date back to 1893 for the Bischler–Napieralski reaction [12] and 1911 for the Pictet–Spengler reaction [13]. Both reactions employ intramolecular electrophilic cyclization strategies to achieve their dihydro- and tetrahydroisoquinolines, respectively. A less common method to access dihydroisoquinoline structures uses the intracyclic Ritter reaction [14,15]. Of direct interest to the current work is the reported cyclization reaction that dihydroisoquinolines compounds undergo with dimethyl acetylenedicarboxylate, affording structures of the type shown in Figure 1 [16]. However, no biological activities have been investigated thus far.

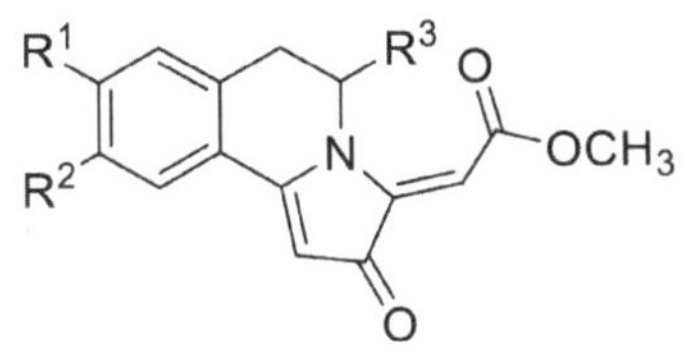

Compound	R^1	R^2	R^3
8a	OCH_3	H	H
8b	H	OCH_3	H
8c	OCH_2O		H
8d	H	H	H
8e	OCH_3	OCH_3	H
8f	OCH_3	OCH_3	CH_3

Figure 1. Structures of the tricyclic isoquinoline compounds investigated in this work.

The unique tricyclic core structure of these compounds was of interest to investigate for potential antibacterial properties against a range of Gram-positive and Gram-negative bacteria. Therefore, we sought to synthesize and evaluate the antibacterial properties of the compounds outlined in Figure 1 and determine the scope of functional groups at R^2 in dihydroisoquinoline precursors. Each of these compounds, except **8e** [16], has not been previously reported.

2. Materials and Methods

2.1. Chemistry

Chemical reagents were purchased from Merck (Darmstadt, Germany) and used without further purification, with the exception of dimethyl acetylenedicarboxylate, which was purchased from TCI chemicals. Compounds prepared using adapted literature procedures have the relevant citations included within the procedures below. Column chromatography used 60-mesh silica gel (Merck). NMR spectra were recorded on an Agilent 500 MHz or a Bruker 400 MHz spectrometer operating at 500 or 400 MHz for ^{1}H NMR, respectively, and 125 or 100 MHz for ^{13}C NMR, respectively. Spectra were referenced to the residual non-deuterated solvent signal using either $CDCl_3$ (^{1}H δ 7.26, ^{13}C δ 77.00) or DMSO-d_6 (^{1}H δ 2.50, ^{13}C δ 39.52). Multiplicity was assigned as singlet (s), doublet (d), triplet (t), quartet (q), quintet (quin), sextet (sxt), or multiplet (m). High-resolution mass spectra (HRMS) were recorded on an Agilent Technologies 6510 Q-TOF MS. Compound purity was confirmed to be >95% prior to biological assays using quantitative NMR spectroscopy (see Supplementary Materials). Schemes 1–3 (see Section 3.1) provide an outline of the synthetic routes described in detail below.

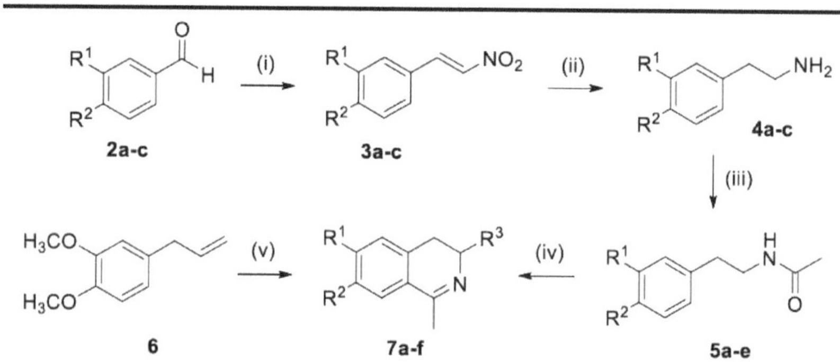

Compound	R¹	R²	R³
7a	OCH_3	H	H
7b	H	OCH_3	H
7c	OCH_2O		H
7d	H	H	H
7e	OCH_3	OCH_3	H
7f	OCH_3	OCH_3	CH_3

Scheme 1. Synthesis of compounds **7a–f**. Conditions: **(i)** CH_3COOH, NH_4CH_3COO (1.1 equiv.), CH_3NO_2 (3 equiv.), reflux, 4 h **(ii)** THF, $LiAlH_4$ (3 equiv.), reflux, 3 h **(iii)** CH_2Cl_2, CH_3COCl (1.5 equiv.), NEt_3 (2 equiv.), r.t., 2 h **(iv)** for **7a–e** only: P_2O_5 (1.5 *w*/*w*), $POCl_3$ (3 equiv.), toluene, 110 °C, 2–16 h. **(v)** for **7f** only: CH_3CN (25 equiv.), H_2SO_4/benzene (3:4), r.t., 3 h.

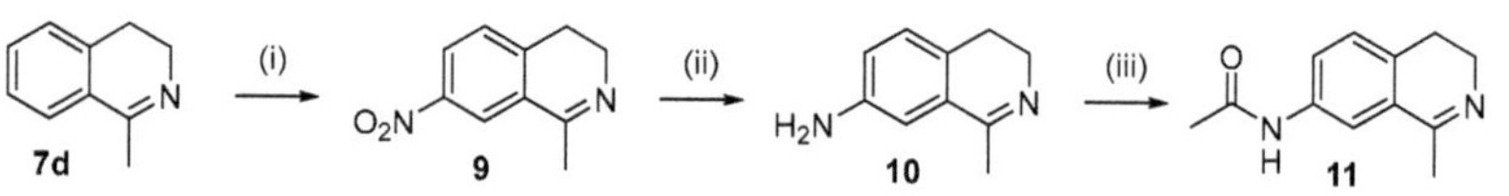

Scheme 2. Synthesis of compounds **9–11**. Conditions: **(i)** KNO_3 (1.1 equiv.), H_2SO_4, 60 °C, 4 h **(ii)** $SnCl_2$ (4 equiv.), HCl (32%), EtOH, 60 °C, 2 h **(iii)** CH_3COOH (1 equiv.), CDI (1.2 equiv.), THF, N_2, r.t., 2 h.

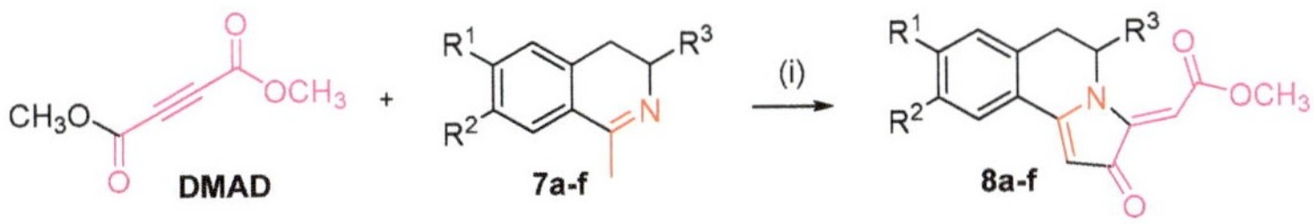

Scheme 3. Synthesis of compounds **8a–f**, and attempted cyclisation reactions with compounds **9** and **11**. Conditions: (i) DMAD (1.5 equiv.), methanol (see Section 2).

2.1.1. Synthesis of 3-Methoxybenzaldehyde (**2**)

3-Hydroxybenzaldehyde (**1**) (5.00 g, 40.9 mmol) was dissolved in anhydrous acetone (30 mL), to which potassium carbonate (11.82 g, 2 equiv.) was added, and stirred at room temperature for five minutes. Methyl iodide (8.72 g, 3.82 mL, 1.5 equiv.) was added, and the reaction was heated under reflux for three hours. Volatiles were removed with reduced pressure, and water (50 mL) was added. The aqueous phase was washed with chloroform (2 × 25 mL), and the combined organic extracts were washed with water (50 mL) and brine (50 mL) and then dried using anhydrous potassium carbonate. The solvent was removed with reduced pressure, affording a light orange liquid that was used without further purification. Spectral data were consistent with reported values [17]. ^{1}H NMR: (500 MHz, $CDCl_3$) δ 9.97 (s, 1H, CHO), 7.44 (m, 2H, H-4, 6), 7.39 (d, *J* = 2.0 Hz, 1H, H-2), 7.17 (dt, *J* = 6.5, 2.5 Hz, 1H, H-5). ^{13}C NMR: (125 MHz, $CDCl_3$) δ 192.1, 160.1, 137.8, 130.0, 123.5, 121.5, 112.0, 55.5.

2.1.2. General Procedure for the Synthesis of Nitrostyrene Compounds (**3a–3c**)

The general procedure for the synthesis of nitrostyrenes from arylaldehydes (**2a–c**) was adapted from a literature procedure [18]. The respective aldehyde was dissolved in glacial acetic acid (5 mL/g), to which nitromethane (3 equiv.) and ammonium acetate (1.1 equiv.) were added. The solution was heated under reflux for three hours, then poured into a beaker of crushed ice (50 g), and then filtered once all the ice had melted. The bright yellow cake was washed with water (2 × 20 mL) and recrystallized from ethanol to afford bright yellow needles.

- 3-Methoxy-β-nitrostyrene (**3a**) (6.80 g, 93%) from 3-hydroxybenzaldehyde over two steps. ^{1}H NMR: (500 MHz, $CDCl_3$) δ 7.97 (d, *J* = 14.0 Hz, 1H, H-2′), 7.57 (d, *J* = 13.5 Hz, 1H, H-1′), 7.36 (dd, *J* = 9.0, 7.5 Hz, 1H, H-5), 7.05–7.03 (m, 2H, 2H-Ar), 3.85 (s, 3H, OCH_3). ^{13}C NMR: (125 MHz, $CDCl_3$) δ 160.1, 139.0, 137.3, 131.3, 130.4, 121.7, 117.9, 114.0, 55.4. Spectral data were consistent with reported values [19].
- 4-Methoxy-β-nitrostyrene (**3b**) (5.21 g, 79%). ^{1}H NMR: (500 MHz, $CDCl_3$) δ 7.90 (d, *J* = 13.5 Hz, 1H, H-2′), 7.51 (d, *J* = 13.5 Hz, 1H, H-1′), 7.49 (d, *J* = 9.0 Hz, 2H, H-2, 6), 6.95 (d, *J* = 9.0 Hz, 2H, H-3, 5), 3.86 (s, 3H, OCH_3). ^{13}C NMR: (125 MHz, $CDCl_3$) δ 162.9, 139.0, 135.0 131.1, 122.5, 114.9, 55.5. Spectral data were consistent with reported values [20].
- 3,4-Dimethoxy-β-nitrostyrene (**3c**) (2.75 g, 71%) ^{1}H NMR: (500 MHz, $CDCl_3$) δ 7.93 (d, *J* = 13.5 Hz, 1H, H-2′), 7.47 (d, *J* = 13.0 Hz, 1H, H-1′), 7.07 (dd, *J* = 8.0, 1.5 Hz, 1H, H-6), 7.00 (d, *J* = 1.0 Hz, 1H, H-2), 6.87 (d, *J* = 8.0 Hz, 1H, H-5), 6.06 (s, 2H, CH_2). ^{13}C NMR: (125 MHz, $CDCl_3$) δ 151.4, 148.8, 139.1, 135.4, 126.6, 124.2, 109.1, 107.0, 102.0. Spectral data were consistent with reported values [21].

2.1.3. General Procedure for the Synthesis of Phenylethylamines (**4a–c**)

The general procedure for synthesizing phenylethylamines from nitrostyrenes (**3a–c**) was adapted from a literature procedure [22]. The respective nitrostyrene was dissolved in anhydrous tetrahydrofuran (c.a. 5 mL/g) and added dropwise at a rate of approximately 1 mL/minute to a stirring slurry of lithium aluminum hydride (3 equiv.) in tetrahydrofuran

(5–10 mL/g) at 0 °C under a nitrogen atmosphere. After the addition, the reaction was heated under reflux for three hours and then cooled to 0 °C. Aqueous hydrochloric acid (1 M) was added slowly with stirring under nitrogen until bubbling subsided and the pH was <1 (typically 20–30 mL). The aqueous phase was then washed with dichloromethane (3 × 30 mL). The aqueous phase was cooled to 0–5 °C with an ice bath and stirred vigorously. Sodium hydroxide pellets were slowly added until the pH was >11 and a white precipitate appeared. The solution was vacuum-filtered, and the filtrate was washed with dichloromethane (3 × 30 mL). The combined organic extracts were washed sequentially with saturated potassium sodium tartrate (50 mL), water (50 mL), and brine (50 mL) and then dried using anhydrous sodium carbonate. The solvent was removed with reduced pressure to afford the desired amine as a yellow oil and was used without further purification.

- 2-(3-Methoxyphenyl)ethylamine (**4a**) Yellow oil (2.00 g, 35%). ^{1}H NMR: (500 MHz, $CDCl_3$): δ 7.22 (t, *J* = 7.5 Hz, 1H H-3), 6.79 (d *J* = 7.5 Hz, H-2), 6.77–6.76 (m, 1H, H-4), 6.75 (s, 1H, H-6), 3.80 (s, 3H, OCH_3), 2.97 (t, *J* = 7.0 Hz, 2H, H-1″), 2.73 (t, *J* = 7.0 Hz, 2H, H-1′). ^{13}C NMR: (125 MHz, $CDCl_3$) δ 159.7, 141.4, 129.4, 121.2, 114.6, 111.4, 55.1, 43.4, 40.0. Spectral data were consistent with reported values [23].
- 2-(4-Methoxyphenyl)ethylamine (**4b**) Yellow oil (2.13 g, 63%). ^{1}H NMR (500 MHz, $CDCl_3$) δ 7.11 (d, *J* = 9.0 Hz, 2H, H-2, 6), 6.84 (d, *J* = 8.5 Hz, 2H, H-3, 5), 3.78 (s, 3H, OCH_3), 2.92 (t, *J* = 7.0 Hz, 2H, H-2′), 2.68 (t, *J* = 7.0 Hz, 2H, H-1′). ^{13}C NMR (125 MHz, $CDCl_3$) δ 158.0, 131.8, 129.7, 133.8, 55.2, 43.7, 39.1. Spectral data were consistent with reported values [24].
- 2-(3,4-Methylenedioxyphenyl)ethylamine (**4c**) Pale yellow oil (0.731 g, 71%). ^{1}H NMR (500 MHz, $CDCl_3$) δ 6.73 (d, *J* = 8.0 Hz, 1H, H-5), 6.68 (d, *J* = 2.0 Hz, 1H, H-2), 6.63 (dd, *J* = 8.0, 1.5 Hz, 1H, H-6), 5.91 (s, 2H, OCH_2O), 2.90 (t, *J* = 7.0 Hz, 2H, H-2′), 2.65 (t, *J* = 7.0 Hz, 2H, H-1′) (NH_2 Not observed). ^{13}C NMR (125 MHz, $CDCl_3$) δ 147.6, 145.8, 133.5 121.6, 109.1, 108.1, 100.7, 43.6, 39.7. Spectral data were consistent with reported values [21].

2.1.4. General Procedure for the Synthesis of *N*-Phenylethylacetamides (**5a–e**)

The respective phenylethylamine* was dissolved in anhydrous chloroform (25 mL/g) and cooled to 0 °C. Acetyl chloride (1.5 equiv.) was added dropwise over one minute with stirring and maintaining temperature. Triethylamine (2 equiv.) was added dropwise with vigorous stirring over five minutes and then warmed slowly to room temperature and stirred for two to three hours. Water (50 mL) was added and stirred vigorously for five minutes. Then, the aqueous layer was removed. The organic phase was washed sequentially with water (2 × 50 mL) and brine (50 mL) and dried using anhydrous sodium carbonate. The solvent was removed with reduced pressure and purified using silica gel column chromatography where specified. (For **5d**, commercially available hydrochloride salt was used. Three equivalents of triethylamine were used to generate the free base in situ).

- *N*-[2-(3-Methoxyphenyl)ethyl]acetamide (**5a**) The dark oil was purified by gradient silica gel column chromatography (50–100% ethyl acetate in hexane) to afford a yellow oil (1.86 g, 74%). ^{1}H NMR: (500 MHz, $CDCl_3$) δ 7.23 (t, *J* = 7.5 Hz, 1H, H-3), 6.77 (d, *J* = 8.0 Hz, 2H, H-2,4) 6.74 (s, 1H, H-6), 5.50 (br s, 1H, NH), 3.80 (s, 3H, OCH_3), 3.51 (q, *J* = 6.5 Hz, 2H, H-1″), 2.79 (t, *J* = 7.0 Hz, 2H, H-1′), 1.94 (s, 3H, $NHCOCH_3$). ^{13}C NMR: (125 MHz, $CDCl_3$) δ 170.0, 159.8, 140.4, 129.6, 121.0, 114.4, 111.8, 55.2, 40.5, 35.6, 23.3. Spectral data were consistent with reported values [23].
- *N*-[2-(4-Methoxyphenyl)ethyl]acetamide (**5b**) The dark oil was purified by gradient silica gel column chromatography (40–100% ethyl acetate in hexane) to afford a light yellow solid (2.08 g, 77%). ^{1}H NMR: (500 MHz, $CDCl_3$) δ 7.11 (dt, *J* = 6.5, 1.5 Hz, 2H,

H-2, 6), 6.85 (dt, J = 7.0 Hz, 1.5 Hz, 2H, H-3, 5), 5.52 (br s, 1H, NH), 3.79 (s, 3H, OCH_3), 3.46 (q, J = 7.0 Hz, 2H, H-2′), 2.75 (t, J = 7.0 Hz, 2H, H-1′), 1.93 (s, 3H, $NHCOCH_3$). ^{13}C NMR: (125 MHz, $CDCl_3$) δ 170.0, 158.2, 130.8, 129.6, 114.0, 55.2, 40.8, 34.7, 23.3. Spectral data were consistent with reported values [25].

- *N*-[2-(3,4-Methylenedioxyphenyl)ethyl]acetamide (**5c**) Pale yellow solid (0.543 g, 96%) was used, requiring no further purification. 1H NMR (500 MHz, $CDCl_3$) δ 6.74 (d, J = 8.0 Hz, 1H, H-5), 6.67 (d, J = 2.0 Hz, 1H, H-2), 6.62 (dd, J = 8.0, 2.0 Hz, 1H, H-6), 5.93 (s, 2H, OCH_2O), 5.50 (br s. 1H, NH), 3.45 (q, J = 7.0 Hz, 2H, H-2′), 2.72 (t, J = 7.0 Hz, 2H, H-1′), 1.95 (s, 3H, CH_3). ^{13}C NMR (125 MHz, $CDCl_3$) δ 170.0, 147.8, 146.2, 132.6, 121.6, 109.0, 108.4, 100.9, 40.8, 35.3, 23.3 Spectral data were consistent with reported values [26].
- *N*-(2-Phenyl)ethylacetamide (**5d**) Brown solid requiring no further purification (2.84 g, >99%). ^{1}H NMR: (500 MHz, $CDCl_3$) δ 7.28 (t, J = 7.5 Hz, 2H, 2Ar), 7.20 (d, J = 7.5 Hz, 1H, H-4), 7.17 (t, J = 7.5 Hz, 2H, 2Ar), 6.09 (br s, 1H, NH), 3.46 (q, J = 7.5 Hz, 2H, H-2′), 2.79 (t, J = 7.5 Hz, 2H, H-1′), 1.90 (s, 3H, CH_3). ^{13}C NMR: (125 MHz, $CDCl_3$) δ 170.0, 138.8, 128.7, 128.6, 126.5, 40.6, 35.6, 23.3. Spectral data were consistent with reported values [27].
- *N*-[2-(3,4-Dimethyoxyphenyl)ethyl]acetamide (**5e**) Yellow solid (3.70 g, >99%). ^{1}H NMR: (500 MHz, $CDCl_3$) δ 6.80–6.79 (m, 1H, H-Ar), 6.72 (m, 2H, 2Ar), 5.56 (br s, 1H, NH), 3.86 (s, 6H, 2 × OCH_3), 3.48 (q, J = 7.0 Hz, 2H, H-1′), 2.75 (t, J = 7.0 Hz, 2H, H-2′), 1.93 (s, 3H, $COCH_3$). ^{13}C NMR: (125 MHz, $CDCl_3$) δ 170.0, 149.0, 147.7, 131.3, 129.6, 111.8, 111.4, 111.3, 55.9, 55.8, 40.7, 35.1, 23.3. Spectral data were consistent with reported values [28].

2.1.5. General Procedure for the Synthesis of 1-Methyl-3,4-dihydroisoquinolines (**6a–e**)

The respective *N*-(1-phenylethyl)acetamide (**5a–d**) was dissolved in anhydrous toluene (10 mL/g) (except **5e**, which was dissolved in anhydrous acetonitrile), to which phosphorus pentoxide (1.5 *w*/*w*) and phosphorus oxychloride (3 equiv.) were added, and heated under reflux for the duration specified below. The solution was cooled to 0 °C, poured onto crushed ice (c.a. 20 g), and stirred vigorously for five minutes. The mixture was transferred to a separatory funnel, and the aqueous layer was removed. Then, the organic phase was washed with an additional portion of water (30 mL). The combined aqueous extracts were washed with dichloromethane (2 × 20 mL), and then the aqueous phase was cooled to 0 °C and stirred vigorously. Sodium hydroxide pellets were slowly added to the solution until the pH was >10, and then the mixture was extracted with dichloromethane (3 × 30 mL). The combined extracts were washed sequentially with water (50 mL) and brine (50 mL) and dried using anhydrous sodium carbonate. Products that required purification were purified by silica gel column chromatography with 0–25% methanol in ethyl acetate (Table 1).

Table 1. Reaction times used in the synthesis of compounds **7a–e** from **5a–e**.

Compound	Reaction Time (h)
7a	4
7b	16
7c	24
7d	3
7e	3

- 6-Methoxy-1-methyl-3,4-dihydroisoquinoline (**7a**) Yellow oil (1.06 g, 65%). ^{1}H NMR: (500 MHz, $CDCl_3$) δ 7.41 (d, J = 8.5 Hz, 1H, H-8), 6.78 (dd, J = 8.5, 2.0 Hz, 1H, H-7), 6.70 (br s, 1H, H-5), 3.83 (s, 3H, OCH_3), 3.62, (t, J = 7.5 Hz, 2H, H-3), 2.68 (t, J = 7.5 Hz,

2H, H-4), 2.34 (br s, 3H, H-1′). ^{13}C NMR: (125 MHz, $CDCl_3$) δ 164.0, 131.1, 139.6, 128.2, 123.1, 112.7, 111.8, 55.3, 46.7, 26.6, 23.2. Spectral data were consistent with reported values [29].

- 7-Methoxy-1-methyl-3,4-dihydroisoquinoline (**7b**) Yellow oil (0.430 g, 24%). ^{1}H NMR: (500 MHz, $CDCl_3$) 7.09 (d, *J* = 8.5 Hz, 1H, H-5), 7.02 (d, *J* = 2.5 Hz, 1H, H-8), 6.90 (dd, *J* = 8.5, 2.5 Hz, 1H, H-7), 3.82 (s, 3H, OCH_3), 3.64 (tq, *J* = 7.5, 1.5 Hz, 2H, H-3), 2.63 (t, *J* = 7.5 Hz, 2H, H-4), 2.37 (t, *J* = 1.5 Hz, 3H, H-1′). ^{13}C NMR: (125 MHz, $CDCl_3$) δ 164.2, 158.4, 130.2, 129.4, 128.1, 115.6, 111.4, 55.4, 47.2, 25.1, 23.2. Spectral data were consistent with reported values [30].
- 6,7-Methylenedioxy-1-methyl-3,4-dihydroisoquinoline (**7c**) Brown solid that was used without further purification (0.450 g, 62%). ^{1}H NMR: (500 MHz, $CDCl_3$) δ 6.97 (s, 1H, H-8), 6.66 (s, 1H, H-5), 5.97 (s, 2H, OCH_2O), 3.59 (td, *J* = 7.5, 1.5 Hz, 2H, H-3), 2.60 (t, *J* = 7.5 Hz, 2H, H-4), 2.32 (t, *J* = 1.5 Hz, 3H, H-1′). ^{13}C NMR: (125 MHz, $CDCl_3$) δ 163.5, 146.3, 132.8, 123.7, 107.8, 106.0, 101.2, 46.9, 26.3, 23.6. Spectral data were consistent with reported values [31].
- Methyl-3,4-dihydroisoquinoline (**7d**) Brown oil required no further purification. (0.635 g, 74%). ^{1}H NMR: (500 MHz, $CDCl_3$) δ 7.47 (dd, *J* = 7.5, 1.5 Hz, 1H, H-Ar) 7.34 (td, *J* = 7.5, 1.5 Hz, 1H, H-Ar), 7.30 (td, *J* = 7.5, 1.5 Hz, 1H, H-Ar), 7.17, (dd, *J* = 7.5, 1.5, 1H, H-Ar), 3.66 (tq, *J* = 7.5, 1.5 Hz, 2H, H-3), 2.70 (t, *J* = 7.5 Hz, 2H, H-4), 2.38 (s, 3H, CH_3). ^{13}C NMR (125 MHz, $CDCl_3$) δ 164.3, 137.4, 130.5, 129.5, 127.4, 126.8, 125.3, 46.9, 26.0, 23.3. Spectral data were consistent with reported values [32].
- 6,7-Dimethoxy-1-methyl-3,4-dihydroisoquinoline (**7e**) Brown oil was used in the next reaction step without further purification (2.40 g, 72%). ^{1}H NMR: (500 MHz, $CDCl_3$) δ 6.98 (s, 1H, H-8), 6.68 (s, 1H, H-5), 3.90 (s, 3H, OCH_3), 3.89 (s, 3H, OCH_3), 3.62 (td, *J* = 7.5, 1.5 Hz, 2H, H-3), 2.63 (t, *J* = 7.5 Hz, 2H, H-4) 2.36 (s, 3H, CH_3). ^{13}C NMR: (125 MHz, $CDCl_3$) δ 163.9, 151.0, 147.5, 131.2, 122.3, 110.2, 109.1, 56.2, 56.0, 46.8, 25.7, 23.3. Spectral data were consistent with reported values [33].

2.1.6. Synthesis of (±)-6,7-Dimethoxy-1,3-dimethyl-3,4-dihydroisoquinoline (**7f**)

Anhydrous acetonitrile (35.6 mL, 25 equiv.) was cooled in a flask to −10 °C using an ice bath with sodium chloride. Sulfuric acid (98%, 15.0 mL) was added dropwise and stirred for two minutes. Then, methyl eugenol (**6**) (4.90 g, 27.5 mmol) was added dropwise as a mixture in benzene (20 mL), and the flask was fitted with a drying tube. The reaction mixture was warmed to room temperature over one hour and stirred for another two hours. Water (20 mL) was then added, and the mixture was washed with diethyl ether (20 mL) twice, allowing ten minutes of vigorous stirring for each wash. The aqueous layer was cooled in an ice bath and basified with aqueous sodium hydroxide (4 M). The mixture was then washed with chloroform (3 × 20 mL), and the combined extracts were washed with brine (50 mL) and dried using anhydrous sodium sulfate. The solvent was then removed in vacuo, and the product was used without further purification. Yellow oil (3.58 g, 16.3 mmol, 59%). ^{1}H NMR: (500 MHz, $CDCl_3$) δ 6.93 (s, 1H, H-5), 6.61 (s, 1H, H-8), 3.86 (s, 3H, -OCH_3), 3.84 (s, 3H, -OCH_3), 3.47–3.42 (m, 1H, H-3), 2.58 (dd, *J* = 5.5, 15.5 Hz, 1H, H-4a), 2.35 (dd, *J* = 13.0, 15.5 Hz, 1H, H-4b), 2.30 (d, *J* = 2.0 Hz, 3H, H-1′), 1.36 (d, *J* = 6.5 Hz, 3H, H-3′). ^{13}C NMR: (100 MHz, $CDCl_3$) δ 162.8, 150.8, 147.4, 130.9, 122.1, 110.3, 108.9, 56.2, 55.9, 51.9, 33.1, 23.3, 21.9. HRMS (ESI/Q-TOF) *m*/*z*: $[M + H]^+$ Calcd for $C_{13}H_{19}NO_2$ 220.1332; found 220.1339.

2.1.7. Synthesis of 7-Nitro-1-methyl-3,4-dihydroisoquinoline (**9**)

This compound was prepared from **5d** using a literature method [34] to afford a pink solid that was used without further purification (0.508 g, 79%). ^{1}H NMR: (500 MHz, $CDCl_3$)

δ 8.32 (d, *J* = 2.0 Hz, 1H, H-8), 8.22 (dd *J* = 8.0, 2.5 Hz, 1H, H-6), 7.37 (t, *J* = 8.5 Hz, 1H, H-5), 3.74 (tq, *J* = 7.5, 1.5 Hz, 2H, H-4), 2.82 (t, 3H, *J* = 8.0 Hz, H-3), 2.47 (t, *J* = 7.5 Hz, 3H, CH_3).^{13}C NMR: (125 MHz, $CDCl_3$) δ 162.5, 144.7, 130.0. 128.5, 125.2, 120.1, 46.3, 26.1, 23.2. Spectral data were consistent with reported values.

2.1.8. Synthesis of 7-Amino-1-methyl-3,4-dihydroisoquinoline (**10**)

This method was adapted from a literature procedure [35]. Compound **9** (0.500 g, 2.65 mmol) was dissolved in ethanol (50 mL), to which tin(II) chloride dihydrate (2.72 g, 4 equiv.) and hydrochloric acid (32%, 5 mL) were added, and stirred at 60 °C for two hours. The reaction mixture was poured onto crushed ice (30 g) and washed with chloroform (2 × 20 mL). The aqueous phase was then basified with chilled aqueous sodium hydroxide (4 M) and extracted with chloroform (2 × 20 mL). The combined extracts were washed with water (50 mL) and brine (50 mL), dried using anhydrous sodium carbonate, and concentrated with reduced pressure to afford a brown solid that was used without further purification. ^{1}H NMR: (500 MHz, $CDCl_3$) δ 6.97 (d, *J* = 7.5 Hz, 1H, H-5), 6.82 (d *J* = 2.5 Hz, 1H, H-8), 6.70 (dd, *J* = 8.0, 2.5 Hz, 1H, H-6), 3.66 (br s, 2H, NH_2), 3.62 (tq, 2H, *J* = 7.5, 1.5 Hz, H-3), 2.58 (t, 8.0 Hz, 2H, H-4) 2.47 (t, *J* = 8.0 Hz, 3H, CH_3). ^{13}C NMR: (125 MHz, $CDCl_3$) δ 164.3, 145.1, 130.2, 128.1, 127.4, 117.2, 112.1, 47.5, 25.1, 23.3. Spectral data were consistent with reported values.

2.1.9. Synthesis of 7-Acetamido-3,4-dihydro-1-methylisoquinoline Synthesis (**11**)

Acetic acid (188 mg, 179 μL, 1 equiv.) was dissolved in dry tetrahydrofuran (10 mL), to which carbonyl diimidazole (607 mg, 1.2 equiv.) was added, and stirred in a nitrogen atmosphere for two hours. The solution was transferred to a mixture of compound **10** (0.500 g, 3.12 mmol, 1 equiv.) in dry tetrahydrofuran (10 mL) and left to stir for 22 h at room temperature under a nitrogen atmosphere. The solvent was removed with reduced pressure, diluted with aqueous hydrochloric acid (20 mL, 0.25 M), and then washed with chloroform (20 mL). The aqueous phase was basified with aqueous sodium hydroxide (1 M) and extracted with chloroform (3 × 20 mL). The extracts were combined and washed sequentially with water (30 mL) and brine (50 mL), dried using anhydrous potassium carbonate, and then concentrated with reduced pressure to afford a dark yellow wax that was used without further purification. ^{1}H NMR: (400 MHz, $CDCl_3$) δ 8.61 (br s, 1H, NH), 7.73 (d, *J* = 1.6 Hz, H-8), 7.45 (dd, *J* = 1.6, 8.0 Hz, 1H, H-6), 7.08 (d, *J* = 8.4 Hz, 1H, H 5), 3.61 (t, *J* = 7.2 Hz, 2H, H-3), 2.63 (t, *J* = 7.6 Hz, 2H, H-4), 2.30 (s, 3H, H-1′), 2.15 (s, 3H, $NHCOCH_3$). ^{13}C NMR: (100 MHz, $CDCl_3$) δ 168.9, 164.4, 137.1, 133.0, 129.6, 127.7, 122.3, 117.4, 46.7, 25.4, 24.2, 23.0. HRMS (ESI/Q-TOF) m/z: $[M + H]^+$ Calcd for $C_{12}H_{16}N_2O_4$ 203.1177; found 203.1179.

2.1.10. Synthesis of Methyl (2Z)-[2-oxo-5,6-dihydropyrrolo[2,1,a]isoquinolin-3-ylidene]-2-ethanoates (**8a–f**)

Compounds **8a-f** were synthesized using an adapted literature procedure [16]. The respective dihydroisoquinoline **7a-f** was dissolved in methanol (10 mL), to which DMAD (1.5 equiv.) was added dropwise, and stirred at room temperature for the duration specified below. Reactions that formed bright red precipitates were filtered, and those that did not were concentrated with reduced pressure. Crude products were purified by silica gel column chromatography, with mobile phases specified for each compound except for **8e**, which was triturated with cold diethyl ether (2 × 5 mL) (Table 2).

Table 2. Reaction times used in the synthesis of compounds **8a–f** from **7a–f**.

Compound	Reaction Time (h)
8a	1
8b	16
8c	12
8d	16
8e	2
8f	16

- Methyl (2Z)-8-methoxy-[2-oxo-5,6-dihydropyrrolo[2,1,a]isoquinolin-3-ylidene]-2-ethanoate (**8a**) (20–100% ethyl acetate in hexane) to afford a bright red/orange solid (0.817 g, 52%). ^{1}H NMR: (500 MHz, $CDCl_3$) δ 7.66 (d, *J* = 8.5 Hz, 1H, H-8), 6.90 (d, *J* = 9.0 Hz, 1H, H-7), 6.80 (br s, 1H, H-5), 6.05 (s, 1H, C=CH), 5.73 (s, 1H, C=CH), 4.31 (t, *J* = 6.5 Hz, 2H, H-3), 3.89 (s, 3H, OCH3), 3.78 (s, 3H, OCH3), 3.10 (t, *J* = 6.5 Hz, 2H, H-4). ^{13}C NMR: (125 MHz, $CDCl_3$) δ 187.0, 166.5, 166.3, 163.2, 143.4, 138.9, 130.4, 117.6, 114.1, 113.3, 98.6, 94.3, 55.6, 51.9, 42.2, 29.0. HRMS (ESI/Q-TOF) *m*/*z*: $[M + H]^+$ Calcd for $C_{16}H_{16}NO_4$ 286.1073; found 286.1110. m.p. 150 °C (decomp.)
- Methyl (2Z)-9-methoxy-[2-oxo-5,6-dihydropyrrolo[2,1,a]isoquinolin-3-ylidene]-2-ethanoate (**8b**) (20–40% ethyl acetate in hexane) to afford the desired product as a bright red solid (0.070 g, 11%). ^{1}H NMR: (500 MHz, $CDCl_3$) δ 7.21 (d, *J* = 8.0 Hz, 1H, H-5), 7.16 (d, *J* = 2.5 Hz, 1H, H-8), 7.05 (dd, *J* = 8.0, 2.5 Hz), 6.05 (s, 1H, C=CH), 5.78 (s, 1H, C=CH), 4.27 (t, *J* = 6.0 Hz, 2H, H-3), 3.85 (s, 3H, OCH_3), 3.78 (s, 3H, OCH_3), 3.04 (t, *J* = 6.0 Hz, 2H, H-4). ^{13}C NMR: (125 MHz, $CDCl_3$) δ 187.5, 166.3, 166.1, 158.7, 142.8, 129.7, 128.8, 125.7, 119.4, 111.9, 99.3, 95.7, 55.5, 51.9, 42.5, 27.9. HRMS (ESI/Q-TOF) *m*/*z*: $[M + H]^+$ Calcd for $C_{16}H_{16}NO_4$ 286.1073; found 286.1099. m.p. 151 °C (decomp.)
- Methyl (2Z)-8,9-methylenedioxy-[2-oxo-5,6-dihydropyrrolo[2,1,a]isoquinolin-3-ylidene]-2-ethanoate (**8c**) (9:1 dichloromethane/ethyl acetate) affording a bright red solid (0.162 g, 24%). ^{1}H NMR: (500 MHz, $CDCl_3$) δ 7.09 (s, 1H, H-8), 6.75 (s, 1H, H-5), 6.06 (s, 2H, OCH_2O), 6.03 (s, 1H, C=CH), 5.67 (s, 1H, C=CH), 4.28 (t, *J* = 6.5 Hz, 2H, H-3), 3.78 (s, 3H, OCH_3), 3.04 (t, *J* = 6.5 Hz, 3H, H-4). 13C NMR: (125 MHz, $CDCl_3$) δ 187.1, 166.5, 166.0, 151.8, 147.4, 147.4, 143.3, 133.0, 118.5, 108.6, 107.1, 102.0, 98.7, 94.9, 51.9, 42.2, 29.0. HRMS (ESI/Q-TOF) *m*/*z*: $[M + H]^+$ Calcd for $C_{16}H_{14}NO_5$ 300.0866; found 300.0850. m.p. 199 °C (decomp.)
- Methyl (2Z)-[2-oxo-5,6-dihydropyrrolo[2,1,a]isoquinolin-3-ylidene]-2-ethanoate (**8d**) (20–40% ethyl acetate in hexane) to afford a bright red solid (0.080 g, 18%). ^{1}H NMR: (500 MHz, $CDCl_3$) δ 7.71 (d, *J* = 7.5 Hz, 1H, H-8), 7.48 (t, *J* = 7.5 Hz, 1H, H-7), 7.37 (t, *J* = 7.0 Hz, 1H, H-6), 7.31 (d, *J* = 7.5 Hz, 1H, H-5), 6.06 (s, 1H, C=CH), 5.82 (s, 1H, C=CH), 4.31 (t, *J* = 6.0 Hz, 2H, H-3), 3.78 (s, 3H, OCH_3), 3.12 (t, *J* = 6.0 Hz, 2H, H-4). ^{13}C NMR: (125 MHz, $CDCl_3$) δ 187.5, 166.4, 166.2, 142.8, 136.4, 132.6, 128.6, 128.1, 127.4, 124.9, 99.3, 95.6, 51.9, 42.3, 28.7. HRMS (ESI/Q-TOF) *m*/*z*: $[M + H]^+$ Calcd $C_{15}H_{14}NO_3$ 256.0968; found 256.1009. m.p. 138 °C (decomp.)
- Methyl (2Z)-8,9-dimethoxy-[2-oxo-5,6-dihydropyrrolo[2,1,a]isoquinolin-3-ylidene]-2-ethanoate (**8e**) Triturated with diethyl ether to afford a bright red solid (0.416 g, 83%). ^{1}H NMR: (500 MHz, $CDCl_3$) δ 7.09 (s, 1H, H-8), 6.76 (s, 1H, H-8), 6.03 (s, 1H, C=CH), 5.70 (s, 1H, C=CH), 4.31 (t, *J* = 6.5 Hz, 2H, H-3), 3.96 (s, 3H, OCH_3), 3.92 (s, 3H, OCH_3), 3.78 (s, 3H, OCH_3), 3.06 (t, *J* = 6.5 Hz, 2H, H-4). ^{13}C NMR: (125 MHz, $CDCl_3$) δ 187.0, 166.5, 166.1, 153.2, 148.5, 143.5, 131.2, 117.1, 110.8, 109.8, 98.5, 94.5, 56.2, 56.2, 51.9, 42.3, 28.4. HRMS (ESI/Q-TOF) *m*/*z*: $[M + H]^+$ Calcd $C_{17}H_{18}NO_5$ 316.1179; found 316.1183. Spectral data were consistent with reported values [16].

- (±)-Methyl (2Z)-8,9-dimethoxy-[2-oxo-5-methyl-5,6-dihydropyrrolo[2,1,a] isoquinolin-3-ylidene]-2-ethanoate (**8f**) Compound **7f** (0.200 g, 1.26 mmol) was dissolved in anhydrous methanol (10 mL) and stirred in a flask at room temperature. DMAD (194 µL, 168 mg, 1.1 equiv.) was added and stirred for 16 h at room temperature. The product precipitated out of the solution as the reaction progressed and was cooled in an ice bath and then filtered. The solid was then washed with cold diethyl ether and dried to give a bright red powder (0.084 g, 28%). ^{1}H NMR: (400 MHz, $CDCl_3$) δ 7.09 (s, 1H, H-6), 6.74 (s, 1H, H-9), 6.05 (s, 1H, H-4″), 5.71 (s, 1H, H-1″), 5.57–5.54 (m, 1H, H-3), 3.96 (s, 3H, -OCH_3), 3.93 (s, 3H, -OCH_3), 3.78 (s, 3H, -OCH_3), 3.42 (dd, *J* = 6.4, 16.4 Hz, 1H, H-4a), 2.72 (dd, *J* = 1.6, 16.4 Hz, 1H, H-4b), 1.09 (d, *J* = 6.8 Hz, 3H, H-3′). ^{13}C NMR: (100 MHz, $CDCl_3$) δ 187.1, 166.4, 164.8, 153.5, 148.3, 142.3, 129.2, 116.6, 111.6, 109.7, 98.6, 94.8, 56.2, 56.1, 51.9, 47.4, 34.4, 18.0 HRMS (ESI/Q-TOF) *m*/*z*: $[M + H]^+$ Calcd $C_{18}H_{21}NO_5$ 330.1336; found 330.1343. m.p. 189 °C (decomp.)

2.2. *Biological Studies*

2.2.1. MIC Evaluation for Other Bacterial Species (Microdilution Assay)

The MIC was determined by broth microdilution according to Clinical Laboratory Standards Institute guidelines and performed in biological triplicate [36]. The respective pathogen was streaked onto an agar plate of cation-adjusted Mueller–Hinton broth (CAMHB) and allowed to incubate at 37 °C for 16 h. A single colony was suspended in 10 mL of CAMHB media (with the exception of *Enterococcus faecium*, where Brain-heart infusion media were used) and incubated at 37 °C for 16 h.

The respective drug concentration (or DMSO solvent control) was added to each well of a 96-well plate in media (95 µL), to which 5 µL of the pathogen suspended in media was added (5×10^5 CFU/mL). The plate was covered with a sealing film and incubated at 37 °C for 24 h. The absorbance of each well was analyzed on a TECAN Spark 10M plate reader at a wavelength of 600 nm (OD_{600}). Growth inhibition was calculated using Microsoft Excel^{TM} relative to uninhibited controls, using the following equation. Compounds that reduced the OD_{600} by ≥90% relative to vehicle controls were considered active for the tested concentration.

$$\text{Bacterial inhibition}(\%) = \frac{\text{Absorbance}_{\text{Drug}} - \text{Absorbance}_{\text{Media control}}}{\text{Absorbance}_{\text{Vehicle control}} - \text{Absorbance}_{\text{Media control}}}$$

2.2.2. Cytotoxicity Assay (MTS)

The MTS assay was performed according to a reported protocol [37]. HEp-2 (human epithelial type 2, ATCC® CCL-23) or McCoy B (mouse fibroblast, ATCC® CRL-1696) cells were grown to 80–100% confluence, and cell density was determined with a hemocytometer. Cells were then diluted to 1×10^5 cells/mL, and 100 µL was added to each well in a 96-well plate, with three wells left as controls. Cells were grown for 24 h, and the media were removed and then replaced with fresh media containing 0.1% DMSO and the drug for testing in triplicate. Drugs and cells were incubated for 24 h, and then 20 µL of MTS and PMS solution was added and incubated for a further four hours. Absorbance was measured at 490 nm on a TECAN M200 Infinite plate reader, and data were processed using Microsoft Excel^{TM} with the following equation to determine cell viability relative to DMSO controls. Then, statistical analysis was performed using GraphPad PRISM version 8.0.1. Statistical significance was determined using Tukey's one-way analysis of variance (ANOVA), followed by Tukey's post hoc test, with $p < 0.05$ considered significant.

$$\text{Cell viability}(\%) = \frac{\text{Absorbance}_{\text{Drug}} - \text{Absorbance}_{\text{Dye}}}{\text{Absorbance}_{\text{DMSO}} - \text{Absorbance}_{\text{Dye}}}$$

3. Results and Discussion

3.1. Design and Synthesis of Tricyclic Isoquinolines

We designed and synthesized a small library of new tricyclic isoquinoline structures using 3,4-dihydro-1-methylisoquinoline compounds as substrates for [2 + 3] cyclization reactions with dimethyl acetylenedicarboxylate (DMAD) [16]. The Bischler–Napieralski reaction was used to access the intermediate dihydroisoquinolines **7a–e**, and an intracyclic Ritter reaction was used to access compound **7f**.

Scheme 1 outlines the synthesis of key compounds in this investigation. Compound **2a** was prepared from methylation of 3-hydroxybenzaldehyde (**1**). Then, aryl-aldehydes (**2a–c**) were reacted with nitromethane to afford the corresponding nitrostyrene compounds (**3a–c**), which were subsequently reduced with lithium aluminum hydride to the corresponding phenylethylamine compounds (**4a–c**). Acylation with acetyl chloride afforded the corresponding amides (**5a–e**), which were cyclized to the dihydroisoquinolines compounds (**7a–7e**) using phosphoryl chloride and phosphorus pentoxide. Compound **7f** was synthesized from the intracyclic Ritter reaction using methyl eugenol (**6**) and acetonitrile under strongly acidic conditions.

Compound **7d** was further derivatized using literature procedures (shown in Scheme 2) to afford the corresponding 7-nitro-3,4-dihydro-1-methylisoquinoline (**9**), and then the nitro group was reduced to 7-amino-3,4-dihydro-1-methylisoquinoline (**10**) and subsequently acylated using an amide coupling reaction to afford compound **11** (*N*-(1-methyl-3,4-dihydroisoquinolin-7-yl)acetamide). These compounds were used to investigate the scope of reactions with DMAD, as the current literature reports only the reaction occurring when aryl-ethers are substituted with $R^1 = R^2 = OCH_3$.

The synthesis of compounds **8a–8f** was performed using a modified literature procedure [16] with dihydroisoquinolines (**7a–7f**), as shown in Scheme 3. The reaction was found to proceed rapidly and exothermically when R^1 acted as electron-donating substituents, affording the desired product in good yield. However, compounds bearing no donating substituent at R^1 produced lower yields. Compound **7d** ($R^1 = R^2 = H$) reacted slowly and afforded compound **8d** in an 18% yield. The reaction of DMAD with compounds **9-11** did not form the desired tricyclic product.

An investigation using compound **8e** as a substrate was undertaken to ascertain potential methods of derivatization. However, attempts to hydrolyze the terminal ester using acidic hydrolysis at room temperature or at reflux in aqueous 1M hydrochloric acid did not afford any conversation from the starting material. Alkaline hydrolysis (1M NaOH/THF (1:3)) led to the formation of multiple indiscernible and inseparable products, likely due to the formation of multiple diastereoisomers from the nucleophilic addition of hydroxide at several α–β unsaturated carbonyl positions. Attempts to transesterify the terminal ester were unsuccessful (conditions: EtOH, p-TsOH (5 mol%), reflux, 4 h).

Alkylation of the nitrogen was attempted using methyl iodide (conditions: CH_3I (5 equiv.), acetone, reflux, 16 h), and a literature method was used to functionalize the nitrogen to the corresponding *N*-oxide [38] (conditions: 30% H_2O_2, MeOH, r.t., 16 h). Both reactions were unsuccessful. However, the reported crystal structure of **8e** highlights that the nitrogen is in a trigonal planar geometry, suggesting that the lone pair is highly delocalized throughout the structure and may not behave as a typical tertiary nitrogen, preventing its use in derivatization reactions [39].

3.2. Cytotoxicity Evaluation of Synthesized Compounds Against Mammalian Cell Lines

The capacity of compounds **8a–f** to reduce the viability of two mammalian cell lines was investigated to determine if the antibacterial activity was selective. Compounds **8a–f** were screened against HEp-2 and McCoy cells, and their effects against cell viability were assessed using the MTS assay. **8a**, **b**, **d**, and **8f** showed mild cytotoxic effects against HEp-2 cells, as shown in Figure 2. Bridging the methoxy group to a methylenedioxy group (**8c** and **8e**) was found to have a small effect on the cytotoxicity at 125 µM (**8c** = 65% vs. **8e** = 76%). All other compounds were found to have larger cytotoxic effects at 125 µM, suggesting that removal of aryl ethers from R^1 or R^2 may enhance these effects (**8e** vs. **8a** = 36% and **8b** = 17%).

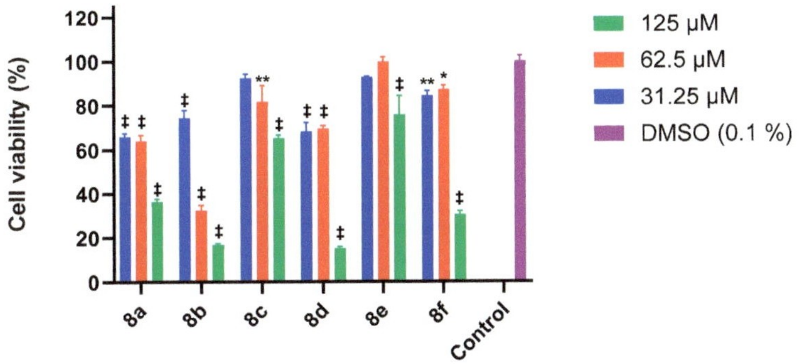

Figure 2. Cytotoxicity of compounds **8a–8f** against HEp-2 cells (24 h; 31.25 µM, 62.5 µM, and 125 µM). Values represent the average ± SEM from three independent experiments. Significance shown is relative to solvent (0.1% DMSO) vehicle controls: (*) $p \leq 0.05$, (**) $p \leq 0.01$, (‡) $p \leq 0.001$ vs. vehicle control (n = 3, one-way ANOVA, Tukey's post hoc).

The cytotoxic properties against McCoy cells were found to correlate strongly with those observed in HEp-2 cells (Figure 3); however, compound **8f** was significantly more cytotoxic to McCoy B cells at all tested concentrations (31.25 µM = 39%). Compounds **8c** and **8e** were not significantly cytotoxic at all assay concentrations, suggesting that the electron-rich nature of the aromatic ring may moderate its biological effects. Further study into various substituents on the ring may provide insight into this. Compounds **8a** (125 µM, 17%), **8b** (125 µM, 10%), and **8d** (125 µM, 12%) were found to have similar cytotoxic properties, suggesting that lipophilicity may be a driving factor for the observed cytotoxicity in McCoy cells. Removing the C-5 methyl group significantly decreased the cytotoxicity (**8e** vs. **8f**). Compounds **8a–f** possess an electron-poor Michael acceptor, which might allow non-specific covalent binding to cellular proteins, leading to non-selective cytotoxicity [40].

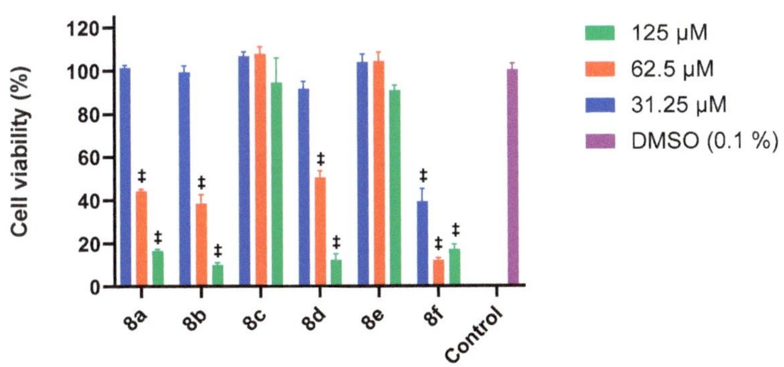

Figure 3. Cytotoxicity of compounds **8a–8f** against McCoy B cells (24 h; 31.25 µM, 62.5 µM, and 125 µM). Values represent the average ± SEM from three independent experiments. Significance shown is relative to solvent (0.1% DMSO) vehicle controls: (‡) $p \leq 0.001$ vs. vehicle control (n = 3, one-way ANOVA, Tukey's post hoc).

3.3. Evaluation of Antibacterial Properties

Compounds **8a–f** were screened against a panel of bacteria, and the results are shown in Table 3.

Table 3. Minimum inhibitory concentration (MIC) of compounds **8a-f** against Gram-positive and Gram-negative pathogens.

Compound	R^1	R^2	R^3	MIC (μg/mL)							
				S. aureus	*E. coli*	*E. coli* Δ*tolC*	*S. pneumoniae*	*E. faecium*	*P. aeruginosa*	*A. baumanii*	*A. baylyi*
8a	OCH_3	H	H	>32	>32	>32	>32	>32	>32	>32	>32
8b	H	OCH_3	H	>32	>32	>32	>32	>32	>32	>32	>32
8c	OCH_2O		H	>32	>32	>32	>32	>32	>32	>32	>32
8d	H	H	H	16	>128	>128	>32	128	>128	>32	>32
8e	OCH_3	OCH_3	H	>32	>32	>32	>32	>32	>32	>32	>32
8f	OCH_3	OCH_3	CH_3	32	>128	>128	32	64	>128	>128	>128
CIP [a] [11]	-	-	-	0.25	≤0.03	≤0.03	1	-	-	-	0.06 [41]
MER [b] [42]	-	-	-	0.12	≤0.06	-	-	-	0.5	0.5	-
VAN [c] [43]	-	-	-	-	-	-	-	1	-	-	-

[a] Ciprofloxacin, [b] meropenem, and [c] vancomycin.

For Gram-positive pathogens, compound **8d** was the most active, as it inhibited the growth of *S. aureus* with an MIC_{90} concentration half that of **8f**, with no observed change in potency relative to other screened pathogens. Since **8d** lacks substitution on the aromatic ring, the result suggests that the dimethoxy substituents in **8f** are not required for activity or lower potency. This finding is supported by the poor activity of the mono-methoxylated isoquinolines **8a** and **8b**, indicating that substitution of the aromatic ring at the R^1 position is not well tolerated but is acceptable at R^2, albeit with a reduction in potency.

Methoxylation of the aryl ring was found to abolish antibacterial activity (**8d** vs. **8a–c** and **8e**). It is proposed that with the presence of several Michael acceptors within the compounds, electron donating groups may weaken their electrophilicity due to their highly conjugated nature. Development of compounds with electron withdrawing substituents, as attempted in the current work, would be of benefit to evaluate the influence of this property on antibacterial activity.

For compounds **8a–b**, poor aqueous solubility prevented testing at higher concentrations. Bridging the dimethoxy analogue to a methylene linker (**8c**) was found not to influence potency, and although slightly less lipophilic than the dimethoxy analogue, solubility was not noticeably improved. The antibacterial properties were found to correlate with the cytotoxic properties of compounds **8d** and **8f**, potentially reducing their efficacy as antibacterial agents without further development to enhance their selectivity toward bacteria over mammalian cell lines.

The mechanism of action of these compounds has not been determined, although there are structural similarities with compounds such as berberine, which is known to inhibit bacterial replication by the inhibition of FtsZ [10]. As cytotoxicity was observed in mammalian cell lines, the mechanism of action may encompass a broader, more conserved mechanism of action, such as interfering with DNA replication. Highly conjugated and planar isoquinoline-based compounds such as HSN584 have been shown to interfere with cell-wall synthesis [8], and they could be explored as a potential mechanism of action of **8d** and **8f**. Future studies may warrant determination of the minimum bactericidal concentration (MBC) to ascertain if active compounds are bacteriostatic or bactericidal and could provide insight into the mechanism of action. Furthermore, it would be of interest to examine if tested compounds exhibit antibacterial activity on a broader range of pathogens, including drug-resistant bacteria pertinent to clinical settings, including methicillin-resistant *S. aureus* and vancomycin-resistant enterococci.

There was no observed inhibitory activity against Gram-negative pathogens (*E. coli*, *E. faecium*, *P. aeruginosa*, *A. baumanii*, and *A. baylyi*). In the case of *E. coli*, screening against the Δ*tolC* mutant suggests the lack of activity is not due to drug efflux. Although the compounds may be benign upon uptake, for Gram-negative pathogens, uptake is one of the most challenging issues facing drug efficacy. As the scope of compatible chemical reactions with compounds **8a–f** was restricted, the use of drug-delivery systems such as nanoformulations including polyphenols [44], silica [45], and metallic nanoparticles [46] may simultaneously address drug uptake issues, as well as improve the limited solubility of these compounds in aqueous systems, opening the potential for assessing their antibacterial properties in vivo.

Moreover, drug uptake in Gram-negative bacteria could be improved by the exploitation of active transport systems. As the cyclisation step of compound **7** with DMAD to compound **8** employs mild reaction conditions, it can be envisioned that alkynes with an array of synthetic handles, or pre-functionalized moieties such as glycosides, peptides, or an ionizable functional group could be installed [47–49]. This may facilitate better transport across Gram-negative cell membranes, and could be utilized to improve the efficacy, and potentially, the selectivity of compounds toward various pathogens tested this work.

4. Conclusions

Access to novel scaffolds that afford unique compounds with antibacterial properties is critical in the development of antibacterial agents. Isoquinoline compounds and their fused-ring derivatives possess a wide range of pharmacological properties, including antibacterial applications. In this work, six methyl (2Z)-[2-oxo-5,6-dihydropyrrolo[2,1,a]-isoquinoline-3-ylidene]-2-ethanoate compounds were synthesized. Five of these were previously unreported. The scope of compatible chemistry for the synthesis of new aryl substituents was investigated and is thus far restricted to an unsubstituted aryl system and aryl ethers. Attempts to derivatize **8e** were unsuccessful, but the attempted reactions provide valuable insight into specific properties of these compounds, including the non-nucleophilic nature of the nitrogen and the limited scope of compatible aryl substituents on 1-methyl-3,4-dihydroisoquinolines that undergo cyclisation with DMAD. It was shown that compounds **9**, **10**, and **11** do not readily react with DMAD to afford tricyclic isoquinolines compounds.

The inhibition of bacterial growth was observed against the Gram-positive bacteria *S. aureus* and *E. faecium* for **8d** and **8f**. Additionally, **8f** weakly exhibited activity against *S. pneumoniae*. Gram-negative pathogens were found to be uninhibited by **8a–f**. As cytotoxicity was observed against mammalian cell lines, the utility of these compounds as antibacterial agents may be limited without further development to enhance the selectivity toward bacterial pathogens. This may be accomplished by functionalizing electron deficient alkynes with glycosides, peptides, or ionizable functional groups prior to cyclization with 1-methyl-3,4-dihydroisoquinolines to facilitate increased drug uptake, or by incorporating nanoencapsulation of active compounds to improve their uptake into pathogens of interest, and mitigate the observed cytotoxicity by improving selectivity.

Supplementary Materials: The following supporting information can be downloaded at https://www.mdpi.com/article/10.3390/biochem5010001/s1, NMR spectra of new compounds and compounds for which data were not previously reported: Table S1: Purity of **8a–f** determined by qNMR. Figure S1: The 500 MHz 1H NMR spectrum of **7f** in CDCl3; Figure S2: The 100 MHz 13C NMR spectrum of **7f** in $CDCl_3$; Figure S3: The 400 MHz ^1H NMR spectrum of **8a** in $CDCl_3$; Figure S4: The 100 MHz ^{13}C NMR spectrum of **8a** in $CDCl_3$; Figure S5: The 400 MHz ^1H NMR spectrum of **8b**

in $CDCl_3$.; Figure S6: The 100 MHz ^{13}C NMR spectrum of **8b** in $CDCl_3$; Figure S7: The 400 MHz ^{1}H NMR spectrum of **8c** in $CDCl_3$; Figure S8: The 100 MHz ^{13}C NMR spectrum of **8c** in $CDCl_3$; Figure S9: The 400 MHz ^{1}H NMR spectrum of **8d** in $CDCl_3$; Figure S10: The 100 MHz ^{13}C NMR spectrum of **8d** in $CDCl_3$; Figure S11: The 400 MHz ^{1}H NMR spectrum of **8f** in $CDCl_3$; Figure S12: The 100 MHz ^{13}C NMR spectrum of **8f** in $CDCl_3$; Figure S13: The 400 MHz ^{1}H NMR spectrum of **11** in $CDCl_3$; Figure S14: The 100 MHz ^{13}C NMR spectrum of **11** in $CDCl_3$; Figure S15: The 400 MHz ^{1}H NMR spectrum of **12** in $CDCl_3$; Figure S16: The 100 MHz ^{13}C NMR spectrum of **12** in $CDCl_3$ [50,51].

Author Contributions: Conceptualization, M.J.A.P., A.T.U. and A.M.M.; methodology, M.J.A.P., A.T.U., E.J.H., J.A. and A.M.M.; software, M.J.A.P.; formal analysis, M.J.A.P.; investigation, M.J.A.P.; resources, E.J.H. (for antibacterial studies); writing—original draft preparation, M.J.A.P. and A.M.M.; writing—review and editing, M.J.A.P., A.T.U., E.J.H., J.A. and A.M.M.; supervision, A.T.U., E.J.H., J.A. and A.M.M. All authors have read and agreed to the published version of the manuscript.

Funding: This project received no external funding.

Data Availability Statement: Data is contained within the article.

Acknowledgments: The authors would like to thank Willa Huston at the University of Technology Sydney for providing access to laboratory facilities and supplying the reagents and cell lines used to perform MTS assays. We would also like to thank Tristan Rawling for helpful discussions. This research was supported by an Australian Government Research Training Program (RTP) Scholarship.

Conflicts of Interest: The authors declare no conflicts of interest.

References

1. Rice, L.B. Federal Funding for the Study of Antimicrobial Resistance in Nosocomial Pathogens: No ESKAPE. *J. Infect. Dis.* **2008**, *197*, 1079–1081. [CrossRef]
2. World Health Organization. *WHO Publishes List of Bacteria for Which New Antibiotics Are Urgently Needed*; World Health Organization: Geneva, Switzerland, 2017.
3. World Health Organization. *2023 Antibacterial Agents in Clinical and Preclinical Development: An Overview and Analysis*; World Health Organization: Geneva, Switzerland, 2023.
4. World Health Organization. *2021 Antibacterial Agents in Clinical and Preclinical Development*; World Health Organization: Geneva, Switzerland, 2022.
5. Brown, E.D.; Wright, G.D. Antibacterial drug discovery in the resistance era. *Nature* **2016**, *529*, 336–343. [CrossRef] [PubMed]
6. Coates AR, M.; Hu, Y.; Holt, J.; Yeh, P. Antibiotic combination therapy against resistant bacterial infections: Synergy, rejuvenation and resistance reduction. *Expert Rev. Anti-Infect. Ther.* **2020**, *18*, 5–15. [CrossRef]
7. Toews, M.L.; Bylund, D.B. Pharmacologic Principles for Combination Therapy. *Proc. Am. Thorac. Soc.* **2005**, *2*, 282–289. [CrossRef]
8. Karanja, C.W.; Naganna, N.; Abutaleb, N.S.; Dayal, N.; Onyedibe, K.I.; Aryal, U.; Seleem, M.N.; Sintim, H.O. Isoquinoline Antimicrobial Agent: Activity against Intracellular Bacteria and Effect on Global Bacterial Proteome. *Molecules* **2022**, *27*, 5085. [CrossRef] [PubMed]
9. Payne, M.; Bottomley, A.L.; Och, A.; Hiscocks, H.G.; Asmara, A.P.; Harry, E.J.; Ung, A.T. Synthesis and biological evaluation of tetrahydroisoquinoline-derived antibacterial compounds. *Bioorg. Med. Chem.* **2022**, *57*, 116648. [CrossRef] [PubMed]
10. Sun, N.; Chan, F.; Lu, Y.; Neves MA, C.; Lui, H.; Wang, Y.; Chow, K.; Chan, K.; Yan, S.; Leung, Y.; et al. Rational Design of Berberine-Based FtsZ Inhibitors with Broad-Spectrum Antibacterial Activity. *PLoS ONE* **2014**, *9*, e97514. [CrossRef] [PubMed]
11. Panchaud, P.; Bruyère, T.; Blumstein, A.; Bur, D.; Chambovey, A.; Ertel, E.A.; Gude, M.; Hubschwerlen, C.; Jacob, L.; Kimmerlin, T.; et al. Discovery and Optimization of Isoquinoline Ethyl Ureas as Antibacterial Agents. *J. Med. Chem.* **2017**, *60*, 3755–3775. [CrossRef]
12. Napieralski, B. *Über Einige Secundäre Diamine der Fettreihe: Zur Kenntniss Einer Neuen Isochinolinsynthese*; Verlag Nicht Ermittelbar: Basel, Switzerland, 1893.
13. Pictet, A.; Spengler, T. Über die Bildung von Isochinolin-derivaten durch Einwirkung von Methylal auf Phenyl-äthylamin, Phenyl-alanin und Tyrosin. *Berichte Der Dtsch. Chem. Ges.* **1911**, *44*, 2030–2036. [CrossRef]
14. Ritter, J.J.; Murphy, F.X. N-Acyl-β-phenethylamines, and a New Isoquinoline Synthesis. *J. Am. Chem. Soc.* **1952**, *74*, 763–765. [CrossRef]
15. Surikova, O.V.; Yusov, A.S.; Makhmudov, R.R.; Mikhailovskii, A.G. Synthesis and Analgesic Activity of 1-substituted 3-methyl-6-methoxy-7(n-butoxy)-3,4-dihydroisoquinolines. *Pharm. Chem. J.* **2017**, *51*, 18–21. [CrossRef]

16. Qinghong, L.; Ball, G.E.; Bishop, R. Ritter reactions. XII. Reappraisal of the reactivity of methyl Schiff bases with dimethyl acetylenedicarboxylate. *Tetrahedron* **1997**, *53*, 10899–10910. [CrossRef]
17. Nagarapu, L.; Vulupala, H.R.; Bantu, R.; Sajja, Y.; Nanubolu, J.B. Efficient synthesis and resolution of novel 2-(hydroxymethyl)-7,8-dihydro-1H-indeno[5,4-b] furan-6(2H)-one by lipase Pseudomonas cepacia. *Tetrahedron Asymmetry* **2014**, *25*, 578–582. [CrossRef]
18. Quan, X.; Ren, Z.; Wang, Y.; Guan, Z. p-Toluenesulfonic Acid Mediated 1,3-Dipolar Cycloaddition of Nitroolefins with NaN_3 for Synthesis of 4-Aryl-*NH*-1,2,3-triazoles. *Org. Lett.* **2014**, *16*, 5728–5731. [CrossRef]
19. Jakubec, P.; Cockfield, D.M.; Hynes, P.S.; Cleator, E.; Dixon Darren, J. Enantio- and diastereoselective Michael additions of C-succinimidyl esters to nitro olefins using cinchonine-derived bifunctional organocatalysts. *Tetrahedron Asymmetry* **2011**, *22*, 1147–1155. [CrossRef]
20. Greger, J.G.; Yoon-Miller SJ, P.; Bechtold, N.R.; Flewelling, S.A.; MacDonald, J.P.; Downey, C.R.; Cohen Eric, A.; Pelkey, E.T. Synthesis of Unsymmetrical 3,4-Diaryl-3-pyrrolin-2-ones Utilizing Pyrrole Weinreb Amides. *J. Org. Chem.* **2011**, *76*, 8203–8214. [CrossRef] [PubMed]
21. Crestey, F.; Jensen, A.A.; Borch, M.; Andreasen, J.T.; Andersen, J.; Balle, T.; Kristensen, J.L. Design, Synthesis, and Biological Evaluation of Erythrina Alkaloid Analogues as Neuronal Nicotinic Acetylcholine Receptor Antagonists. *J. Med. Chem.* **2013**, *56*, 9673–9682. [CrossRef]
22. Smith, R.A.; White, R.L.; Krantz, A. Stereoisomers of allenic amines as inactivators of monoamine oxidase type B. Stereochemical probes of the active site. *J. Med. Chem.* **1988**, *31*, 1558–1566. [CrossRef] [PubMed]
23. Henry, M.C.; Senn, H.M.; Sutherland, A. Synthesis of Functionalized Indolines and Dihydrobenzofurans by Iron and Copper Catalyzed Aryl C–N and C–O Bond Formation. *J. Org. Chem.* **2019**, *84*, 346–364. [CrossRef] [PubMed]
24. Haddenham, D.; Pasumansky, L.; DeSoto, J.; Eagon, S.; Singaram, B. Reductions of Aliphatic and Aromatic Nitriles to Primary Amines with Diisopropylaminoborane. *J. Org. Chem.* **2009**, *74*, 1964–1970. [CrossRef]
25. Wright, A.E.; Roth, G.P.; Hoffman, J.K.; Divlianska, D.B.; Pechter, D.; Sennett, S.H.; Guzmán, E.A.; Linley, P.; McCarthy, P.J.; Pitts, T.P.; et al. Isolation, Synthesis, and Biological Activity of Aphrocallistin, an Adenine-Substituted Bromotyramine Metabolite from the Hexactinellida Sponge Aphrocallistes beatrix. *J. Nat. Prod.* **2009**, *72*, 1178–1183. [CrossRef]
26. An, R.; Gu, Z.; Sun, H.; Hu, Y.; Yan, R.; Ye, D.; Liu, H. Self-assembly of Fluorescent Dehydroberberine Enhances Mitochondria-Dependent Antitumor Efficacy. *Chem. A Eur. J.* **2018**, *24*, 9812–9819. [CrossRef]
27. Alalla, A.; Merabet-Khelassi, M.; Aribi-Zouioueche, L.; Riant, O. Green Synthesis of Benzamides in Solvent- and Activation-Free Conditions. *Synth. Commun.* **2014**, *44*, 2364–2376. [CrossRef]
28. Johannes, M.; Altmann, K. A Ring-Closing Metathesis-Based Approach to the Synthesis of (+)-Tetrabenazine. *Org. Lett.* **2012**, *14*, 3752–3755. [CrossRef] [PubMed]
29. Haraszti, F.; Igricz, T.; Keglevich, G.; Milen, M.; Abranyi-Balogh, P. Synthetic and Mechanistic Study on the Microwave Assisted Fries Rearrangement of 1-Methylidene-3,4-dihydroisoquinoline-2(*1H*)-yl-methanones. *Curr. Org. Chem.* **2018**, *22*, 912–922. [CrossRef]
30. Václavík, J.; Pecháček, J.; Vilhanová, B.; Šot, P.; Januščák, J.; Matoušek, V.; Přech, J.; Bártová, S.; Kuzma, M.; Kačer, P. Molecular Structure Effects in the Asymmetric Transfer Hydrogenation of Functionalized Dihydroisoquinolines on (S,S)-[RuCl(η6-p-cymene)TsDPEN]. *Catal. Lett.* **2013**, *143*, 555–562. [CrossRef]
31. Kusama, H.; Yamashita, Y.; Uchiyama, K.; Narasaka, K. ChemInform Abstract: Transformation of Oximes of Phenethyl Ketone Derivatives to Quinolines and Azaspirotrienones Catalyzed by Tetrabutylammonium Perrhenate and Trifluoromethanesulfonic Acid. *ChemInform* **1997**, *28*, 965–975.
32. Muraca AC, A.; Perecim, G.; Rodrigues, A.; Raminelli, C. Convergent Total Synthesis of (±)-Apomorphine via Benzyne Chemistry: Insights into the Mechanisms Involved in the Key Step. *Synthesis* **2017**, *49*, 3546–3557.
33. Gibson, H.W.; Berg MA, G.; Dickson, J.C.; Lecavalier, P.R.; Wang, H.; Merola, J.S. Diastereomeric Reissert Compounds of Isoquinoline and 6,7-Dimethoxy-3,4-dihydroisoquinoline in Stereoselective Synthesis. *J. Org. Chem.* **2007**, *72*, 5759–5770. [CrossRef]
34. Heer, J.P.; Harling, J.D.; Thompson, M. Preparation of 1,7-disubstituted-1,2,3,4-tetrahydroisoquinolines. *Synth. Commun.* **2002**, *32*, 2555–2563. [CrossRef]
35. Harling, J.D.; Orlek, B.S.; Thompson, M. Anti-Convulsant Isoquinolyl-Benzamide Derivatives. Patent US6277861B1, 16 March 1998.
36. The Clinical and Laboratory Standards Institute. *Methods for Dilution Antimicrobial Susceptibility Tests for Bacteria That Grow Aerobically*, 11th ed.; The Clinical and Laboratory Standards Institute: Wayne, PA, USA, 2018.
37. Riss, T.L.; Moravec, R.A.; Niles, A.L.; Duellman, S.; Benink, H.A.; Worzella, T.J.; Minor, L. *Cell Viability Assays*; Eli Lilly & Company and the National Center for Advancing Translational Sciences: Bethesda, MD, USA, 2013; pp. 1–23.
38. Cullen, D.R.; Pengon, J.; Rattanajak, R.; Chaplin, J.; Kamchonwongpaisan, S.; Mocerino, M. Scoping Studies into the Structure-Activity Relationship (SAR) of Phenylephrine-Derived Analogues as Inhibitors of Trypanosoma brucei rhodesiense. *ChemistrySelect* **2016**, *1*, 4533–4538. [CrossRef]

39. Lin, Q.; Djaidi, D.; Bishop, R.; Craig, D.C.; Scudder, M.L. Ritter Reactions. XIII Reactivity of Methyl Schiff Bases with Dimethyl Acetylenedicarboxylate and Mercaptoacetic Acid. *Aust. J. Chem.* **1998**, *51*, 799–806. [CrossRef]
40. Jackson, P.A.; Widen, J.C.; Harki, D.A.; Brummond, K.M. Covalent Modifiers: A Chemical Perspective on the Reactivity of α,β-Unsaturated Carbonyls with Thiols via Hetero-Michael Addition Reactions. *J. Med. Chem.* **2017**, *60*, 839–885. [CrossRef] [PubMed]
41. Chen, T.-L.; Siu, L.-K.; Lee, Y.-T.; Chen, C.-P.; Huang, L.-Y.; Wu Roy, C.-C.; Cho, W.-L.; Fung, C.-P. Acinetobacter baylyi as a Pathogen for Opportunistic Infection. *J. Clin. Microbiol.* **2008**, *46*, 2938–2944. [CrossRef] [PubMed]
42. Zampaloni, C.; Mattei, P.; Bleicher, K.; Winther, L.; Thäte, C.; Bucher, C.; Adam, J.-M.; Alanine, A.; Amrein, K.E.; Baidin, V.; et al. A novel antibiotic class targeting the lipopolysaccharide transporter. *Nature* **2024**, *625*, 566–571. [CrossRef] [PubMed]
43. Rodriguez Carlos, A.; Agudelo, M.; Gonzalez Javier, M.; Vesga, O.; Zuluaga Andres, F. An Optimized Mouse Thigh Infection Model for Enterococci and Its Impact on Antimicrobial Pharmacodynamics. *Antimicrob. Agents Chemother.* **2014**, *59*, 233–238. [CrossRef] [PubMed]
44. Wahnou, H.; Liagre, B.; Sol, V.; El Attar, H.; Attar, R.; Oudghiri, M.; Duval, R.E.; Limami, Y. Polyphenol-Based Nanoparticles: A Promising Frontier for Enhanced Colorectal Cancer Treatment. *Cancers* **2023**, *15*, 3826. [CrossRef] [PubMed]
45. Şen Karaman, D.; Kettiger, H. Chapter 1—Silica-based nanoparticles as drug delivery systems: Chances and challenges. In *Inorganic Frameworks as Smart Nanomedicines*; Grumezescu, A.M., Ed.; William Andrew Publishing: Norwich, NY, USA, 2018; pp. 1–40.
46. Ramalingam, V. Multifunctionality of gold nanoparticles: Plausible and convincing properties. *Adv. Colloid Interface Sci.* **2019**, *271*, 101989. [CrossRef]
47. Delcour, A.H. Outer membrane permeability and antibiotic resistance. *Biochim. Biophys. Acta (BBA)—Proteins Proteom.* **2009**, *1794*, 808–816. [CrossRef] [PubMed]
48. Saxena, D.; Maitra, R.; Bormon, R.; Czekanska, M.; Meiers, J.; Titz, A.; Verma, S.; Chopra, S. Tackling the outer membrane: Facilitating compound entry into Gram-negative bacterial pathogens. *npj Antimicrob. Resist.* **2023**, *1*, 17. [CrossRef]
49. Zhang, R.; Qin, X.; Kong, F.; Chen, P.; Pan, G. Improving cellular uptake of therapeutic entities through interaction with components of cell membrane. *Drug Deliv.* **2019**, *26*, 328–342. [CrossRef]
50. Bharti, S.K.; Roy, R. Quantitative 1H NMR spectroscopy. *TrAC Trends Anal. Chem.* **2012**, *35*, 5–26. [CrossRef]
51. Cushman, M.; Georg, G.I.; Holzgrabe, U.; Wang, S. Absolute Quantitative 1H NMR Spectroscopy for Compound Purity Determination. *J. Med. Chem.* **2014**, *57*, 9219. [CrossRef] [PubMed]

BioChem MDPI

Article

Overproduction of Phenolic Compounds in *Pseudomonas putida* KT2440 Through Endogen Deregulation of the Shikimate Pathway

William Merre, Ricardo Andrade, Cyril Perot, Alexia Chandor-Proust * and Caroline Ranquet

BGene Genetics, 7 Rue des Arts et Métiers, 38000 Grenoble, France
* Correspondence: alexia.chandor@bgene-genetics.com

Abstract: Metabolic engineering of the shikimate pathway offers a promising strategy for enhancing the production of aromatic compounds in microbial hosts. However, feedback inhibition of key enzymes, such as the 3-deoxy-D-arabino-heptulosonate 7-phosphate synthase (DAHP synthase), often limits the yield of target products. In this study, we focused on the DAHP synthase (AroF-I) from *Pseudomonas putida*. Through computational modeling and experimental validation, we identified specific amino-acid residues responsible for tyrosine-mediated feedback inhibition. By targeted mutagenesis, we engineered DAHP synthase variants that exhibit reduced sensitivity to feedback inhibition. The introduction of these engineered enzymes into a metabolically engineered *Pseudomonas putida* strain resulted in significantly increased production of p-coumaric acid. Our findings provide valuable insights into the regulation of the shikimate pathway and demonstrate the potential of protein engineering to improve microbial production of aromatic compounds.

Keywords: feedback resistance; DAHP synthase; *Pseudomonas putida*; enzyme engineering; shikimate pathway; p-coumaric acid

1. Introduction

Synthetic biology is the engineering of microbes for the creation of new biologically based parts and novel devices and systems, as well as redesigning existing, natural biological systems (Kitney and Freemont [1]). One of its main applications is the production of high-interest molecules originally obtained by other means, such as chemical synthesis or plant extraction, using microorganisms as cell factories. For example, the insertion of a heterologous pathway from plants will allow the synthesis of a plant molecule into a microbe. Several studies from this last decade have shown successful examples of microbe engineering (Martău et al. [2], Wang et al. [3], Dudnik et al. [4], Peralta-Yahya et al. [5], Aggarwal et al. [6]). Nevertheless, this process is not straightforward and needs fine-tuning, particularly in the central metabolism. Indeed, efficient carbon fluxes through the new pathway for an interesting biosynthesis of the final product are needed, as well as precursor availability (Li et al. [7], Maurya et al. [8], Leonard et al. [9]).

One of the best examples is the use of aromatic amino acids synthesized by microorganisms (bacteria or yeasts) to produce aromatic compounds of high interest, such as phenylpropanoids (Suzuki et al. [10], Limem et al. [11], Wang et al. [12], Averesch and Krömer [13]). These molecules are derived from L-Phe and L-Tyr and are produced mainly in plants. The phenylpropanoid acids cinnamic acid (CA) and p-coumaric acid (pCA) are important building blocks allowing the biosynthesis of a great variety of secondary metabolites and natural products widely used in industry for flavoring, pharmaceuticals, and cosmetics (Vargas-Tah and Gosset [14]).

The shikimate pathway is the central metabolic route leading to the formation of L-Trp, L-Phe, and L-Tyr from simple carbon sources in microorganisms and plants (Herrmann and Weaver [15]). This metabolic pathway has been extensively studied: it is highly regulated as the accumulation of aromatic amino acids may be deleterious for the cells (Averesch and Krömer [13]). An illustration can be seen in Figure 1. In the Appendix A Section, we provide a more detailed illustration (Figure A1) based on KeGG pathways.

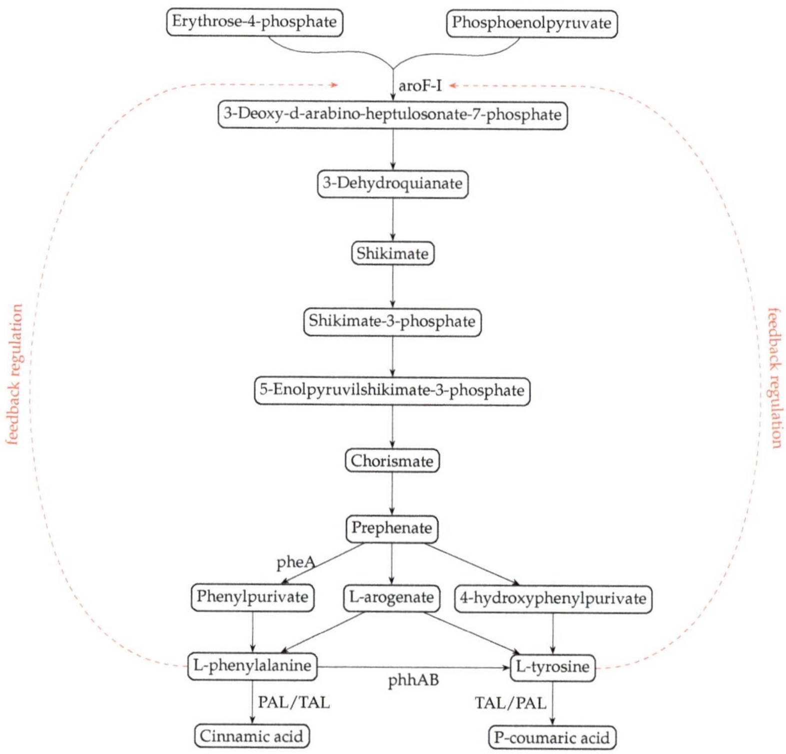

Figure 1. An illustration of the reactions from the shikimate pathway to the production of phenolic compounds cinnamic acid and p-coumaric acid. The red arrows indicate common negative feedback regulation of DAHP synthase by aromatic amino acids.

Among the 15 enzymes that are implicated in this pathway, some, like the DAHP synthases (3-deoxy-D-arabino-heptulosonate 7-phosphate synthase), are negatively repressed by the aromatic amino acid. This phenomenon is called feedback inhibition. This regulation is widespread among microorganisms and tends to be a major bottleneck for the bioproduction of aromatic amino acids derived compounds (Jayaraman et al. [16], Wu et al. [17], Luttik et al. [18], Bilal et al. [19]). Obtaining feedback-inhibition resistant (fbr) enzymes is one of the best strategies to generate L-Phe or L-Tyr overproducer strains, and it was conducted with success in the two well-known chassis for bioproduction, *Escherichia coli* and *Saccharomyces cerevisiae* (Lütke-Eversloh and Stephanopoulos [20], Kang et al. [21], Santos et al. [22], Juminaga et al. [23], Rodriguez et al. [24]). Nevertheless, these microbial chassis are not the best adapted to produce phenylpropanoids, as these aromatic molecules are highly toxic for cells. On the contrary, bacteria of the genus *Pseudomonas* are useful hosts for synthetic biology as they have high-stress tolerance and an interesting metabolic versatility (Nikel et al. [25], Ankenbauer et al. [26], Schwanemann et al. [27]). On top of that, they have an efficient aromatic catabolic pathway, and tolerance towards p-coumaric acid was recently shown (Molina-Santiago et al. [28], Calero et al. [29], Mohamed et al. [30]). In their work, Calero et al. [29] clearly showed that *Pseudomonas putida* KT2440 was twice as tolerant to p-coumaric acid as *E. coli*, highlighting its suitability for the production of phenolic compounds. Previous works demonstrated that production of phenol or pCA with a correct titer could be obtained in some *Pseudomonas* strains, often by transforming these strains with fbr genes from *Escherichia coli* (aroGfbr and tyrAfbr) (Nijkamp et al. [31], Wierckx et al. [32], Calero et al. [33], Wynands et al. [34], Otto et al. [35]). Random mutagenesis experiments were also performed in these strains, and some mutations in genes from the central metabolism and/or the shikimate pathway could improve aromatic molecule production. The overproduction of the *Pseudomonas* DAHP synthase AroF-I gene could also improve this production (Wierckx et al. [32]), showing that this enzyme could be a good target to obtain efficient fbr enzymes from *Pseudomonas*. AroF-I is identified as a phospho-2-dehydro-3-deoxyheptonate aldolase (UniProt Q88KG6).

Historically, most known feedback-resistant mutants in the shikimate pathway were obtained by random mutagenesis approaches, using transposons or highly toxic chemicals, like fluoro-analogs of tyrosine or phenylalanine, or ethyl methane sulfonate, (Weaver and Herrmann [36], Lütke-Eversloh and Stephanopoulos [37]) and rational designs only appeared recently (Jayaraman et al. [16], Liu et al. [38]). While random mutagenesis is a straightforward strategy to obtain mutants with a desired phenotype, it must be associated with a laborious and time-consuming screening phase (Chen et al. [39], Rowlands [40]). Also, reverse engineering is frequently used following these methods to identify impacting mutations and introduce them in a final strain (Zhao et al. [41]). On the contrary, a rational design drastically reduces the time dedicated to screening and alleviates the need for reverse engineering.

The aim of this study was to rationally identify and evaluate the fbr mutant of the AroF-I DAHP synthase from *Pseudomonas putida* KT2440. First, the sensitivity of AroF-I toward the three aromatic amino acids L-Phe, L-Trp, and L-Tyr was tested in vitro, showing a negative regulation of the enzyme in the presence of L-Tyr. Then, bioinformatics studies were carried out to determine putative amino acids responsible for the negative feedback regulation of AroF-I by L-Tyr. Thus, several potential deregulated mutants of AroF-I were constructed and tested in vitro. A double mutant was found to be fully active and resistant to feedback inhibition by tyrosine. This is the first feedback-resistant DAHP enzyme designed, described, and characterized for *Pseudomonas putida*. A mutant strain harboring this deregulated AroF-I shows promising results for high-titer bioproduction of p-coumaric acid.

2. Results

2.1. AroF-I from Pseudomonas putida Is Highly Sensitive to and Inhibited by Tyrosine

To first evaluate if AroF-I is a feedback-inhibited enzyme, the corresponding gene (PP_2324) was cloned into the pET28 expression vector, overexpressed into *E. coli* and the His-tagged protein was purified (see Appendix A Figure A2). The activity of AroF-I was tested in vitro and was found to be fully active in the tested conditions (Figure 2). All the phosphoenolpyruvate was consumed in 1 h or less. Either phenylalanine, tyrosine, or tryptophane were added to the in vitro test, along with the purified AroF-I enzyme. Phenylalanine and tryptophane seemed to slightly inhibit AroF-I, whereas the inhibition by tyrosine was dramatic (Figure 2). We concluded that AroF-I is highly sensitive to tyrosine, which exerts a negative feedback inhibition on this enzyme.

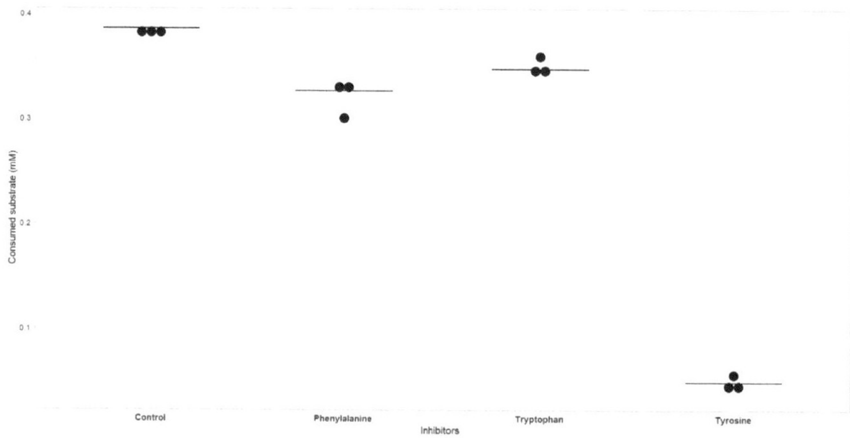

Figure 2. Activity of AroF-I in the absence or presence of inhibitors. Dots represent each replicate as a raw value of consumed substrate (mM), and the mean is represented as a horizontal black bar.

2.2. *Bioinformatics Helps in the Prediction of Allosteric Inhibition of the AroF-I Enzyme*

The 3D structure of AroF-I from *Pseudomonas putida* KT2440 was reconstructed from its amino-acid sequence through homology modeling by SWISS-MODEL (Waterhouse et al. [42]) and used for docking with tyrosine. The protein region that is responsible for the allosteric regulation was identified by comparison with a homolog protein, AroG, a phospho-2-dehydro-3-deoxyheptonate aldolase (UniProt ID P0AB91) from *Escherichia coli* strain K12, known to be regulated by phenylalanine. Through a 3D structure alignment, we could identify the same structure in the AroF-I model. Analyzing different solutions obtained for the docking of AroF-I with tyrosine, we could identify a few residues that seem close enough to the molecule to interact with it. These residues are the following: 160 (P160), 164 (Q164), 190 (S190), 191 (G191), 193 (S193), and 225 (I225) (numbering corresponds to the position of the amino acid in the FASTA sequence of AroF-I).

For each one of these residues, we identified possible substitutions based on the literature and the alignment against naturally tyrosine feedback-resistant AroF enzymes. These approaches are described in the Material and Methods Section, and the results can be seen in Table 1.

Table 1. AroF-I variant positions and amino-acid substitutions assayed in vitro.

Original Amino Acid (AroF-I)	Position	Amino Acids Substitutes
P	160	L
Q	164	A
S	190	A
G	191	K
S	193	A
I	225	P

2.3. *Effect of Point Mutations on AroF-I Activity in the Presence or Absence of an Inhibitor*

As described in the previous section, multiple potential mutation sites were predicted by the in silico study. Some of the best candidates were chosen, and mutations were introduced into the AroF-I enzyme. It included P160L, Q164A, S190A, G191K and S193A. As for the wildtype, the corresponding genes were cloned into the pET28a(+) expression vector, overexpressed into *E. coli* and the mutated His-tagged proteins were purified. Some single mutations (Q164A, S190A, and I225P) prevented the correct synthesis and/or purification of the AroF-I enzyme or shown strictly no activity, thus stopping the study at this step. The in vitro analysis was conducted on the three remaining active single mutants of AroF-I (P160L, G191K, and S193A). First, the activity in the absence of an inhibitor was assessed. As shown in Figure 3, the introduction of point mutation G191K almost completely abolished AroF-I activity. Then, this mutant was no longer considered to be an AroF-I fbr candidate. Results were contrasted for mutant AroF-I S193A, both in the absence and presence of tyrosine. Enzymatic activity without tyrosine was comparable to the WT, whereas the activity in the presence of tyrosine seemed to be slightly higher. The most interesting mutation was the AroF-I P160L. This mutant had a significantly higher activity in the presence of L-Tyr than the WT enzyme (Figure 4). Furthermore, the activity of mutant P160L without inhibitor was comparable to the WT, making AroF-I P160L the best candidate for further engineering.

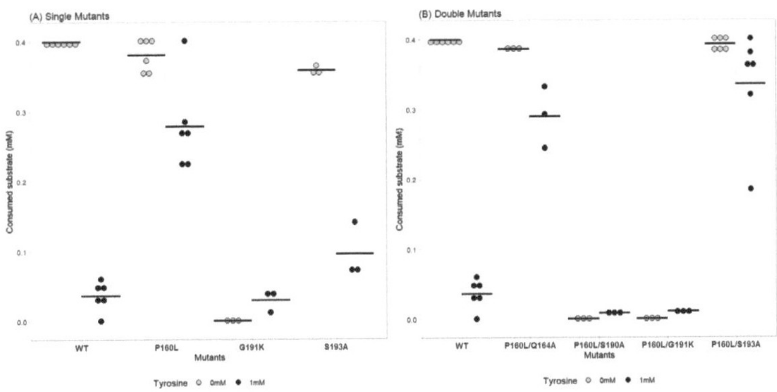

Figure 3. Comparison of the consumed substrate (mM) for WT AroF-I vs. single mutants (**A**) and double mutants (**B**) in the absence and presence of tyrosine. Gray dots represent raw values of replicates in the absence of tyrosine; Black dots represent raw values of replicates in the presence of 1 mM tyrosine. Horizontal black bars represent the mean value of all replicates for each condition.

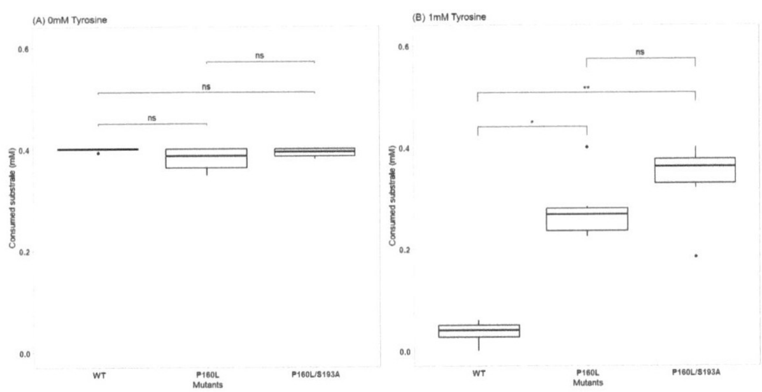

Figure 4. Comparison of the consumed substrate (mM) for WT AroF-I vs. promising mutants in the absence (**A**) and presence (**B**) of 1 mM tyrosine. Boxplot with $n = 6$ for each tested condition. For statistical analysis, a Kruskal–Wallis non-parametric test with Dunn's post hoc analysis was used. p-value adjusted using Holm's method (significance = ns: p-value > 0.05; *: p-value < 0.05; **: p-value < 0.01) . Cohen's d test values for effect size evaluation are: WT vs. P160L = 0.78; WT vs. P160L/S193A = 1.03; P160L vs. P160L/S193A = 0.52. Values show significant differences between the compared data.

2.4. A Double Mutation Improves the Tyrosine Resistance of AroF-I and Its Enzymatic Activity

The promising results observed with AroF-I P160L led us to further work on this mutant. We thought to introduce a second mutation from the list described in the bioinformatic analysis section. All combinations were considered with no a priori as it would be interesting to observe potential synergistic effects of mutations, which are usually hard to predict (Mildvan [43], Chaparro-Riggers et al. [44]). Following the same procedure as above, we constructed several plasmids with mutated genes of AroF-I with the aim of expressing and purifying double mutants of the AroF-I enzyme. We obtained four combinations of mutations, all including P160L: P160L/Q164A; P160L/S190A; P160L/G191K; P160L/S193A). Interestingly, when compared to the single mutants, all the combinations were correctly expressed and purified (see Appendix A). The in vitro characterization was then pursued on the four double mutants obtained. First, the activity of these mutants was assayed in the absence of any inhibitor. Two out of four enzymes showed no activity (AroF-I P160L/S190A; AroF-I P160L/G191K) and were no longer studied (Figure 3). For the P160L/Q164A variant, the activity with and without tyrosine was similar to the single mutant P160L. The characteristics of this mutant are then mostly linked to the P160L mutation, and the second introduced mutation has a null or low impact on the measured activity. Regarding double mutant AroF-I P160L/S193A, the activity in the presence of tyrosine was significantly higher than the WT protein (Figure 4). Surprisingly, the combination of mutations P160L and S193A seemed to improve the resistance to tyrosine compared to the P160L single mutation. Thus, the double mutant AroF-I P160L/S193A was elected AroF-I fbr and selected to study its in vivo impact on p-coumaric acid production.

2.5. Construction of *Pseudomonas putida* Strains Overproducing Aromatic Amino Acids

To evaluate the potential of the AroF-I P160L/S193A enzyme for p-coumaric acid bioproduction, the gene coding for this mutant was inserted under the control of a strong constitutive promoter in the chromosome of a pCA-producing strain of *Pseudomonas putida* KT2440 (called Ppu201). This strain overexpresses a tyrosine ammonia lyase (TAL, EC 4.3.1.23) (integrated into the chromosome), as well as aroH, one of the three DAHP synthases found in *P. putida*'s chromosome. The resulting strain was named Ppu235. To assess the producing capacities of both strains, they were cultured for 24 h in a minimal medium containing glucose as the sole carbon source, and an HPLC analysis was performed on the supernatant fractions to detect and quantify the produced phenolic compounds. As shown in Figure 5, the strain harboring the AroF-I mutant showed a higher production capability, being approximately 0.66 mM Phenolics/1 OD600 nm, compared to Ppu201 producing 0.53 mM Phenolics/1 OD600 nm. The two major products detected are cinnamic acid and p-coumaric acid. Indeed, the TAL enzyme, responsible for the conversion of tyrosine to p-coumaric acid, is also able to use phenylalanine as a substrate to produce cinnamic acid. These results indicated that the Ppu235 strain produces more tyrosine and more phenylalanine than its homolog without the AroF-I mutant, indicating a deregulation of the shikimate pathway. It is worth noting that Ppu235 showed a final OD600 nm comparable with Ppu201, suggesting that the introduction of the feedback-resistant mutant did not seem to alter the overall fitness of Ppu235.

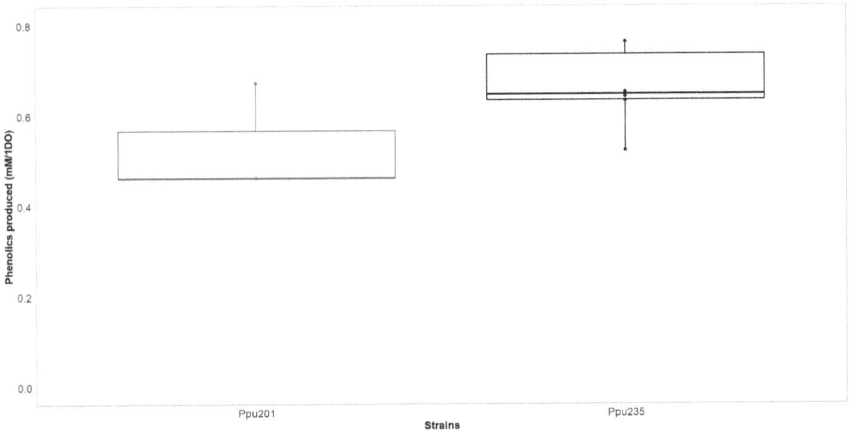

Figure 5. In vivo production of phenolic compounds by *P. putida* strains expressing (Ppu235) or not (Ppu201) AroF-I fbr variant (P160L/S193A).

2.6. A Synergistic Effect of AroF-I Fbr and phhAB Improves Production of p-Coumaric Acid

As stated earlier, a *Pseudomonas putida* KT2440 strain with a deregulated shikimate pathway was constructed by introducing a double mutant P160L/S193A of the AroF-I enzyme into the chromosome of a previously constructed p-coumaric acid-producing strain. It is suggested that in *Pseudomonads*, the majority of tyrosine is produced through the conversion of phenylanine (Otto et al. [35], Wierckx et al. [45]). Therefore, we thought to overexpress the two genes phhA (PP_4490) and phhB (PP_4491) coding for the enzymes responsible for the synthesis of tyrosine from phenylalanine. The phhAB operon was cloned into a pBBR1 vector under the control of the arabinose inducible promoter araC/Pbad, creating pC2F327. This plasmid was transformed into Ppu201 and Ppu235, and the resulting strains were cultured in minimal medium MPpu with glucose as the sole carbon source, supplemented or not with 0.5% L-arabinose. A plasmid expressing eGFP under the control of the arabinose inducible promoter was used as a control (plasmid pPpu226). On the one hand, as depicted in Figure 6, Ppu235 showed a slight increase in total phenolics produced when the control plasmid was used. Furthermore, the repartition between cinnamic acid and p-coumaric was lower in Ppu235 than in Ppu201, which is in accordance with the results observed for both strains without bearing any plasmid. Induction of eGFP did not change the conclusions except that fewer phenolics were produced, which is not surprising considering the metabolic burden created by gene overexpression on plasmid (Glick [46]). On the other hand, results obtained with the expression of phhAB were very promising. When the expression of the phhAB operon was not induced, both strains produced an amount of phenolics comparable with the control plasmid. On the contrary, when arabinose was supplemented in the medium, a sharp increase in total phenolics produced was measured for Ppu235 bearing pC2F327, reaching 1.9 mM phenolics/1 OD600 nm, which is 3 times higher than with the control plasmid. In addition, pCA represented 97% of the total phenolics produced in this condition, improving by 10% the amount of pCA in total phenolics produced. However, this enhanced pCA production is correlated with a decrease in growth rate (see Appendix A Table A1). While *P. putida* KT2440 is known to tolerate high pCA concentrations, ruling out pCA toxicity. The observed growth reduction may be attributed to L-phenylalanine depletion and/or broader metabolic imbalances resulting from phhAB overexpression (Calero et al. [29]). Nevertheless, the strain Ppu235 expressing phhAB is a promising strain towards a high-titer and high-yield industrial production of p-coumaric acid.

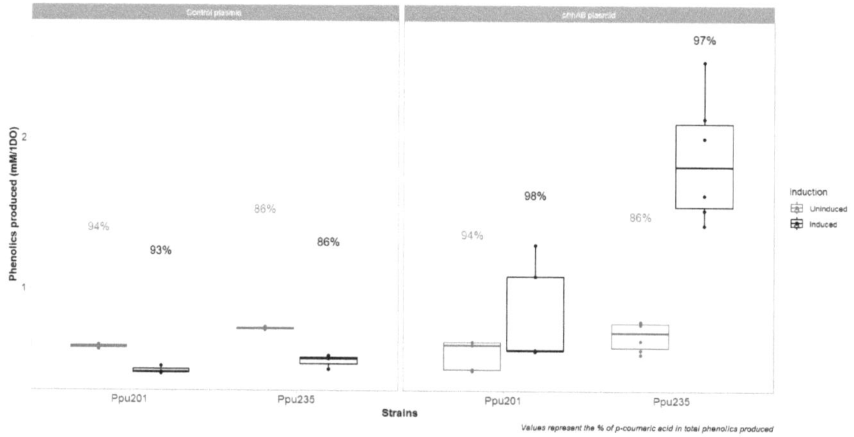

Figure 6. In vivo production of phenolic compounds by *P. putida* strains expressing (Ppu235) or not (Ppu201) AroF-I fbr in combination with the overexpression of phhAB genes. Bold box plots are when phhAB is induced, gray ones when phhAB is not induced by Arabinose.

3. Materials and Methods

3.1. Bacterial Strains, Plasmids, and Growth Conditions

The bacterial strains and plasmids used in this study are listed in Table 2. *P. putida* strains were grown at 30 °C and *E. coli* strains at 37 °C either in liquid LB medium (Bertani [47]) or on solid agar LB plates (additionally containing 1.5% (*w*/*v*) agar). For the selection of plasmid transformants, 50 µg/mL kanamycin (Kan) was added. After mating procedures, recombinant *Pseudomonas* were isolated on Yeast Extract Tryptone (YT) plates supplemented with Ampicillin (100 mg/mL) and kanamycin (50 mg/mL). The loss of the suicide plasmid was selected on YT plates containing 10% sucrose.

Table 2. Bacterial strains and plasmids used in this study.

Strains and Plasmids	Genotype or Markers; Characteristics and Uses	Source or References
Strains		
Escherichia coli BL21(DE3)	$F^{-} ompT, gal, dcm, lon, hsdSB(r_B^- m_B^-), \lambda(DE3\ [lacI, lacUV5 - T7, p07\ ind1, sam7, nin5]), [malB^{+}]_{K-12}(\lambda^S)$	Novagen
Escherichia coli S17.1	$recA, pro, hsdR,$ $RP4\text{-}2\text{-}Tc::Mu\text{-}Km::Tn7$	Simon et al. [48]
Pseudomonas putida KT2440	WT	NBRC100650
Pseudomonas putida $Ppu201$	p-coumaric productive strain with aroH under strong constitutive promoter	This study
Pseudomonas putida $Ppu235$	$Ppu201, aroH::aroFI\ P160L/S193A$	This study
Plasmids		
pBBR1MCS-2	Broad-host-range cloning vector, mobilizable, Kan^R	Kovach et al. [49]
pUC-araC/pBAD	pUC containing the araC/pBAD promoter, *ampR*	SEVA plasmid collection
pBBR1-araC/pBAD	pBBR1MCS-2 containing the araC/pBAD promoter	This study

Table 2. *Cont.*

Strains and Plasmids	Genotype or Markers; Characteristics and Uses	Source or References
pC2F387	pBBR1-araC/pBAD with the phhA/B genes cloned downstream of the araC/pBAD promoter	This study
pPpu226	pBBR1-araC/pBAD with the eGFP gene cloned downstream of the araC/pBAD promoter	This study
pET28a(+)		Novagen
pET28-AroF-I WT		This study
pET28-AroF-I G191K		This study
pET28-AroF-I P160L		This study
pET28-AroF-I S193A		This study
pET28-AroF-I P160L/G191K		This study
pET28-AroF-I P160L/Q164A		This study
pET28-AroF-I P160L/S190A		This study
pET28-AroF-I P160L/S193A		This study
pK18mobsacB	Widely used gene modifications suicide vector, Kan^R and *sacB* for counter-selection	Schäfer et al. [50]
pK18-aroH::AroF-I P160L/S193A	pK18mobsacB containing the homologous arms of *aroH* and the *aroF-I* P160L/S193A gene to be inserted, Kan^R	This study

For enzyme production and purification, *Escherichia coli* strains were grown in a ZYM auto-inducer medium (Studier [51]). For p-coumaric acid production assay, *P. putida* KT2440 strains were grown in a minimal medium containing 4.77 g/L Na_2HPO_4, 2.02 g/L KH_2PO_4, 2.36 g/L $(NH_4)_2SO_4$, 0.24 g/L $MgSO_4$ and 5 g/L glucose, with the following trace elements: 13.9 mg/L $FeSO_4{\cdot}7H_2O$, 14.7 mg/L $CaCl_2{\cdot}2H_2O$, 8.6 mg/L $ZnSO_4{\cdot}7H_2O$. To induce gene expression using the araC/Pbad promoter, 0.5% (*m*/*v*) arabinose were added to culture when necessary. The chemicals used in this work were obtained from Carl Roth (Karlsruhe, Germany), Merck (Darmstadt, Germany), or Sigma–Aldrich (Saint Louis, MO, USA).

3.2. Vector Construction and Strain Engineering

For plasmid constructions, DNA fragments were PCR amplified using the Platinum SuperFi high-fidelity polymerase (Thermo Fisher Scientific, Waltham, MA, USA) with corresponding overhangs to enable subsequent Gibson assembly (Gibson [52]) with the NEBuilder HiFi DNA Assembly mix (New England Biolabs, New Ipswich, NH, USA). Heat shock was used for the transformation of DNA assemblies and purified plasmids into *E. coli*. Plasmids were introduced in *Pseudomonas putida* KT2440 by conjugational transfer of mobilizable plasmids. Mating procedures were performed as follows. Overnight cultures of the *E. coli* donor, S17.1 strain bearing the desired plasmid, and the recipient *Pseudomonas putida* strain, were washed twice in $MgSO_4$ 10 mM. Equal volumes of the cultures (50 µL) were mixed and incubated for 4 to 5 h at 30 °C without agitation. The mix was then harvested and resuspended in 100 µL of $MgSO_4$ 10 mM before plating on a selective medium. Plates were incubated overnight at 30 °C. Replicative pBBR1-MCS2-based plasmids (Kovach et al. [49]) were used for gene expression, pET28a(+)-based plasmids (Novagen) were used for protein production and purification, and pK18mobsacB (Schäfer et al. [50])-based plasmids were used for chromosomal modification of *P. putida* strains. The pK18mobsacB vector is suicidal in *Pseudomonas putida* and can be used for gene insertion into a definite chromosomal target. When this plasmid was used for chromosomic modifications in *Pseudomonas*, the mating mixture for conjugational transfer was incubated overnight before following the procedure as described earlier. Eventually, for chromosomal modifications of *Pseudomonas putida* strains, antibiotics-resistant clones were incubated overnight at 30 °C in LB medium without selection pressure. A streak of this culture was then performed on YT sucrose 10%.

E. coli or *Pseudomonas putida* transformants, and chromosomally engineered *Pseudomonas* were screened by colony PCRs using the DreamTaq 2X Master Mix PCR (Thermo Fisher Scientific, Waltham, MA, USA). Primers were ordered as unmodified DNA oligonucleotides from Eurofins Genomics (Ebersberg, Germany). Plasmid inserts and gene deletions or insertions were confirmed by Sanger sequencing performed by Microsynth AG (Balgach, Switzerland). AroF-I mutated genes were ordered as synthetic DNA fragments from Eurofins Genomics (Ebersberg, Germany). PCR products and plasmids were purified using NucleoSpin Gel and a PCR cleanup purification kit from Macherey–Nagel (Düren, Germany) according to the manufacturer's instructions.

3.3. Production and Purification of the AroF-I Proteins

All DNA fragments corresponding to mutated variants of AroF-I genes were amplified with a flanking overhang and cloned into a pET28a(+) backbone. Assembly mixtures were transformed into a BL21(DE3) strain and the culture was plated on LB+Agar medium with appropriate antibiotic. Positive clones were PCR-checked and sequenced to confirm successful assembly.

Auto-inducible ZYM medium was inoculated at an initial OD of 0.05 with an overnight pre-seed culture and incubated for 24 h. For the first 5 h, the temperature was set to 25 °C and then lowered to 19 °C for the remaining incubation time. Next, protein purification was performed using the Protino Ni-TED purification kit for His-tagged proteins following the manufacturer's instructions (Macherey–Nagel, Düren, Germany). Briefly, cells were separated from the medium by cold centrifugation for 10 min at 5000× *g*. The pellet was resuspended in an adapted buffer with the addition of lysozyme (final concentration 1 mg/mL) and kept on ice for 30 min. Then, cells were disrupted using sonication at 30% power, pulse set to 40, and a total time of 8 min (Omni Sonic Ruptor 400, Omni International (Kennesaw, GA, USA)). Streptomycine was added to a final concentration of 0.22% (m/v), and samples were cold-centrifugated for 30 min at 15,000× *g*. The supernatants were kept, and the enzymes were purified. After purification, proteins were concentrated approximately 10 times in Tris-HCl 0.1 M pH 7.5 + 10% glycerol using Amicon Ultra-4 30 kD from Merck (Darmstadt, Germany). Purified proteins were checked by SDS-PAGE electrophoresis and stored at −80 °C.

3.4. Enzymatic Assay for Determination of AroF-I Activity

Enzymatic activity was assessed by measurement of substrate consumption using HPLC detection. Assays were performed in a final volume of 200 µL containing 40 mM phosphate buffer pH 7, 300 µM phosphoenolpyruvate (PEP), 20 µg of purified enzyme and 3 µM HCl. For inhibition tests, 1 mM of aromatic acid was added to the reaction mixture. The final volume was adjusted using ultrapure water. After a 2 min pre-incubation at 30 °C, reaction was started by adding 300 µM of erythrose-4-phosphate (E4P) and incubated at 30 °C for 60 min. To stop the reaction, the mixture was heated to 80 °C for 5 min. Samples were then centrifugated for 10 min at 15,000× *g*. The supernatant was kept and filtrated on a 0.22 µm membrane before subsequent HPLC analysis. The remaining PEP concentration was determined using the Agilent 1100 system (Agilent Technologies, Santa Clara, CA, USA) with a Luna OMEGA polar C18 column (Phenomenex, Torrance, CA, USA). Injection of 15 μL was analyzed for each sample. The mobile phase was a 1 mM phosphoric acid solution. The rate flow was set to 0.3 mL/min, and detection occurred via UV detection at 220 nm. The total time of analysis for each sample was 15 min.

3.5. Replacement of aroH by Mutated AroF-I

The gene AroF-I P160L/S193A was amplified from the synthesized DNA fragment and cloned into the vector pK18mobsacB. The pK18mobsacB vector is suicidal in *Pseudomonas putida* and can be used for gene insertion into a definite chromosomal target. To perform chromosomal modification, the suicide plasmid was constructed based on pK18mobsacB bearing two homology arms of 800 bp around aroH (PP_1866) upstream and downstream the gene coding for the mutated version of AroF-I. Using this plasmid, chromosomal modification was achieved to obtain strain Ppu235 from strain Ppu201, a p-coumaric acid-producing strain obtained in our laboratory in which aroH is overexpressed under the control of a strong constitutive promoter and a tyrosine ammonia lyase is inserted into the chromosome.

3.6. Overexpression of phhAB Genes

The two genes phhA (PP_4490) and phhB (PP_4491) coding, respectively, for a Phe-4-monooxygenase and a Pterin-4-alpha-carbinolamine dehydratase were cloned into a pBBR1-MCS2 expression plasmid as an operon under the control of the inducible araC/Pbad promoter. The WT genes were amplified from purified genomic DNA with a flanking overhang to allow insertion into the plasmid backbone. The assembly mixture was then transformed into an S17.1 strain and plated LB+Agar with the appropriate antibiotic. Positive clones were PCR-checked, and amplicons were sequenced to confirm successful assembly. The final plasmid was transferred to the recipient *P. putida* KT2440 strain by conjugation.

3.7. Assessment of pCA Production in Batch Culture

To evaluate the production of p-coumaric acid by genetically modified *Pseudomonas putida* strains, batch culture in Erlenmeyer flasks was performed. First, strains were plated on an LB agar plate containing appropriate antibiotics when needed. One colony was PCR-checked and cultured overnight in 5 mL minimal medium supplemented with 5 g/L glucose at 30 °C with agitation. The next day, the pre-seed culture was used to inoculate 25 mL of minimal medium supplemented with 5 g/L glucose and appropriate antibiotic in an Erlenmeyer flask. When needed, arabinose was used to induce gene expression. Growth was followed by measuring OD at 600 nm, and p-coumaric acid production was assessed by HPLC measurement at the end of the batch culture. The analytical method used to identify and quantify p-coumaric acid and cinnamic acid was as follows: Kinetex 5 μm F5 100 A 150 × 4.6 mM column (Phenomenex, Torrance, CA, USA), formic acid 0.1% and acetonitrile as mobile phases with variating gradient (0–5 min = 100% formic acid; 5–25 min = 75% formic acid + 25% acetonitrile; 25–30 min = 62% formic acid + 38% acetonitrile; 30–35 min = 100% formic acid), flow rate of 1 mL/min, oven at 40 °C, injection volume of 10 μL and UV detection at 315 nm for p-coumaric acid and 28 nm for cinnamic acid.

3.8. Statistical Analysis

A Shapiro–Wilk normality test (Shapiro and Wilk [53]) and a Bartlett variance equality test (Jones et al. [54]) were first performed on data to be analyzed. As data did not meet the criteria to be analyzed via a parametric statistical analysis, a non-parametric Kruskal–Wallis test was conducted (Kruskal and Wallis [55]) followed by a Dunn's post hoc analysis with Holm's adjustment method for the p-value (Dunn [56]). To confirm results obtained using the Kruskal–Wallis test, a Cohen's d measure for effect size evaluation was performed (Cohen [57]). The results of these two tests are shown in Figure 4.

3.9. Bioinformatics Protocols

Our primary objective was to identify point mutations in our target, AroF-I, that may reduce the feedback regulation of tyrosine. To achieve this, we performed a docking analysis of tyrosine within the AroF-I structure to pinpoint amino acids that are likely to play a significant role in the molecule's binding. We then aimed to substitute these residues with amino acids exhibiting different properties, thereby blocking or diminishing the likelihood of tyrosine binding.

The PDB structure for the UniProt entry Q88KG6 (AroF-I from *Pseudomonas putida* KT 2440) was not available at the moment of this study, so we reconstructed it using the Swiss-Model approach (Waterhouse et al. [42]). Our structure was built with ProMod3 (version 2.0) using 1of6.4.G structure as a template. We obtained a Global Model Quality Estimate (GMQE) of 0.77, which is considered reasonable.

We used this structure to perform docking against tyrosine. We used Vina (Trott and Olson [58]) as our docking software, and we considered a large region for the docking, as we can see in Figure 7. The center of the docking box is located at the coordinates (17.214, 39.976, 59.293), with dimensions of 44, 48, and 32 ångströms along the x, y, and z axes, respectively. This choice was based on the analysis of AroG, a phospho-2-dehydro-3-deoxyheptonate aldolase (UniProt ID P0AB91) from *Escherichia coli* K12 (crystal structure 1kfl from PDBe), where a phenylalanine is docked in the protein. This protein is known to be regulated by phenylalanine and, based on studies by Shumilin et al. [59] and Cui et al. [60] on *Escherichia coli* DAHP synthases, we assumed the docking region for the tyrosine would be similar. Analyzing the docking solutions, we could identify six residues in interaction with at least one atom of the docked tyrosine: P160, Q164, S190, G191, S193, and I225. In Figure 8, we present our solutions overlapped within the protein pocket, highlighting the identified residues and some of the hypothetical bonds. Since the identified residues present putative interactions with tyrosine, we supposed that altering these residues could, in theory, inhibit the docking of this molecule.

Figure 7. Docking region used in our docking studies with the AroF-I from *Pseudomonas putida* KT2440. The docking box is located at the coordinates (17.214, 39.976, 59.293), with dimensions of 44, 48, and 32 along the x, y, and z axes, respectively.

To identify possible replacement residues for this region, we consulted the literature on known phenylalanine feedback-resistant mutations for phospho-2-dehydro-3-deoxyheptonate aldolases in *Escherichia coli* (Cui et al. [60], Ding et al. [61]). Being neutral and non-polar, alanine is frequently used as a replacement residue, but we also found other candidates. We employed an alternative approach based on the structural and sequence alignment of our target enzyme, AroF-I, with other enzymes known to exhibit natural feedback resistance to tyrosine. The aim was to determine which natural residues occupy the positions of those identified in our docking study. We considered UniProt entries Q9YEJ7 (AroG from *Aeropyrum pernix*), A0A5C0XSG1 (aroF from *Pyrococcus furiosus*) and Q9V1I0 (AroG from *Pyrococcus abyssi*) that were aligned with AroF-I and we identified different amino acids present at positions corresponding to the target residues identified in the docking approach. Finally, based primarily on amino-acid properties, we selected one candidate mutation for each position. For each target residue, we chose a replacement amino acid that was either neutral or exhibited opposing physicochemical properties compared to the original in AroF-I. The chosen mutations can be seen in Table 1.

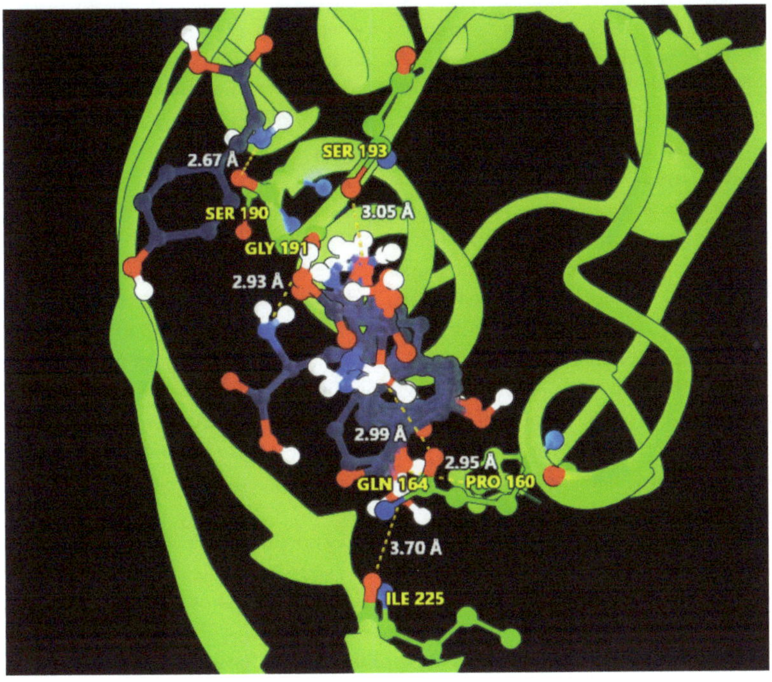

Figure 8. Overlapped solutions of the docking of tyrosine into the AroF-I from *Pseudomonas putida* KT2440. In yellow we see the name and number of the residues that form a bond with at least one solution. These residues are depicted with green carbons. The dashed lines highlight the found interactions, and in white, we have the distance of the bonds.

4. Discussion

Cellular metabolic pathways, from the simple to the complex, are highly regulated in response to external nutrient changes, internal metabolite accumulation, or stress responses (Shimizu [62], Chubukov et al. [63]. Different strategies are adopted by the cell for these tight regulations: genes transcriptional and/or translational regulations, through regulatory proteins or small RNAs, and fine-tuning of enzymatic activities (Shimizu and Matsuoka [64], Nudler [65]).

One good example of tight metabolic regulation is found in fungi, bacteria, and plants, with the production of aromatic amino acids (L-tyrosine, L-phenylalanine, L-tryptophane) via the shikimate pathway. The shikimate pathway and how to optimize the metabolic fluxes to overproduce certain intermediates were extensively studied in several microbial chassis, especially since aromatic intermediates are widely used as raw materials in the medical, chemical, food, and other industries (Wu et al. [66], Jiang and Zhang [67], Cao et al. [68]).

In *S. cerevisiae*, the initial step of the shikimate pathway is catalyzed by two DAHP synthase isozymes, encoded by ARO4 and ARO3 and regulated through feedback inhibition by L-tyrosine and L-phenylalanine, respectively (Helmstaedt et al. [69]). In *Escherichia coli* three DAHP synthase isozymes exist (aroF, aroG, aroH), each one being feedback-inhibited by one of the three aromatic amino acids (Jayaraman et al. [16], Jossek et al. [70]). Regulation of DAHP synthases, the entry point in the shikimate pathway, and the mechanisms involved are well described and conserved among microorganisms (Mir et al. [71]). Some studies were already done regarding the shikimate pathway in *P. putida* KT2440, but most of them involve mutated enzymes from *E. coli* (Wierckx et al. [32], Calero et al. [33], Otto et al. [35]). To our knowledge, no evidence of such a regulation was available for DAHP synthases from *P. putida* even if a putative fbr mutant of AroF-I was described for *P. putida* S12 (Wierckx et al. [32]). Knowing that *P. putida* KT2440 is well suited for the industrial production of aromatic amino-acid derivatives. We thought to highlight the regulation of the shikimate pathway in this strain and engineer a DAHP synthase to obtain a phenolypropanoids overproducing strain.

Feedback-resistant mutants have initially been found using random mutagenesis and screening of the best aromatic amino-acid producers (Ger et al. [72], Ray et al. [73], Kikuchi et al. [74]). However, this process is time-consuming because promising mutants can be hard to find among all the diversity created by random mutagenesis. Nevertheless, over recent decades, the development of bioinformatics tools, combined with knowledge of protein structure, functions, and regulations, have paved the way for the rational engineering of protein (Jayaraman et al. [16], Lutz [75], Chen et al. [76], Nixon and Firestine [77]) and allows us to create variants of genes purposefully as to obtain feedback-resistant forms of DAHP synthases. We chose to apply these principles to the interesting AroF-I DAHP synthase from *P. putida* KT2440, with the aim of having a reasonable set of rationally designed candidates to obtain a feedback-resistant mutant of this enzyme. Using protein structure modeling, ligand interaction modeling, and knowledge from the literature (Cui et al. [60], Ding et al. [61]), several mutants of AroF-I were synthesized, then cloned and expressed in *E. coli*. We were confident in the results obtained using bioinformatic tools, as we successfully identified residues that had previously been described in similar DAHP synthases. Not surprisingly, some mutations proved critical for protein expression and activity, likely altering the protein's secondary structure and preventing proper folding (Faisca et al. [78], Sen et al. [79]). Regarding these issues in protein variant expression, it may be worthwhile to conduct site-saturation experiments, as alternative amino-acid substitutions could influence the overall shape and activity of the enzyme. (Chronopoulou and Labrou [80]). Another option would be to use $\Delta\Delta G$ energy prediction between mutated protein and wildtype as it is shown that energy variation can correlate with protein stability(Marabotti et al. [81]). Even if predictions are not perfect and do not always correlate with experimental values, these tools can help select candidates for experimental testing. Fortunately, the prediction of mutation stability has advanced significantly in recent years, and the development of new artificial intelligence models and algorithms will further enhance the reliability of these predictions. (Pucci et al. [82], Fang [83], Cheng et al. [84]).

Since we explored just a small set of mutations and experimentally tested them all, we did not perform this type of prediction.

Among the few candidates we ordered, two single mutants showed better resistance to tyrosine inhibition in vitro (P160L and S193A) without the need for intensive screening. Pursuing our work further, we combined single mutations to create double mutants of AroF-I. In the end, we demonstrated that the AroF-I P160L/S193A was a fully active and fully tyrosine-resistant mutant. To our knowledge, despite a few studies of DAHP synthases in other *Pseudomonas* strains (Wynands et al. [34]), this is the first rationally designed and characterized mutant of a variant DAHP synthase for *P. putida*.

Single mutants capable of rendering a DAHP synthase feedback-resistant have never been finely enzymatically characterized. Some mutations are known to impact the allosteric regulation by tyrosine and phenylalanine, as discussed previously in this manuscript, and the binding site of the *E. coli* DAHP synthase is known and composed of residues from two adjacent subunits of a dimer (Shumilin et al. [59], Cui et al. [60]). However, the allosteric mechanism is not fully understood, and the possible synergic effects of double mutants have not yet, to our knowledge, been investigated. It is known that the synergistic effect of mutations may render a protein more stable, even if identifying or understanding these synergies is not straightforward (Hamborg et al. [85]). The same difficulty can be translated into the study of allosteric regulation. One way of approaching this problem may be through the aid of normal mode analysis methods. Normal mode analysis can help in understanding allostery in proteins by exposing the dynamic relationships between structural changes and functional outcomes. By analyzing the vibrational modes of proteins, we can identify allosteric sites and predict how conformational changes propagate through the protein structure (Greener and Sternberg [86], Yamato and Laprévote [87], Kolossváry [88]). Surely, these methods do not exclude in vitro or in vivo validations, but it may be a way to reduce the highly combinatorial solution space of mutations to explore. Another possibility could be the exploration of deep learning methods and protein language models that are literally revolutionizing the understanding of proteins (Ruffolo and Madani [89]). These models can extract structural and functional insights from amino-acid sequences or protein structures. This can help us to assess functional properties, and identify structural features, thereby enhancing our comprehension of protein dynamics. However, pre-trained protein language models alone are not enough since their accuracy and interpretability remain limited. However, an approach based on few-shots (few or scarce data points) used to retrain protein language models has been proven recently useful for this kind of task (Schmirler et al. [90], Zhou et al. [91], Biswas et al. [92]).

We then decided to incorporate the promising variant into a *P. putida* strain, already engineered to overproduce p-coumaric acid and screen the effect of a plasmid copy of phhAB. Plasmid-driven gene expression is the easiest and most common strategy for recombinant protein production, rapid screening, and strain engineering. It has been extensively used in recent decades to study enzyme activities, namely in the well-known T7 system in *E. coli* (Chen [93]). This system was used in this study to express, purify, and assay the catalytic parameters of the AroF-I variants and compare them to each other.

However, studying the proteins in their bacterial environment has limitations when expressed from a plasmid. It is thus preferable to opt for a chromosomal-driven strategy (Goussé et al. [94]), explaining why we adopted this strategy to express our AroF-I variant. For example, a too-high translation rate can lead to inclusion bodies or proteolytic degradation and thus a weak production (Kachroo et al. [95].). We can also be confronted with metabolic burden and/or plasmid instability. Moreover, inducing gene expression from a multi-copy plasmid is known to be a source of variability between batches when compared to chromosomal integration or single-copy plasmid (Zucca et al. [96]). This is true if considering the pCA production with our Ppu235 strain bearing the phhAB plasmid, which shows considerable variations between batches. Regarding the designed AroF-I fbr, as the aim is to express a variant of a native enzyme, we chose the native chromosomal locus. However, the choice of a different locus can also be interesting in the fine-tuning of expression level or inserting an additional copy of the gene in the future (Tyo et al. [97]). Moreover, when constructing a strain for secondary metabolite production, it is preferable to adopt a strategy for the long-term stability of the strain. The technology we developed allows the insertion of genes in the chromosome without scar or marker (e.g., antibiotic resistance (Patent [98]). Several modifications can be performed in the genome at different loci without the necessity of accumulating antibiotics or thinking about compatible plasmids. This strategy will be performed as well for the chromosomal overexpression of phhAB instead of a plasmid-driven expression.

Reviews on several microbial hosts showed that the deregulation of DHAP synthase is not sufficient and that multiple engineering strategies have to be tested to obtain a high-titer and high-yield production of aromatic amino acids and their derivatives (Bilal et al. [19], Wu et al. [66], Li et al. [99]). The shikimate pathway is highly regulated, interconnected, and central in bacterial metabolism. Thus, maximizing the fluxes towards the production of L-Phe and L-Tyr needs efforts in cutting unnecessary reactions and boosting key reactions (Bentley and Haslam [100], Liu et al. [101]). Further work will be needed on the obtained Ppu235 strain to continue improving titer and yield and make it suitable for industrial production of p-coumaric acid. For example, the derivation point around chorismate in the shikimate pathway is known to be regulated to ensure a balance between branched pathways (Kroll et al. [102], Brown and Dawes [103], Zhang et al. [104]). In *Pseudomonas putida*, a deregulated variant of the chorismate mutase-prephenate dehydratase pheA (PP_1769, EC 5.4.99.5), catalyzing the conversion of chorismate to prephenate, was found and used in production strains (Molina-Santiago et al. [28], Otto et al. [35]). Using this pheA fbr variant in our Ppu235 strain could enhance the production of L-Tyr and p-coumaric acid.

From an industrial perspective, an internal techno-economic study sets a target pCA titer of at least 65 mM in large-scale bioreactors. Efficient sugar conversion is essential to minimize both operating (OPEX) and capital (CAPEX) expenditures. Extrapolating from our shake-flask data, and assuming typical *P. putida* bioreactor densities (OD600 of 30–50) (Mokwatlo et al. [105]), our engineered strains are projected to achieve 55–92 mM pCA without any supplementation. This represents a significant improvement over previous efforts, such as the *P. putida* S12 strain requiring L-phenylalanine supplementation (10.6 mM) and the highly engineered yeast strain achieving 76 mM (Flourat et al. [106]). Unlike these approaches, our method combines rational enzyme engineering for feedback resistance with stable chromosomal overexpression to enhance tyrosine flux. This strategy offers a more robust and industrially relevant platform for pCA production, with the potential to meet industrial demands and elevate bioproduction into the spotlight.

Also, thinking about product purification at the step of strain construction seems important. To this extent, reducing or suppressing cinnamic acid production is crucial to the development of an efficient process. With the maximization of fluxes towards L-Phe and L-Tyr, we could end up with an accumulation of these amino acids. On the one hand, an accumulation of L-Phe would lead to a greater production of cinnamic acid. On the contrary, an over-accumulation of L-Tyr could lead to an incomplete conversion to p-coumaric acid. To avoid the first hypothesis, chromosomal expression of phhAB is a promising strategy, as described earlier in this work. Also, the use of bioinformatics tools to identify or design highly active TAL enzymes unable to use L-Phe as a substrate to form cinnamic acid seems of high priority. Several studies have reviewed works done on this enzyme to elucidate product specificity (Louie et al. [107], Watts et al. [108], Jendresen et al. [109]). Finding a highly active TAL would also have the side effect of avoiding L-Tyr accumulation and maximizing its conversion to p-coumaric acid.

In this study, we investigated the regulation of the shikimate pathway in *Pseudomonas putida* KT2440 with the aim of enhancing p-coumaric acid production. We successfully engineered an L-tyrosine feedback-resistant mutant of the DAHP synthase AroF-I (P160L/S193A) using rational mutagenesis and computational modeling. This approach proved advantageous over traditional random mutagenesis, leading to the identification of key residues involved in tyrosine-mediated feedback inhibition. The resulting DAHP synthase variant was incorporated into a *P. putida* strain previously pre-wired for p-coumaric acid production. While plasmid-based overexpression of the phhAB gene was initially used for rapid prototyping, future work will focus on chromosomal integration for long-term strain stability. In parallel, further metabolic engineering strategies are planned to maximize p-coumaric acid production. These include incorporating a deregulated chorismate mutase-prephenate dehydratase and exploring the use of highly active, L-tyrosine-specific tyrosine ammonia lyase (TAL) enzymes. By optimizing the metabolic pathway, especially with the help of computer science, we aim to develop a robust strain for industrial p-coumaric acid production.

5. Patents

A patent filed under the number WO2022238645A1 is the result of the work presented in this manuscript.

Author Contributions: Conceptualization, W.M., R.A., C.P., A.C.-P. and C.R.; methodology, W.M.; bioinformatics, R.A. and C.P.; validation, A.C.-P. and C.R.; formal analysis, W.M.; data curation, W.M. and R.A.; writing—original draft preparation, C.R.; writing—review and editing, all; visualization, W.M.; supervision, C.R.; project administration, A.C.-P. All authors have read and agreed to the published version of the manuscript.

Funding: This research received no external funding.

Data Availability Statement: No new datasets were created.

Conflicts of Interest: The authors declare no conflicts of interest.

Appendix A

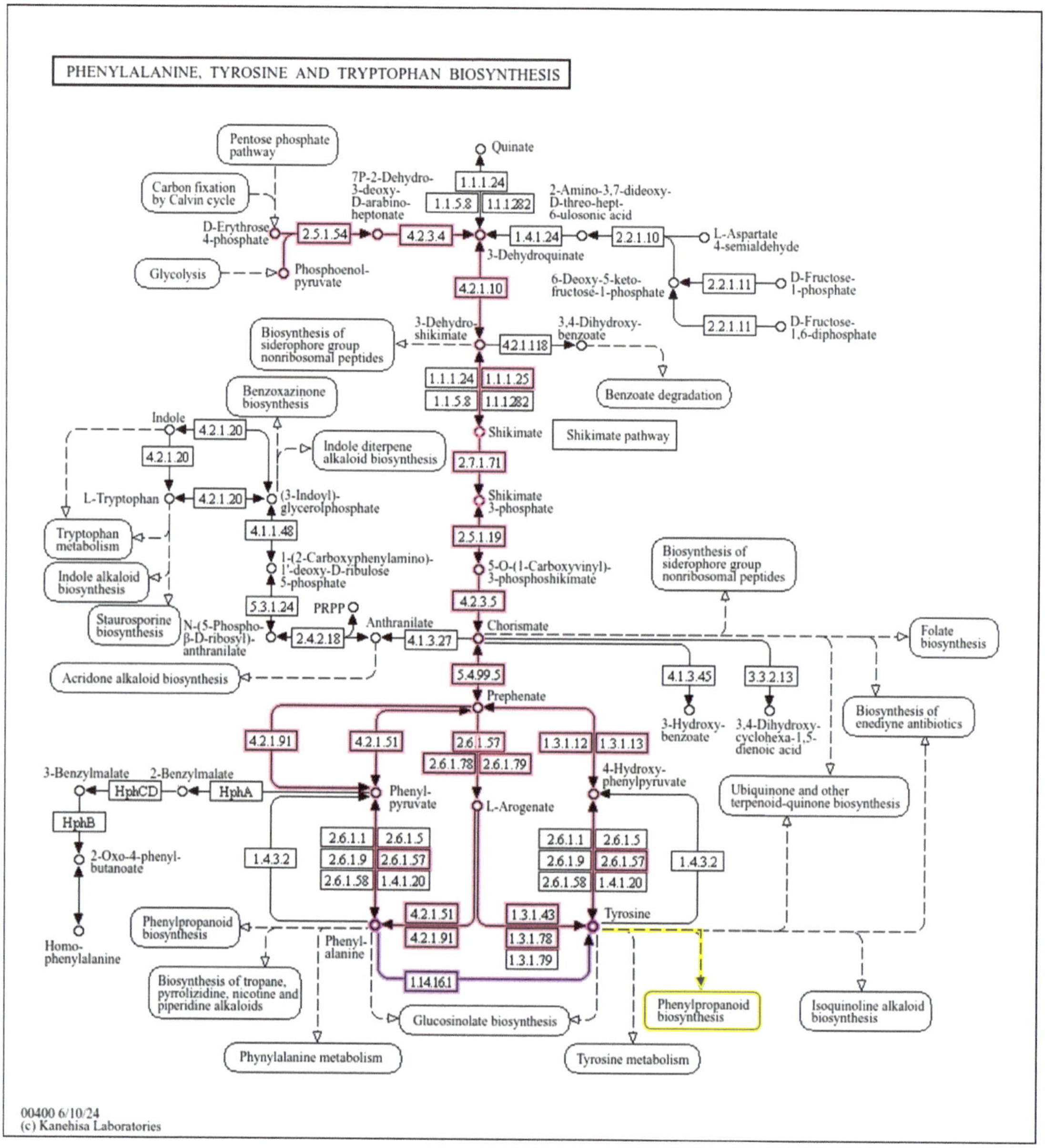

Figure A1. An illustration of the shikimate pathway and related routes based on KeGG data [110]. In red, we have the shikimate pathway (upper part up to chorismate). Also, in red (after chorismate) and magenta, we have the phenylalanine and tyrosine biosynthesis routes. In yellow we can see the entry point to the biosynthesis of phenylpropanoids.

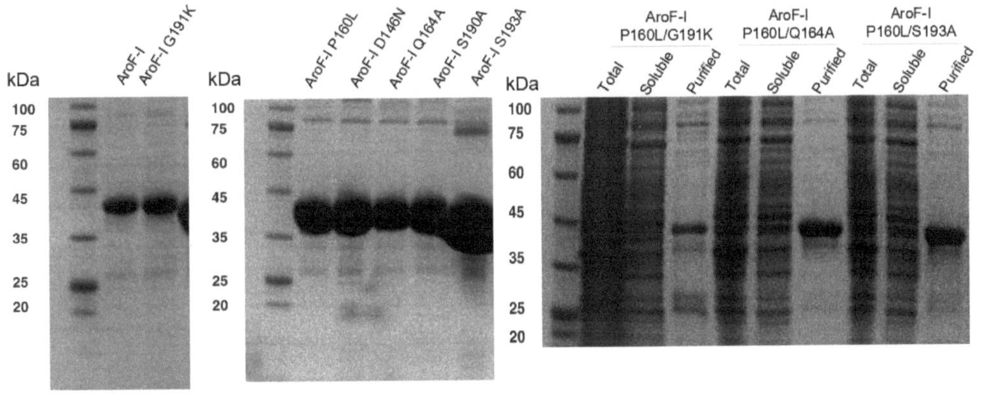

Figure A2. SDS-PAGE gels showing purified proteins used in this study. The expected size is 37 kDa.

Table A1. Strain growth (OD) and concentration of phenolic compounds, concentrations normalized by OD, and total phenolic compounds normalized by the OD. Grayed lines correspond to uninduced and blank lines to induced conditions. CP stands for control plasmid, and PP for phhAB plasmid. CA stands for cinnamic acid, and pCA for coumaric acid.

Strain	OD600	[pCA] (mM)	[CA] (mM)	pCA/1DO	CA/1DO	Total Phenolics/1DO
Ppu201	5.4	0.28	0.14	0.05	0.003	0.08
Ppu201/CP	3.4	1.95	0.12	0.57	0.04	0.61
Ppu201/PP	3.4	1.99	0.13	0.59	0.04	0.62
Ppu235	4.1	1.94	0.66	0.47	0.16	0.63
Ppu235/CP	3.8	2.38	0.38	0.63	0.10	0.72
Ppu235/PP	3.7	2.46	0.37	0.66	0.10	0.76
Ppu201	4.6	0.28	0.14	0.06	0.03	0.09
Ppu201/CP	4.6	1.83	0.15	0.40	0.03	0.43
Ppu201/PP	3.1	1.80	0.03	0.58	0.01	0.59
Ppu235	3.4	1.92	0.66	0.56	0.20	0.76
Ppu235/CP	5.1	2.07	0.30	0.41	0.06	0.46
Ppu235/PP	1.6	2.21	0.07	1.38	0.04	1.42
Ppu201	5.1	0.28	0.14	0.06	0.03	0.08
Ppu201/CP	3.3	1.92	0.13	0.58	0.04	0.62
Ppu201/PP	3.4	2.06	0.13	0.61	0.04	0.64
Ppu235	3.9	1.93	0.65	0.49	0.17	0.66
Ppu235/CP	3.7	2.37	0.37	0.64	0.10	0.74
Ppu235/PP	3.6	2.39	0.37	0.66	0.10	0.77
Ppu201	4.3	0.27	0.14	0.06	0.03	0.09
Ppu201/CP	4.1	1.82	0.14	0.44	0.03	0.48
Ppu201/PP	3.1	1.76	0.03	0.57	0.01	0.58
Ppu235	3.8	1.83	0.65	0.48	0.17	0.65
Ppu235/CP	4.6	2.19	0.36	0.48	0.08	0.55
Ppu235/PP	1.1	2.26	0.06	2.05	0.05	2.10
Ppu201	5	0.28	0.14	0.06	0.03	0.09
Ppu201/CP	3.5	1.94	0.13	0.55	0.04	0.59
Ppu201/PP	3.4	2.04	0.13	0.60	0.04	0.64
Ppu235	3.9	2.00	0.64	0.51	0.17	0.68
Ppu235/CP	3.8	2.40	0.37	0.63	0.10	0.73
Ppu235/PP	3.6	2.44	0.36	0.68	0.10	0.78

Table A1. *Cont.*

Strain	OD600	[pCA] (mM)	[CA] (mM)	pCA/1DO	CA/1DO	Total Phenolics/1DO
Ppu201	4.4	0.27	0.14	0.06	0.03	0.09
Ppu201/CP	4.7	1.90	0.15	0.41	0.03	0.44
Ppu201/PP	3.2	1.85	0.03	0.58	0.01	0.59
Ppu235	3.5	1.98	0.68	0.56	0.20	0.76
Ppu235/CP	5.4	2.45	0.39	0.45	0.07	0.53
Ppu235/PP	1.6	2.37	0.07	1.48	0.04	1.52
Ppu201	3.7	1.5	0.30	0.41	0.08	0.49
Ppu235	4	1.3	0.80	0.32	0.20	0.52
Ppu201/PP	3.3	1.4	0.10	0.42	0.03	0.45
Ppu235/PP	3.6	1.8	0.36	0.50	0.10	0.60
Ppu201	3.8	1.4	0.30	0.38	0.08	0.46
Ppu235	3.9	1.3	0.74	0.33	0.19	0.52
Ppu201/PP	0.4	0.5	0.01	1.26	0.03	1.29
Ppu235/PP	0.4	0.8	0.02	2.08	0.05	2.13
Ppu201	4	1.5	0.28	0.38	0.07	0.45
Ppu235/PP	4.2	1.4	0.76	0.34	0.18	0.52
Ppu201/PP	3.6	1.4	0.11	0.40	0.03	0.43
Ppu235/PP	4	1.8	0.32	0.46	0.08	0.54
Ppu201	4	1.5	0.32	0.38	0.08	0.46
Ppu235	4.1	1.5	0.74	0.36	0.18	0.54
Ppu201/PP	0.5	0.5	0.01	0.99	0.02	1.01
Ppu235/PP	0.8	1.1	0.02	1.40	0.03	1.43
Ppu201	4.5	2.2	0.32	0.49	0.07	0.56
Ppu235	4.6	2.1	0.74	0.45	0.16	0.61
Ppu235/PP	4	2.2	0.36	0.56	0.09	0.65
Ppu201	5.1	3.0	0.46	0.58	0.09	0.67
Ppu235	5.4	2.5	0.86	0.47	0.16	0.63
Ppu235/PP	0.5	1.2	0.04	2.43	0.08	2.51

References

1. Kitney, R.; Freemont, P. Synthetic biology—The state of play. *FEBS Lett.* **2012**, *586*, 2029–2036. [CrossRef] [PubMed]
2. Martău, G.A.; Călinoiu, L.F.; Vodnar, D.C. Bio-vanillin: Towards a sustainable industrial production. *Trends Food Sci. Technol.* **2021**, *109*, 579–592. [CrossRef]
3. Wang, J.; Shen, X.; Rey, J.; Yuan, Q.; Yan, Y. Recent advances in microbial production of aromatic natural products and their derivatives. *Appl. Microbiol. Biotechnol.* **2018**, *102*, 47–61. [CrossRef] [PubMed]
4. Dudnik, A.; Gaspar, P.; Neves, A.R.; Forster, J. Engineering of Microbial Cell Factories for the Production of Plant Polyphenols with Health-Beneficial Properties. *Curr. Pharm. Des.* **2018**, *24*, 2208–2225. [CrossRef] [PubMed]
5. Peralta-Yahya, P.P.; Zhang, F.; del Cardayre, S.B.; Keasling, J.D. Microbial engineering for the production of advanced biofuels. *Nature* **2012**, *488*, 320–328. [CrossRef]
6. Aggarwal, N.; Pham, H.L.; Ranjan, B.; Saini, M.; Liang, Y.; Hossain, G.S.; Ling, H.; Foo, J.L.; Chang, M.W. Microbial engineering strategies to utilize waste feedstock for sustainable bioproduction. *Nat. Rev. Bioeng.* **2023**, *2*, 155–174. [CrossRef]
7. Li, B.; Zhang, B.; Wang, P.; Cai, X.; Chen, Y.Y.; Yang, Y.F.; Liu, Z.Q.; Zheng, Y.G. Rerouting Fluxes of the Central Carbon Metabolism and Relieving Mechanism-Based Inactivation of L-Aspartate-α-decarboxylase for Fermentative Production of β-Alanine in *Escherichia coli*. *Acs Synth. Biol.* **2022**, *11*, 1908–1918. [CrossRef]
8. Maurya, R.; Gohil, N.; Nixon, S.; Kumar, N.; Noronha, S.B.; Dhali, D.; Trabelsi, H.; Alzahrani, K.J.; Reshamwala, S.M.; Awasthi, M.K.; et al. Rewiring of metabolic pathways in yeasts for sustainable production of biofuels. *Bioresour. Technol.* **2023**, *372*, 128668. [CrossRef]
9. Leonard, E.; Lim, K.H.; Saw, P.N.; Koffas, M.A.G. Engineering Central Metabolic Pathways for High-Level Flavonoid Production in *Escherichia coli*. *Appl. Environ. Microbiol.* **2007**, *73*, 3877–3886. [CrossRef]
10. Suzuki, S.; Koeduka, T.; Sugiyama, A.; Yazaki, K.; Umezawa, T. Microbial production of plant specialized metabolites. *Plant Biotechnol.* **2014**, *31*, 465–482. [CrossRef]

11. Limem, I.; Guedon, E.; Hehn, A.; Bourgaud, F.; Chekir Ghedira, L.; Engasser, J.M.; Ghoul, M. Production of phenylpropanoid compounds by recombinant microorganisms expressing plant-specific biosynthesis genes. *Process Biochem.* **2008**, *43*, 463–479. [CrossRef]
12. Wang, S.; Zhang, S.; Xiao, A.; Rasmussen, M.; Skidmore, C.; Zhan, J. Metabolic engineering of *Escherichia coli* for the biosynthesis of various phenylpropanoid derivatives. *Metab. Eng.* **2015**, *29*, 153–159. [CrossRef] [PubMed]
13. Averesch, N.J.H.; Krömer, J.O. Metabolic Engineering of the Shikimate Pathway for Production of Aromatics and Derived Compounds—Present and Future Strain Construction Strategies. *Front. Bioeng. Biotechnol.* **2018**, *6*, 32. [CrossRef]
14. Vargas-Tah, A.; Gosset, G. Production of Cinnamic and p-Hydroxycinnamic Acids in Engineered Microbes. *Front. Bioeng. Biotechnol.* **2015**, *3*, 116. [CrossRef]
15. Herrmann, K.M.; Weaver, L.M. The Shikimate Pathway. *Annu. Rev. Plant Physiol. Plant Mol. Biol.* **1999**, *50*, 473–503. [CrossRef]
16. Jayaraman, K.; Trachtmann, N.; Sprenger, G.A.; Gohlke, H. Protein engineering for feedback resistance in 3-deoxy-D-arabino-heptulosonate 7-phosphate synthase. *Appl. Microbiol. Biotechnol.* **2022**, *106*, 6505–6517. [CrossRef]
17. Wu, J.; Howe, D.L.; Woodard, R.W. Thermotoga maritima 3-Deoxy-D-arabino-heptulosonate 7-Phosphate (DAHP) Synthase. *J. Biol. Chem.* **2003**, *278*, 27525–27531. [CrossRef]
18. Luttik, M.; Vuralhan, Z.; Suir, E.; Braus, G.; Pronk, J.; Daran, J. Alleviation of feedback inhibition in *Saccharomyces cerevisiae* aromatic amino acid biosynthesis: Quantification of metabolic impact. *Metab. Eng.* **2008**, *10*, 141–153. [CrossRef]
19. Bilal, M.; Wang, S.; Iqbal, H.M.N.; Zhao, Y.; Hu, H.; Wang, W.; Zhang, X. Metabolic engineering strategies for enhanced shikimate biosynthesis: current scenario and future developments. *Appl. Microbiol. Biotechnol.* **2018**, *102*, 7759–7773. [CrossRef]
20. Lütke-Eversloh, T.; Stephanopoulos, G. L-Tyrosine production by deregulated strains of *Escherichia coli*. *Appl. Microbiol. Biotechnol.* **2007**, *75*, 103–110. [CrossRef]
21. Kang, S.Y.; Choi, O.; Lee, J.K.; Hwang, B.Y.; Uhm, T.B.; Hong, Y.S. Artificial biosynthesis of phenylpropanoic acids in a tyrosine overproducing *Escherichia coli* strain. *Microb. Cell Factories* **2012**, *11*, 153. [CrossRef] [PubMed]
22. Santos, C.N.S.; Koffas, M.; Stephanopoulos, G. Optimization of a heterologous pathway for the production of flavonoids from glucose. *Metab. Eng.* **2011**, *13*, 392–400. [CrossRef] [PubMed]
23. Juminaga, D.; Baidoo, E.E.K.; Redding-Johanson, A.M.; Batth, T.S.; Burd, H.; Mukhopadhyay, A.; Petzold, C.J.; Keasling, J.D. Modular Engineering of L-Tyrosine Production in *Escherichia coli*. *Appl. Environ. Microbiol.* **2012**, *78*, 89–98. [CrossRef] [PubMed]
24. Rodriguez, A.; Kildegaard, K.R.; Li, M.; Borodina, I.; Nielsen, J. Establishment of a yeast platform strain for production of p-coumaric acid through metabolic engineering of aromatic amino acid biosynthesis. *Metab. Eng.* **2015**, *31*, 181–188. [CrossRef]
25. Nikel, P.I.; Martínez-García, E.; de Lorenzo, V. Biotechnological domestication of pseudomonads using synthetic biology. *Nat. Rev. Microbiol.* **2014**, *12*, 368–379. [CrossRef]
26. Ankenbauer, A.; Schäfer, R.A.; Viegas, S.C.; Pobre, V.; Voß, B.; Arraiano, C.M.; Takors, R. *Pseudomonas putida* KT2440 is naturally endowed to withstand industrial-scale stress conditions. *Microb. Biotechnol.* **2020**, *13*, 1145–1161. [CrossRef]
27. Schwanemann, T.; Otto, M.; Wierckx, N.; Wynands, B. Pseudomonas as Versatile Aromatics Cell Factory. *Biotechnol. J.* **2020**, *15*, 1900569. [CrossRef]
28. Molina-Santiago, C.; Cordero, B.F.; Daddaoua, A.; Udaondo, Z.; Manzano, J.; Valdivia, M.; Segura, A.; Ramos, J.L.; Duque, E. *Pseudomonas putida* as a platform for the synthesis of aromatic compounds. *Microbiology* **2016**, *162*, 1535–1543. [CrossRef]
29. Calero, P.; Jensen, S.I.; Bojanovič, K.; Lennen, R.M.; Koza, A.; Nielsen, A.T. Genome-wide identification of tolerance mechanisms toward p-coumaric acid in *Pseudomonas putida*. *Biotechnol. Bioeng.* **2017**, *115*, 762–774. [CrossRef]
30. Mohamed, E.T.; Werner, A.Z.; Salvachúa, D.; Singer, C.A.; Szostkiewicz, K.; Rafael Jiménez-Díaz, M.; Eng, T.; Radi, M.S.; Simmons, B.A.; Mukhopadhyay, A.; et al. Adaptive laboratory evolution of *Pseudomonas putida* KT2440 improves p-coumaric and ferulic acid catabolism and tolerance. *Metab. Eng. Commun.* **2020**, *11*, e00143. [CrossRef]
31. Nijkamp, K.; Westerhof, R.G.M.; Ballerstedt, H.; de Bont, J.A.M.; Wery, J. Optimization of the solvent-tolerant *Pseudomonas putida* S12 as host for the production of p-coumarate from glucose. *Appl. Microbiol. Biotechnol.* **2007**, *74*, 617–624. [CrossRef]
32. Wierckx, N.J.P.; Ballerstedt, H.; de Bont, J.A.M.; Wery, J. Engineering of Solvent-Tolerant *Pseudomonas putida* S12 for Bioproduction of Phenol from Glucose. *Appl. Environ. Microbiol.* **2005**, *71*, 8221–8227. [CrossRef] [PubMed]
33. Calero, P.; Jensen, S.I.; Nielsen, A.T. Broad-Host-Range ProUSER Vectors Enable Fast Characterization of Inducible Promoters and Optimization of p-Coumaric Acid Production in *Pseudomonas putida* KT2440. *ACS Synth. Biol.* **2016**, *5*, 741–753. [CrossRef] [PubMed]
34. Wynands, B.; Lenzen, C.; Otto, M.; Koch, F.; Blank, L.M.; Wierckx, N. Metabolic engineering of Pseudomonas taiwanensis VLB120 with minimal genomic modifications for high-yield phenol production. *Metab. Eng.* **2018**, *47*, 121–133. [CrossRef]
35. Otto, M.; Wynands, B.; Lenzen, C.; Filbig, M.; Blank, L.M.; Wierckx, N. Rational Engineering of Phenylalanine Accumulation in Pseudomonas taiwanensis to Enable High-Yield Production of Trans-Cinnamate. *Front. Bioeng. Biotechnol.* **2019**, *7*. [CrossRef]
36. Weaver, L.M.; Herrmann, K.M. Cloning of an aroF allele encoding a tyrosine-insensitive 3-deoxy-D-arabino-heptulosonate 7-phosphate synthase. *J. Bacteriol.* **1990**, *172*, 6581–6584. [CrossRef]

37. Lütke-Eversloh, T.; Stephanopoulos, G. Feedback inhibition of chorismate mutase/prephenate dehydrogenase (TyrA) of *Escherichia coli*: Generation and characterization of tyrosine-insensitive mutants. *Appl. Environ. Microbiol.* **2005**, *71*, 7224–7228. [CrossRef]
38. Liu, X.; Liu, J.; Liu, Z.; Qiao, Q.; Ni, X.; Yang, J.; Sun, G.; Li, F.; Zhou, W.; Guo, X.; et al. Engineering allosteric inhibition of homoserine dehydrogenase by semi-rational saturation mutagenesis screening. *Front. Bioeng. Biotechnol.* **2024**, *11*, 1336215. [CrossRef]
39. Chen, M.M.; Snow, C.D.; Vizcarra, C.L.; Mayo, S.L.; Arnold, F.H. Comparison of random mutagenesis and semi-rational designed libraries for improved cytochrome P450 BM3-catalyzed hydroxylation of small alkanes. *Protein Eng. Des. Sel.* **2012**, *25*, 171–178. [CrossRef]
40. Rowlands, R. Industrial strain improvement: Mutagenesis and random screening procedures. *Enzym. Microb. Technol.* **1984**, *6*, 3–10. [CrossRef]
41. Zhao, S.; Tan, M.Z.; Wang, R.X.; Ye, F.T.; Chen, Y.P.; Luo, X.M.; Feng, J.X. Combination of genetic engineering and random mutagenesis for improving production of raw-starch-degrading enzymes in *Penicillium oxalicum*. *Microb. Cell Factories* **2022**, *21*, 272. [CrossRef] [PubMed]
42. Waterhouse, A.; Bertoni, M.; Bienert, S.; Studer, G.; Tauriello, G.; Gumienny, R.; Heer, F.T.; de Beer, T.A.P.; Rempfer, C.; Bordoli, L.; et al. SWISS-MODEL: Homology modelling of protein structures and complexes. *Nucleic Acids Res.* **2018**, *46*, W296–W303. [CrossRef] [PubMed]
43. Mildvan, A.S. Inverse thinking about double mutants of enzymes. *Biochemistry* **2004**, *43*, 14517–14520. [CrossRef] [PubMed]
44. Chaparro-Riggers, J.F.; Polizzi, K.M.; Bommarius, A.S. Better library design: Data-driven protein engineering. *Biotechnol. J. Healthc. Nutr. Technol.* **2007**, *2*, 180–191. [CrossRef]
45. Wierckx, N.; Ruijssenaars, H.J.; de Winde, J.H.; Schmid, A.; Blank, L.M. Metabolic flux analysis of a phenol producing mutant of Pseudomonas putida S12: Verification and complementation of hypotheses derived from transcriptomics. *J. Biotechnol.* **2009**, *143*, 124–129. [CrossRef]
46. Glick, B.R. Metabolic load and heterologous gene expression. *Biotechnol. Adv.* **1995**, *13*, 247–261. [CrossRef]
47. Bertani, G. STUDIES ON LYSOGENESIS I: The Mode of Phage Liberation by Lysogenic *Escherichia coli*. *J. Bacteriol.* **1951**, *62*, 293–300. [CrossRef]
48. Simon, R.; Priefer, U.; Pühler, A. A broad host range mobilization system for in vivo genetic engineering: Transposon mutagenesis in gram negative bacteria. *Bio/Technology* **1983**, *1*, 784–791. [CrossRef]
49. Kovach, M.E.; Elzer, P.H.; Hill, D.S.; Robertson, G.T.; Farris, M.A.; Roop II, R.M.; Peterson, K.M. Four new derivatives of the broad-host-range cloning vector pBBR1MCS, carrying different antibiotic-resistance cassettes. *Gene* **1995**, *166*, 175–176. [CrossRef]
50. Schäfer, A.; Tauch, A.; Jäger, W.; Kalinowski, J.; Thierbach, G.; Pühler, A. Small mobilizable multi-purpose cloning vectors derived from the *Escherichia coli* plasmids pK18 and pK19: Selection of defined deletions in the chromosome of *Corynebacterium glutamicum*. *Gene* **1994**, *145*, 69–73. [CrossRef]
51. Studier, F.W. Protein production by auto-induction in high-density shaking cultures. *Protein Expr. Purif.* **2005**, *41*, 207–234. [CrossRef] [PubMed]
52. Gibson, D.G. Synthesis of DNA fragments in yeast by one-step assembly of overlapping oligonucleotides. *Nucleic Acids Res.* **2009**, *37*, 6984–6990. [CrossRef] [PubMed]
53. Shapiro, S.S.; Wilk, M.B. An analysis of variance test for normality (complete samples). *Biometrika* **1965**, *52*, 591–611. [CrossRef]
54. Jones, D.H.; Snedecor, G.W.; Cochran, W.G. Statistical Methods. *J. Educ. Behav. Stat.* **1994**, *19*, 304. [CrossRef]
55. Kruskal, W.H.; Wallis, W.A. Use of Ranks in One-Criterion Variance Analysis. *J. Am. Stat. Assoc.* **1952**, *47*, 583–621. [CrossRef]
56. Dunn, O.J. Multiple Comparisons Using Rank Sums. *Technometrics* **1964**, *6*, 241–252. [CrossRef]
57. Cohen, J. *Statistical Power Analysis for the Behavioral Sciences*; Routledge: London, UK, 2013.
58. Trott, O.; Olson, A.J. AutoDock Vina: Improving the speed and accuracy of docking with a new scoring function, efficient optimization, and multithreading. *J. Comput. Chem.* **2010**, *31*, 455–461. [CrossRef]
59. Shumilin, I.A.; Zhao, C.; Bauerle, R.; Kretsinger, R.H. Allosteric Inhibition of 3-Deoxy-d-arabino-heptulosonate-7-phosphate Synthase Alters the Coordination of Both Substrates. *J. Mol. Biol.* **2002**, *320*, 1147–1156. [CrossRef]
60. Cui, D.; Deng, A.; Bai, H.; Yang, Z.; Liang, Y.; Liu, Z.; Qiu, Q.; Wang, L.; Liu, S.; Zhang, Y.; et al. Molecular basis for feedback inhibition of tyrosine-regulated 3-deoxy-d-arabino-heptulosonate-7-phosphate synthase from *Escherichia coli*. *J. Struct. Biol.* **2019**, *206*, 322–334. [CrossRef]
61. Ding, R.; Liu, L.; Chen, X.; Cui, Z.; Zhang, A.; Ren, D.; Zhang, L. Introduction of two mutations into AroG increases phenylalanine production in *Escherichia coli*. *Biotechnol. Lett.* **2014**, *36*, 2103–2108. [CrossRef]
62. Shimizu, K. Metabolic Regulation of a Bacterial Cell System with Emphasis on *Escherichia coli* Metabolism. *ISRN Biochem.* **2013**, *2013*, 1–47. [CrossRef] [PubMed]
63. Chubukov, V.; Gerosa, L.; Kochanowski, K.; Sauer, U. Coordination of microbial metabolism. *Nat. Rev. Microbiol.* **2014**, *12*, 327–340. [CrossRef] [PubMed]

64. Shimizu, K.; Matsuoka, Y. Feedback regulation and coordination of the main metabolism for bacterial growth and metabolic engineering for amino acid fermentation. *Biotechnol. Adv.* **2022**, *55*, 107887. [CrossRef] [PubMed]
65. Nudler, E. The riboswitch control of bacterial metabolism. *Trends Biochem. Sci.* **2004**, *29*, 11–17. [CrossRef]
66. Wu, S.; Chen, W.; Lu, S.; Zhang, H.; Yin, L. Metabolic Engineering of Shikimic Acid Biosynthesis Pathway for the Production of Shikimic Acid and Its Branched Products in Microorganisms: Advances and Prospects. *Molecules* **2022**, *27*, 4779. [CrossRef]
67. Jiang, M.; Zhang, H. Engineering the shikimate pathway for biosynthesis of molecules with pharmaceutical activities in *E. coli*. *Curr. Opin. Biotechnol.* **2016**, *42*, 1–6. [CrossRef]
68. Cao, M.; Gao, M.; Suástegui, M.; Mei, Y.; Shao, Z. Building microbial factories for the production of aromatic amino acid pathway derivatives: From commodity chemicals to plant-sourced natural products. *Metab. Eng.* **2020**, *58*, 94–132. [CrossRef]
69. Helmstaedt, K.; Strittmatter, A.; Lipscomb, W.N.; Braus, G.H. Evolution of 3-deoxy-d-arabino-heptulosonate-7-phosphate synthase-encoding genes in the yeast *Saccharomyces cerevisiae*. *Proc. Natl. Acad. Sci. USA* **2005**, *102*, 9784–9789. [CrossRef]
70. Jossek, R.; Bongaerts, J.; Sprenger, G.A. Characterization of a new feedback-resistant 3-deoxy-d-arabino-heptulosonate 7-phosphate synthase AroF of *Escherichia coli*. *FEMS Microbiol. Lett.* **2001**, *202*, 145–148. [CrossRef]
71. Mir, R.; Jallu, S.; Singh, T.P. The shikimate pathway: Review of amino acid sequence, function and three-dimensional structures of the enzymes. *Crit. Rev. Microbiol.* **2013**, *41*, 172–189. [CrossRef]
72. Ger, Y.M.; Chen, S.L.; Chiang, H.J.; Shiuan, D. A Single Ser-180 Mutation Desensitizes Feedback Inhibition of the Phenylalanine-Sensitive 3-Deoxy-D-Arabino-Heptulosonate 7-Phosphate (DAHP) Synthetase in *Escherichia coli*. *J. Biochem.* **1994**, *116*, 986–990. [CrossRef] [PubMed]
73. Ray, J.M.; Yanofsky, C.; Bauerle, R. Mutational analysis of the catalytic and feedback sites of the tryptophan-sensitive 3-deoxy-D-arabino-heptulosonate-7-phosphate synthase of *Escherichia coli*. *J. Bacteriol.* **1988**, *170*, 5500–5506. [CrossRef] [PubMed]
74. Kikuchi, Y.; Tsujimoto, K.; Kurahashi, O. Mutational analysis of the feedback sites of phenylalanine-sensitive 3-deoxy-D-arabino-heptulosonate-7-phosphate synthase of *Escherichia coli*. *Appl. Environ. Microbiol.* **1997**, *63*, 761–762. [CrossRef] [PubMed]
75. Lutz, S. Beyond directed evolution—Semi-rational protein engineering and design. *Curr. Opin. Biotechnol.* **2010**, *21*, 734–743. [CrossRef]
76. Chen, H.; Ma, L.; Dai, H.; Fu, Y.; Wang, H.; Zhang, Y. Advances in Rational Protein Engineering toward Functional Architectures and Their Applications in Food Science. *J. Agric. Food Chem.* **2022**, *70*, 4522–4533. [CrossRef]
77. Nixon, A.E.; Firestine, S.M. Rational and "Irrational" Design of Proteins and Their Use in Biotechnology. *IUBMB Life* **2000**, *49*, 181–187. [CrossRef]
78. Faisca, P.; Travasso, R.; Ball, R.; Shakhnovich, E.I. Identifying critical residues in protein folding: Insights from ϕ-value and Pfold analysis. *J. Chem. Phys.* **2008**, *129*, 095108. [CrossRef]
79. Sen, C.; Logashree, V.; Makde, R.D.; Ghosh, B. Amino acid propensities for secondary structures and its variation across protein structures using exhaustive PDB data. *Comput. Biol. Chem.* **2024**, *110*, 108083. [CrossRef]
80. Chronopoulou, E.G.; Labrou, N.E. Site-saturation mutagenesis: A powerful tool for structure-based design of combinatorial mutation libraries. *Curr. Protoc. Protein Sci.* **2011**, *63*. [CrossRef]
81. Marabotti, A.; Scafuri, B.; Facchiano, A. Predicting the stability of mutant proteins by computational approaches: An overview. *Briefings Bioinform.* **2021**, *22*, bbaa074. [CrossRef]
82. Pucci, F.; Schwersensky, M.; Rooman, M. Artificial intelligence challenges for predicting the impact of mutations on protein stability. *Curr. Opin. Struct. Biol.* **2022**, *72*, 161–168. [CrossRef] [PubMed]
83. Fang, J. A critical review of five machine learning-based algorithms for predicting protein stability changes upon mutation. *Briefings Bioinform.* **2019**, *21*, 1285–1292. [CrossRef] [PubMed]
84. Cheng, P.; Mao, C.; Tang, J.; Yang, S.; Cheng, Y.; Wang, W.; Gu, Q.; Han, W.; Chen, H.; Li, S.; et al. Zero-shot prediction of mutation effects with multimodal deep representation learning guides protein engineering. *Cell Res.* **2024**, *34*, 630–647. [CrossRef] [PubMed]
85. Hamborg, L.; Granata, D.; Olsen, J.G.; Roche, J.V.; Pedersen, L.E.; Nielsen, A.T.; Lindorff-Larsen, K.; Teilum, K. Synergistic stabilization of a double mutant in chymotrypsin inhibitor 2 from a library screen in *E. coli*. *Commun. Biol.* **2021**, *4*, 980. [CrossRef]
86. Greener, J.G.; Sternberg, M.J. AlloPred: Prediction of allosteric pockets on proteins using normal mode perturbation analysis. *BMC Bioinform.* **2015**, *16*, 335. [CrossRef]
87. Yamato, T.; Laprévote, O. Normal mode analysis and beyond. *Biophys. Physicobiol.* **2019**, *16*, 322–327. [CrossRef]
88. Kolossváry, I. A Fresh Look at the Normal Mode Analysis of Proteins: Introducing Allosteric Co-Vibrational Modes. *JACS Au* **2024**, *4*, 1303–1309. [CrossRef]
89. Ruffolo, J.A.; Madani, A. Designing proteins with language models. *Nat. Biotechnol.* **2024**, *42*, 200–202. [CrossRef]
90. Schmirler, R.; Heinzinger, M.; Rost, B. Fine-tuning protein language models boosts predictions across diverse tasks. *Nat. Commun.* **2024**, *15*, 7407. [CrossRef]
91. Zhou, Z.; Zhang, L.; Yu, Y.; Wu, B.; Li, M.; Hong, L.; Tan, P. Enhancing efficiency of protein language models with minimal wet-lab data through few-shot learning. *Nat. Commun.* **2024**, *15*, 5566. [CrossRef]

92. Biswas, S.; Khimulya, G.; Alley, E.C.; Esvelt, K.M.; Church, G.M. Low-N protein engineering with data-efficient deep learning. *Nat. Methods* **2021**, *18*, 389–396. [CrossRef] [PubMed]
93. Chen, R. Bacterial expression systems for recombinant protein production: *E. coli* and beyond. *Biotechnol. Adv.* **2012**, *30*, 1102–1107. [CrossRef] [PubMed]
94. Goussé, M.; Dell'Aglio, E.; Curien, G.; Borland, S.; Renoud, S.; Ranquet, C.; Chandor-Proust, A. *E. coli* chromosomal-driven expression of NADK2 from *A. thaliana*: A preferable alternative to plasmid-driven expression for challenging proteins. *Protein Expr. Purif.* **2022**, *195–196*, 106090. [CrossRef] [PubMed]
95. Kachroo, A.H.; Jayaram, M.; Rowley, P.A. Metabolic engineering without plasmids. *Nat. Biotechnol.* **2009**, *27*, 729–731. [CrossRef]
96. Zucca, S.; Pasotti, L.; Mazzini, G.; Cusella De Angelis, M.G.; Magni, P. Characterization of an inducible promoter in different DNA copy number conditions. *BMC Bioinform.* **2012**, *13*, S11. [CrossRef]
97. Tyo, K.E.J.; Ajikumar, P.K.; Stephanopoulos, G. Stabilized gene duplication enables long-term selection-free heterologous pathway expression. *Nat. Biotechnol.* **2009**, *27*, 760–765. [CrossRef]
98. Schild, F.; Chandor-Proust, A.; Ranquet, C. *Method of Mutagenesis Without Scar*; Institut National de la Propriété' Industrielle: Paris, France, 2017.
99. Li, Z.; Wang, H.; Ding, D.; Liu, Y.; Fang, H.; Chang, Z.; Chen, T.; Zhang, D. Metabolic engineering of *Escherichia coli* for production of chemicals derived from the shikimate pathway. *J. Ind. Microbiol. Biotechnol.* **2020**, *47*, 525–535. [CrossRef]
100. Bentley, R.; Haslam, E. The Shikimate Pathway—A Metabolic Tree with Many Branche. *Crit. Rev. Biochem. Mol. Biol.* **1990**, *25*, 307–384. [CrossRef]
101. Liu, Q.; Yu, T.; Li, X.; Chen, Y.; Campbell, K.; Nielsen, J.; Chen, Y. Rewiring carbon metabolism in yeast for high level production of aromatic chemicals. *Nat. Commun.* **2019**, *10*, 4976. [CrossRef]
102. Kroll, K.; Holland, C.K.; Starks, C.M.; Jez, J.M. Evolution of allosteric regulation in chorismate mutases from early plants. *Biochem. J.* **2017**, *474*, 3705–3717. [CrossRef]
103. Brown, J.F.; Dawes, I.W. Regulation of chorismate mutase in *Saccharomyces cerevisiae*. *Mol. Gen. Genet. MGG* **1990**, *220*, 283–288. [CrossRef] [PubMed]
104. Zhang, S.; Pohnert, G.; Kongsaeree, P.; Wilson, D.B.; Clardy, J.; Ganem, B. Chorismate Mutase-Prephenate Dehydratase from *Escherichia coli*. *J. Biol. Chem.* **1998**, *273*, 6248–6253. [CrossRef] [PubMed]
105. Mokwatlo, S.C.; Klein, B.C.; Benavides, P.T.; Tan, E.C.; Kneucker, C.M.; Ling, C.; Singer, C.A.; Lyons, R.; i Nogué, V.S.; Hestmark, K.V.; et al. Bioprocess development and scale-up for cis, cis-muconic acid production from glucose and xylose by *Pseudomonas putida*. *Green Chem.* **2024**, *26*, 10152–10167. [CrossRef]
106. Flourat, A.L.; Combes, J.; Bailly-Maitre-Grand, C.; Magnien, K.; Haudrechy, A.; Renault, J.H.; Allais, F. Accessing p-Hydroxycinnamic Acids: Chemical Synthesis, Biomass Recovery, or Engineered Microbial Production? *ChemSusChem* **2021**, *14*, 118–129. [CrossRef]
107. Louie, G.V.; Bowman, M.E.; Moffitt, M.C.; Baiga, T.J.; Moore, B.S.; Noel, J.P. Structural Determinants and Modulation of Substrate Specificity in Phenylalanine-Tyrosine Ammonia-Lyases. *Chem. Biol.* **2006**, *13*, 1327–1338. [CrossRef]
108. Watts, K.T.; Mijts, B.N.; Lee, P.C.; Manning, A.J.; Schmidt-Dannert, C. Discovery of a Substrate Selectivity Switch in Tyrosine Ammonia-Lyase, a Member of the Aromatic Amino Acid Lyase Family. *Chem. Biol.* **2006**, *13*, 1317–1326. [CrossRef]
109. Jendresen, C.B.; Stahlhut, S.G.; Li, M.; Gaspar, P.; Siedler, S.; Förster, J.; Maury, J.; Borodina, I.; Nielsen, A.T. Highly Active and Specific Tyrosine Ammonia-Lyases from Diverse Origins Enable Enhanced Production of Aromatic Compounds in Bacteria and *Saccharomyces cerevisiae*. *Appl. Environ. Microbiol.* **2015**, *81*, 4458–4476. [CrossRef]
110. KeGG Kyoto Encyclopedia of Genes and Genomes. Available online: https://www.kegg.jp/kegg/ (accessed on 15 November 2024).

BioChem

MDPI

Article

Hydrogels Made with Tilapia Fish Skin Increase Collagen Production and Have an Effect on MMP-2/MMP-9 Enzymes in Burn Treatment

Berkay Baydogan [1], Aslihan Kucuk [2], Bensu Kozan [2], Merve Erdal [2], Burcin Irem Abas [1] and Ozge Cevik [1,*]

1 Department of Medicinal Biochemistry, School of Medicine, Aydin Adnan Menderes University, Aydin 09100, Turkey; dr.berkaybaydogan@gmail.com (B.B.); burcin.abas@adu.edu.tr (B.I.A.)
2 Department of Molecular Biotechnology, Graduate School of Health Sciences, Aydin Adnan Menderes University, Aydin 09100, Turkey; kucukasli.98@gmail.com (A.K.); bensukozan@gmail.com (B.K.); merverdal93@gmail.com (M.E.)
* Correspondence: ozge.cevik@adu.edu.tr

Abstract: Background/Objectives: Burns are a prevalent health concern that manifest on the skin's surface or within organs due to various traumas and necessitate prompt intervention. The healing process of the skin involves a sequence of time-dependent events, commencing with the activation of growth factors and culminating in the expression of various genes. To expedite the healing process of burn wounds, there is a need to develop biodegradable materials and new technologies that are compatible with the skin. **Methods**: In this study, the roles of tilapia (TL, *Oreochromis niloticus*) fish skin in burn wound treatment processes were investigated. TL or TL-alginate hydrogels (AGTL) were applied to a burn wound created in Sprague Dawley rats for 7 and 14 days. Following the administration of treatment, the levels of hydroxyproline, a critical element in tissue reorganization, along with the gene expression levels of COL1A1, COL3A1, MMP-2, and MMP-9, and the protein expression levels of MMP-2 and MMP-9 were evaluated. **Results**: Wound closure processes were faster in AGTL-groups compared to TL-groups, and hydroxyproline levels were found to be higher. While the increase in MMP-2 levels was less, the increase in MMP-9 gene and protein levels was greater in the AGTL-group. Concurrently, COL1A1 levels decreased over 14 days, while COL3A1 levels increased in the AGTL-group. **Conclusions**: Consequently, it was determined that the biological substances in the TL structure, in conjunction with alginate, were effective in the healing and reorganization of the wound tissue. This finding suggests that tilapia may provide a valuable source of insights for future studies aimed at developing effective wound dressings for wound tissues.

Keywords: burn; fish skin; MMP-2; MMP-9; hydroxyproline

1. Introduction

Burns are the most common type of injury in everyday life, and they are caused primarily by heat or contact with chemicals, electricity, friction, and radioactivity on the outer surface tissue of organs located on or under the skin. Burns are critical and costly problems that require care depending on the type and extent of the material in contact. In cases of burns, depending on the degree of severity, they can be fatal depending on the complications that occur. On the other hand, non-fatal burns are one of the leading causes of morbidity, including long-term hospitalization, deformity and disability, as well as stigma and rejection [1,2]. In recent years, there has been an increasing tendency to utilize biological resources in the development of dressing materials with the objective of accelerating the healing process of patients and reducing the duration of hospital stay for burn treatment. These resources encompass both plant-based compounds and proteins derived from animal sources, which are engineered into forms compatible with various types of wounds [3,4].

Tilapia (*Oreochromis niloticus*) is a freshwater fish species belonging to the Cichlidae family. It is notable for its high economic and nutritional value. Tilapia is able to tolerate elevated water temperatures and wide pH ranges, which makes it a resilient species. The skin of tilapia fish has been utilized in burn treatment due to its reported wound healing properties [5,6]. These properties are attributed to the presence of unique biological components in the healing process. Recent studies have demonstrated a clear increase in the utilization of biological materials in treatment methods. Fish skin grafts, in particular, have been shown to accelerate wound healing due to their content. These grafts have demonstrated promising results in the treatment of diabetic foot ulcers and venous leg ulcers, suggesting their potential for use in the treatment of various other acute and chronic wounds [7]. Collagen, a protein that has the potential to support wound healing, is found in large amounts in fish skin [8]. Research findings have demonstrated that collagen, when dissolved using a tilapia fish source, has enhanced efficacy in promoting wound healing processes [9,10]. Tilapia skin has been reported to contain a collagen content exceeding 40% on a dry weight basis, and it is also used in the production of collagen peptides [11,12]. An analysis of the amino acid composition of collagen obtained by means of enzymatic hydrolysis from tilapia revealed that glycine, proline, and hydroxyproline, as the predominant amino acids, accounted for 20.92%, 11.32%, and 10.28%, respectively [9]. Research has demonstrated that collagen in tilapia comprises the following same four subunits: β chains, γ chains, and two α chains ($\alpha 1$ and $\alpha 2$). The fibrillar structure of collagen is responsible for its significant enhancement of cell adhesion capacities [13]. Therefore, it has been suggested that it is safe to use in the production of collagen-containing biomaterials and as a dressing material. Tilapia has been reported to be biosafe in studies conducted on patients, with accelerated wound healing being a notable outcome [14]. It has been documented that the addition of silver nanoparticles to tilapia skins via polyaniline coating enhances antibiofilm activity. Moreover, it is postulated that this process may play a pivotal role in combating contamination by Staphylococcus aureus [15,16]. Additionally, the potential of silver nanoparticles to mitigate collagen loss during the sterilization of tilapia skin grafts has been underscored. The development of modified tilapia skin grafts that exhibit antimicrobial properties is also recommended [17].

The process of wound healing is accompanied by a series of molecular mechanisms in the physiological process, which consists of four stages. Collagen, a structural protein, plays a critical role in wound healing; however, its production and reorganization are subject to rigorous regulation by enzymes such as matrix metalloproteinases (MMPs). Among the MMPs, MMP-2 and MMP-9 are of particular significance, as they play a pivotal role in the degradation of collagen, particularly type I and type III collagen. MMP-2 interacts with basement membrane structures, such as type IV and type V collagen, while MMP-9 targets collagens located more in the connective tissue and outer cell matrix [18]. In the post-burn healing process, the excessive activation or inhibition of these MMPs can determine the speed and quality of healing. Therefore, the balanced activities of MMP-2 and MMP-9 in burn healing are important for both optimal wound healing and the prevention of scar formation [19,20]. Collagen's function is most significant in the remodeling stage, which is the third stage of the process [21,22]. The remodeling stage is to restore normal tissue structure and ensure the formation of scar tissue with collagen [23]. The function of collagen in this stage is particularly important, as it acts as a biodegradable protein for cellular migration and capillary growth, which are essential for the repair process [24].

In this study, a hydrogel form of tilapia skin was developed for use in burn cases, and its role in wound healing processes in rats was investigated. The aim was to determine changes in markers such as hydroxyproline, MMP-2, MMP-9, COL1A1, and COL3A1, which play important roles in collagen production and destruction, in tilapia skin hydrogels in burn wounds created in rats and to investigate whether they could be a dressing material that could be used in the clinic.

2. Materials and Methods

2.1. Preparation of Hydrogels from Tilapia and Alginate

Oreochromis niloticus, commonly known as the tilapia (TL) fish, were procured from a cultivation facility (Konya, Turkey). The skins of these fish were meticulously separated through mechanical means on ice using a sharp knife. The process of sterilization entailed the meticulous removal of the tilapia skin, which was then immersed in a solution of povidone iodine for a period of 10 min. This was followed by a thorough cleansing with NaCI (0.1%), which was repeated three times. For the decellularization process [25], the skin sample was subjected to an incubation in a phosphate-buffered saline (PBS) solution comprising 0.5% Triton X and 0.02% sodium azide at a rotary speed of 100 rpm for a duration of 1 h at ambient temperature. Subsequently, the sample underwent two washes with phosphate-buffered saline (PBS), each lasting 5 min and conducted at 100 rpm at room temperature. Subsequently, the decellularized skin sample was subjected to incubation with a trypsin solution (0.05 g/mL trypsin in 1 M Tris-HCl, pH 8.5) for a duration of 1 h at room temperature. Following this, the sample was thoroughly washed with PBS to ensure the removal of any residual enzymes. Subsequently, a sterilization solution was formulated containing 50% glycerol, 1% penicillin streptomycin, and an antifungal agent (2.5 μg/mL Amphotericin B), with the objective of ensuring long-term sterilization. Sodium alginate (AG W201502, Sigma-Aldrich Saint Louis, MO, USA,) was dissolved (0.1 g/mL) in sterile PBS and filtered through a 0.45 micron sterile filter. Sterile calcium chloride solutions (2% *w*/*v*) were prepared to reduce the agent. Then, the solution was prepared at a concentration of 1.25% AG. After the tilapia skin (1 g) was homogenized in the KCL solution (10 mL), it was mixed with alginate in equal proportions (10 mL homogenate and 10 mL alginate solution), and the calcium solution (2 mL) was added and kept for 5 min for the formation of the alginate complex. Then, TL-alginate hydrogels (AGTL) were washed with a sterilization solution (50% glycerol, 1% penicillin streptomycin, and 2.5 μg/mL Amphotericin B). Alginate was selected due to its non-toxic composition and substantial water retention capabilities. The decision was made to employ alginate in its hydrogel form to facilitate the adhesion of tilapia skin to the wound surface. The resultant AGTL hydrogel was transferred into a sterile PBS container and subsequently stored in capped falcons. The alginate hydrogels were meticulously shaped into dimensions commensurate with the size of the wound. While AGLT was being placed on the wound surface, it was placed on a dressing to prevent slipping, and the back area of the animals was covered with a plaster. Tilapia hydrogel production stages and experimental design for applications are included in Figure 1.

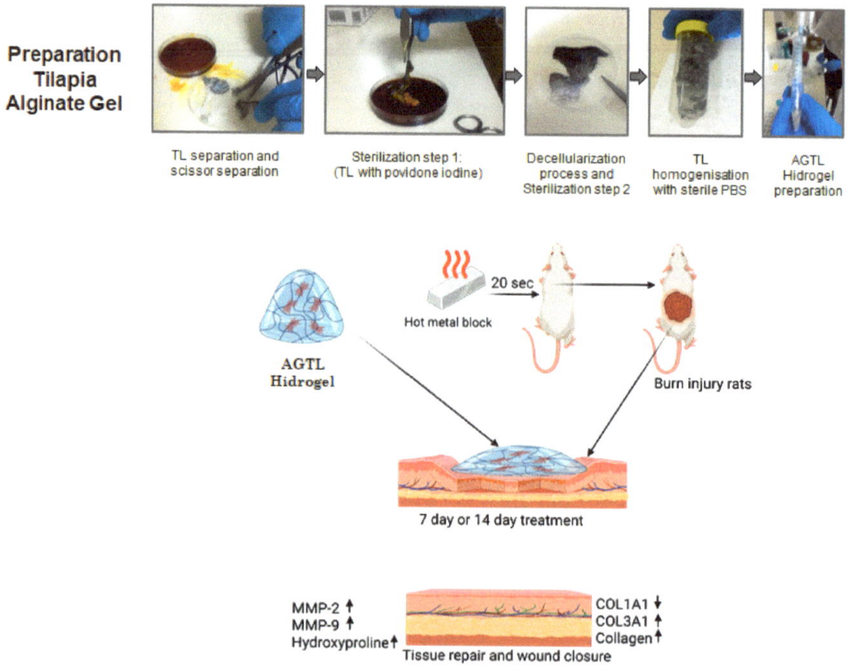

Figure 1. Tilapia hydrogel production stages and experimental design.

2.2. Characterization of AGLT Hydrogels

The rheological behaviors of the hydrogels were measured using a rheometer (HAAKE MARS 40 Rheometer, Thermo Fisher Scientific Inc., Waltham, MA, USA). AG and AGTL viscosity and shear rate were measured every 10 s for 1 min.

Since the wound surface is acidic, the swelling properties of AG or AGLT-containing hydrogel samples were investigated at different pH values. For swelling tests, hydrogel samples were dried in a vacuum oven at 40 °C, and their weights were measured. The dried hydrogels were placed in PBS (pH: 7.4) or acetate buffer (acetic acid/sodium acetate, pH: 4.5) at 25 °C, and the time of placement was taken as t = 0 min. At specified time intervals, the hydrogel sample was removed from the buffer, weighed, and returned to the buffer medium. This process continued until the hydrogel reached its equilibrium swelling value. The swelling values of the hydrogels were calculated by subtracting the mass of the swollen hydrogel from the initial mass of the dry hydrogel.

2.3. Burn Model Creation and Treatment Groups

This study was approved by the Ethical Committee for Animal Experiments at the Aydin Adnan Menderes University, approval number (2023/75). All experiments were conducted according to the Guide for the Care and Use of Laboratory Animals of the National Institutes of Health [26], and the experiments were reported according to the ARRIVE guidelines [27], Aydin Experimental. The burn model was created by anesthetizing the Sprague Dawley Rats (250–300 g, 3 months old) with 75–90 mg/kg of ketamine (5%; Pfizer Inc., New York, NY, USA) and 5–8 mg/kg of xylazine (2%; Bayer Health Care AG, Leverkusen, Germany) intraperitoneally, and then catheterization was performed from the lateral tail vein with a 24 Gauge intravenous catheter. The dorsal areas of the anesthetized rats were shaved and cleansed with povidone iodine for 10 min [28]. A 4 cm × 4 cm metal plate was heated by immersing it in boiling water (100 °C) for 15 s. Then, the heated metal plate was placed on the shaved back of the rat without applying pressure and held for 20 s with a stopwatch, and the same amount of time was applied to all rats. In the experimental group, 6 were used in each group. The following experiments were conducted: A 7-day or 14-day group was created for the burn models, then and a group with no treatment for 7 or 14 days was created. A TL-7-day or TL-14-day burn model was created, and then tilapia skin was applied for 7 or 14 days. An AGTL-7-day or AGTL-14-day burn model was created and then tilapia-containing hydrogel skin was applied for 7 or 14 days. Animals were placed in each cage individually. Infection monitoring and dressing monitoring were performed daily, and no infection was observed during the experiment. During the creation of the burn model and the treatment period, no rats were lost. The experiment concluded with the survival of all rats. The wound size was measured with a caliper and evaluated through the documentation of photographs. At the end of the experiment, rats were euthanized, and burn tissues were excised and utilized for biochemical, molecular, and histological analyses.

2.4. Gene Expression with qPCR

The isolation of RNA was conducted in accordance with the stipulated recommendations of the total RNA purification kit (GeneAll, Hybrid-RTM 305-101, Seoul, Republic of Korea). To this end, 100 mg of skin tissue was homogenized with the RiboEx RNA cell lysis buffer, which was included in the aforementioned kit. Thereafter, the homogenate was subjected to incubation for a period of five minutes at ambient temperature. The mixture was then subjected to a centrifugation process at 12,000× g for 10 min at a temperature of 4 °C. The resultant pellet was transferred to a fresh tube and utilized for RNA isolation. The quantity of RNA was subsequently quantified using nanodrops. To synthesize cDNA from the RNA samples, a commercial kit containing a reverse transcriptase enzyme was employed, following the manufacturer's recommended protocol (Applied Biosystems Inc., High Capacity cDNA Reverse Transcription Kit, Catalog no. 4368813, Vilnius, Lithuania). Subsequently, a mixture of random primer and RNA samples was prepared and maintained at 25 °C for a duration of 10 min. Subsequently, the samples were subjected to incubation at 37 °C for two hours and at 85 °C for five minutes to facilitate the synthesis of cDNA. The quantity of cDNA obtained was then measured using nanodrops. To perform quantitative PCR (qPCR) reactions, 100 ng of cDNA was utilized, along with primers specific to MMP2 and MMP9 genes. GAPDH was selected as the housekeeping gene for normalization purposes. Primer sequences were used for MMP-2 Forward 5′-AGCAAGTAGACGCTGCCTTT-3′, Reverse 5′-ACCTGGGATCCCCTTACCTC-3′; MMP-9 Forward 5′-CTTCTCTTTCGTAGGGCGCA-3′, Reverse 5′-TAGGGCTCCTCCTACTGGTG-3′; COL1A1 Forward 5′-ATGGATCAGGCCAATGGCAA-3′, Reverse 5′-TGTTACTTACAG-TGCAGCCA-3′; COL3A1 Forward 5′-GCGAAGGCAACAGTCGATTC-3′, Reverse 5′-GGACCTGGTCTGGGGATACT-3′; and GAPDH Forward 5′-TTCCACCTTTGATGCTGGG-G-3′, Reverse 5′-AGAGGGCACCAAACCTTCAG-3′. The cDNA was subsequently diluted for qPCR, and the reaction mixture was prepared using a sybr green master mix (Applied Biosystem Inc., Vilnius, Lithuania, Catalog no. 4309155). The prepared mixture was configured to yield a total reaction volume of 20 µL per well. It was then subjected to the following thermal cycling conditions: 95 °C for 10 min (1 cycle), 95 °C for 15 s, and qPCR experiments were subsequently conducted on the qPCR device (Applied Biosystems StepOnePlus Real-Time PCR System) in accordance with the following program: 57 °C for 20 s, 72 °C for 30 s (40 cycles), 95 °C for 2 min (1 cycle).

2.5. Protein Expression with Western Blotting

Skin tissues were lysed in homogenization buffer containing 1% Triton X-100 and NP40, along with a protease inhibitor cocktail. The samples were then kept on ice to allow cell lysis. Subsequently, the samples were centrifugated at 12,000× g at 4 °C for 10 min, after which Western blot studies were performed on the upper phase. Samples were boiled at 95 °C for 5 min with sample loading buffer prepared at a 1:1 ratio. Some 12% SDS-PAGE gels were prepared, and samples containing 25 µg/mL protein were loaded into each well. Immunoblotting was performed with a semi-dry system to transfer proteins in the gel to the PVDF membrane. Following this, the PVDF membrane was blocked with 3% BSA for 2 h at room temperature and then incubated with primary antibody (1:500 dilution, MMP-2 sc-10736, MMP-9 sc-10737, GAPDH sc-32233) overnight. After this step, the membrane was washed 3 times with TBST and treated with a secondary antibody (1:2000 dilution, anti-rabbit-HRP sc-2357, and anti-mouse-HRP sc-2060). Then, it was incubated with a solution containing chemiluminescence substrate (Santa Cruz, Dallas, TX, USA, Western Blotting Luminol Reagent: sc-2048) for 1 min in the dark. The bands obtained in the imaging system (Syngene GBox Chemi XRQ) were evaluated by densitometric analysis, with GAPDH serving as the reference protein in the densitometric analysis and calculations to eliminate changes in the samples. Densitometric analysis of the bands was performed in the ImageJ 1.54 software (NIH, Bethesda, MD, USA) program.

2.6. Hydroxyproline Assay

Hydroxyproline (4-hydroxyproline) is a non-proteinogenic amino acid that is formed through the process of the post-translational hydroxylation of proline. Hydroxyproline is the predominant component of collagen and is employed as an indicator of collagen content due to its ability to stabilize the helical structure. To ascertain the impact of tilapia hydrogels on wound healing, the levels of hydroxyproline in the tissue were quantified by ELISA, a technique that utilizes a commercial kit for homogenized skin tissues. The Hydroxyproline Quantification Kit (Sigma-Aldrich MAK569, Saint Louis, MO, USA) was utilized in accordance with the protocol established by the kit's manufacturer. The measurement of colorimetric color changes at 560 nm in the tissue sample was performed.

2.7. Histological Staining

To examine cell activities, hematoxylin-eosin (HE) staining was performed on biopsy samples taken from the rat burn model [29]. To observe the collagen factor during the healing process, Masson trichrome staining was performed [30]. After cervical dislocation, the skin tissues of the animals in the groups were placed in 4% paraformaldehyde solution and kept at 4 °C for histological examination. Tissue tracking was performed with a tissue tracking device, and the tissues were embedded in paraffin. This process includes the following steps: tissue collection, fixation, dehydration, transparency, impregnation, and embedding. Each paraffin block was sectioned at a thickness of 5 µm, and HE staining was performed on a number of sections for light microscope (Zeiss Axio Vert.A1, Jena, Germany).

2.8. Statistical Analysis

The data are presented as the means ± standard deviations. Statistical analyses were conducted using GraphPad Prism 7. p values less than 0.05 were considered to be statistically significant. The normality of distributions between groups was analyzed using one-way analysis of variance (ANOVA). Additionally, the Bonferroni and Tukey post-hoc tests were employed to further explore the data.

3. Results

3.1. AG and AGTL Hydrogel Viscosity and Swelling Rate

The viscosity was measured as a function of shear rate at 25 °C using a rheometer setup. The hydrogel exhibited shear-thinning behavior, with viscosity decreasing as shear rate increased, indicating its suitability for extrusion-based applications, such as dressing material for injury. AG and AGTL viscosity were measured in a shear rate range of 1–1000 s^{-1} by increasing the shear rate every 10 s for one minute. The resultant shear viscosity values for 1.25% alginate at 25 °C were found to be 381 mPa s, while AGTL shear viscosity exhibited a value of 241 mPa s at the same temperature. In the rheological analysis of hydrogels, an examination was conducted of the relationship between shear rate and viscosity. It was observed that, as the shear rate increased, viscosity decreased. It was determined that the gel exhibited a certain degree of strength at shear rates ranging from 250 to 500 s per minute, as indicated by the shear stress, indicating that the gel was not highly fluid (see Figure 2a).

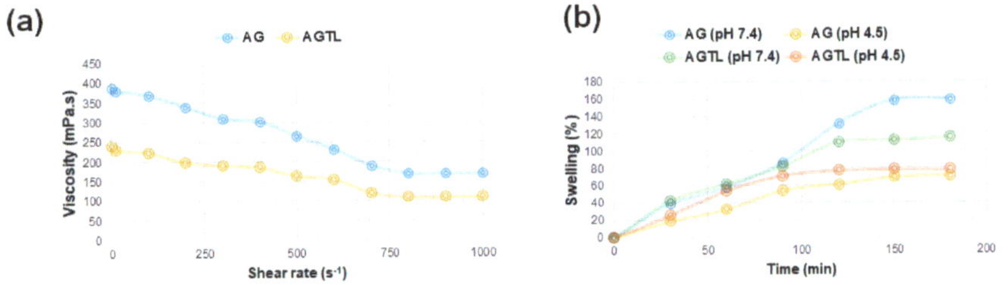

Figure 2. (**a**) Rheological characterization of the alginate hydrogel with a shear rate vs. viscosity curve. (**b**) Swelling behavior of alginate hydrogel in buffer solutions with a pH of 7.4 and a pH of 4.5 over time.

The swelling ratio (%) was measured at regular time intervals to evaluate the hydrogel's responsiveness to different pH conditions. The percentage swelling values of hydrogels in pH 7.4 and pH 4.5 buffer environments are illustrated in Figure 2b. It was determined that the water retention capacity of hydrogels decreased in AG at a pH of 4.5; however, this decrease was less in AGTL. The buffering effect of collagen and proteins present in the AGTL structure contributes to its resistance against changes in acidic environments, thereby minimizing water loss. This behavior highlights the pH-dependent nature of alginate hydrogels, which is relevant for applications in dressing material for burn injury.

3.2. TL and AGTL Treatment Effects on Wound Closure

As illustrated in Figure 3a, images of the burn wounds were captured on the 7th and 14th days post-treatment in rats. Figure 3b presents the calculated wound closure rates, indicating a statistically significant difference ($p < 0.001$) in the burn wound closure between the untreated group on the 14th day and the 7th day. The 7-day TL treatment administered to the burn wound exhibited a significant increase in comparison to the 7-day untreated group ($p < 0.05$). Similarly, the 14-day TL treatment demonstrated a significant increase in comparison to the 14-day untreated group ($p < 0.05$). Furthermore, the 7-day AGTL treatment administered to the burn wound exhibited a significant increase in comparison to the 7-day untreated group ($p < 0.01$). Similarly, the 14-day AGTL treatment demonstrated a significant increase in comparison to the 14-day untreated group ($p < 0.01$).

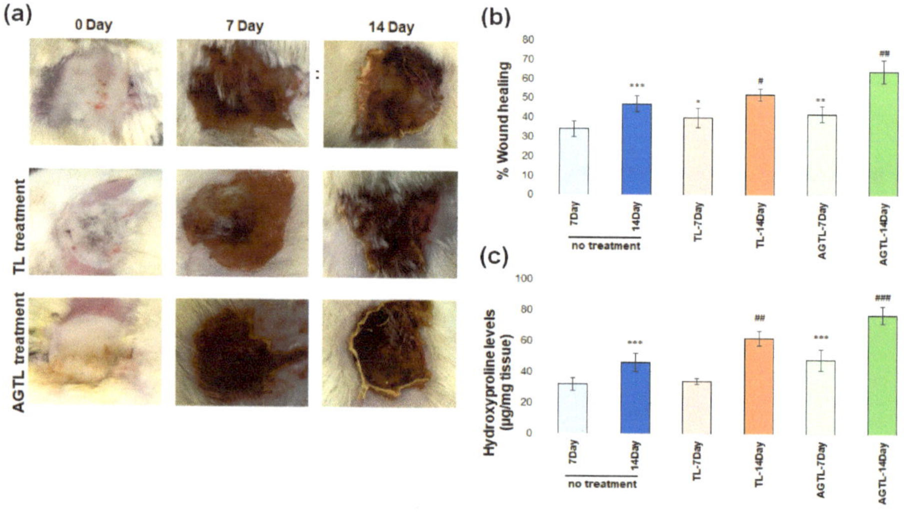

Figure 3. (**a**) Macroscopic wound images of TL and AGTL treatment in rats on day 0, day 7, and day 14. (**b**) The wound closure percentages. (**c**) Hydroxyproline levels of TL and AGTL treatment on day 7 and day 14 (* $p < 0.05$, ** $p < 0.01$,*** $p < 0.001$ day 7, # $p < 0.05$, ## $p < 0.01$, ### $p < 0.001$ day 14 compared with no treatment group).

3.3. TL and AGTL Treatment Effects on Hydroxyproline

Changes in hydroxyproline levels, a critical indicator of wound healing, are demonstrated in Figure 3c. A significant increase in hydroxyproline levels was observed in the untreated group on day 14 compared to the untreated group on day 7 ($p < 0.001$). However, no significant change was detected when TL treatment applied to the burn wound on day 7 was compared to the untreated group on day 7. A similar outcome was observed when TL treatment for 14 days was compared to the untreated group on day 14 ($p < 0.01$). AGTL treatment applied to the burn wound on day 7 demonstrated a significant increase in hydroxyproline levels compared to the untreated group on day 7 ($p < 0.001$). Conversely, no significant increase in hydroxyproline levels was observed when 14 days of AGTL treatment was compared to the 14-day untreated group ($p < 0.001$).

3.4. TL and AGTL Treatment Effects on MMP-2 and MMP-9 Gene and Protein Expression

Western blot gel images of the groups belonging to MMP-2 are shown in Figure 4a. No significant change was observed in the pro-MMP-2 protein expression levels in burn wounds on rats on day 7 and 14 (Figure 4b). Changes in active MMP-2 protein expression levels after TL and AGTL application on day 7 and 14 in burn wounds on rats are shown in Figure 4c. On day 14, the untreated group exhibited a substantial increase in active-MMP-2 protein expression levels in comparison to the untreated group on day 7 ($p < 0.001$). Similarly, TL treatment administered on day 7 resulted in a significant increase in MMP-2 protein expression levels when compared to the untreated group on the same day ($p < 0.05$). No significant change was observed when the 14 days of TL treatment group was compared with the 14 days of untreated group. A significant increase was observed when group with 7 days of AGTL treatment applied to the burn wound was compared with the 7 days untreated group ($p < 0.001$). No significant change was observed in active-MMP-2 protein levels when 14 days of AGTL treatment was compared to the 14 days of untreated group.

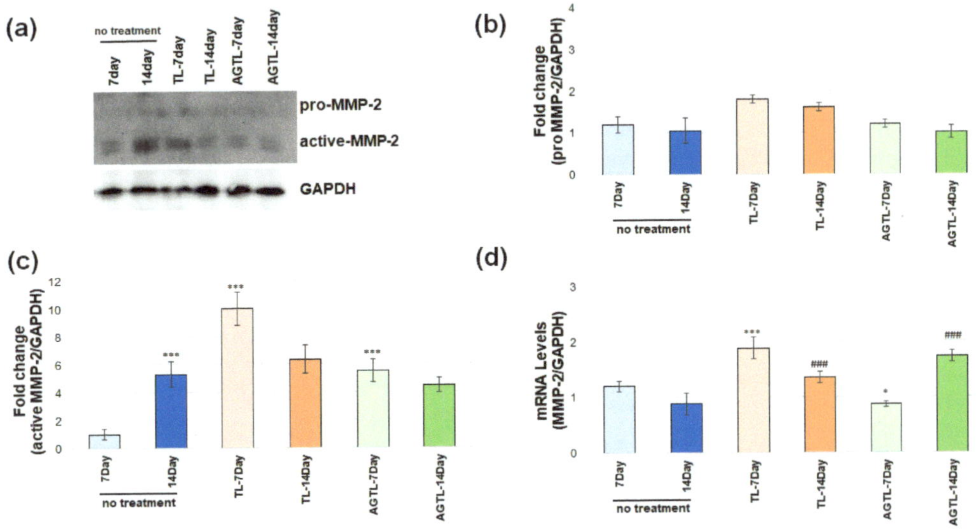

Figure 4. (**a**) MMP-2 western blot image of TL and AGTL treatment on day 7 and day 14, TL and AGTL treatment on day 7 and day 14 (**b**) pro-MMP-2 band and (**c**) active MMP-2 band (**d**) MMP-2 gene expression analyses of TL and AGTL treatment on day 7 and day 14 (* $p < 0.05$, *** $p < 0.001$ on day 7, ### $p < 0.001$ on day 14 compared with no treatment group).

As illustrated in Figure 4d, the present study examined alterations in active MMP-2 gene expression levels following TL and AGTL application on the 7th and 14th days in burn wounds in rats. The investigation revealed no statistically significant change in MMP-2 gene expression levels on the 14th day in the untreated group of the burn wound compared to the 7-day untreated group. A significant increase was observed in the 7-day TL treatment applied to the burn wound compared to the 7-day untreated group ($p < 0.001$). However, no significant change was observed in the 14-day TL treatment compared to the 14-day untreated group. A significant increase was observed in the 7-day AGTL treatment applied to the burn wound compared to the 7-day untreated group ($p < 0.05$). However, no significant increase was observed in the MMP-2 gene expression levels when compared to the 14-day AGTL treatment compared to the 14-day untreated group ($p < 0.001$).

Western blot gel images of the groups belonging to MMP-9 are shown in Figure 5a. No statistically significant alterations were observed in the protein levels of pro-MMP-9 (see Figure 5b). The alterations in active MMP-9 protein expression levels subsequent to TL and AGTL application on the 7th and 14th days in burn wounds in rats are presented in Figure 5c. A significant decrease in active-MMP-9 protein expression levels was observed on the 14th day in the untreated group of the burn wound compared to the 7th day in the untreated group ($p < 0.01$). A significant decrease was also observed when the 7-day TL treatment was applied to the burn wound compared to the 7-day untreated group ($p < 0.01$). However, no significant change was observed when 14 days of TL treatment was compared to the 14-day untreated group ($p < 0.05$). A significant decrease was observed when the 7-day AGTL treatment was compared to the 7-day untreated group ($p < 0.001$). However, no significant change was observed in active-MMP-9 protein levels when the 14-days AGTL treatment was compared to the 14-day untreated group. As illustrated in Figure 5d, the application of TL and AGTL on days 7 and 14 post-burn in rat models resulted in alterations in active MMP-9 gene expression levels. A notable increase in MMP-9 gene expression levels was observed in the untreated group on day 14 compared to the untreated group on day 7 ($p < 0.001$). A significant increase was observed in the 7-day TL treatment applied to the burn wound compared to the 7-day untreated group ($p < 0.05$). No significant change was observed in the 14-day TL treatment compared to the 14-day untreated group. A significant increase was observed in the 7-day AGTL treatment applied to the burn wound compared to the 7-day untreated group ($p < 0.01$). However, no significant increase was observed in the MMP-9 gene expression levels when compared to the 14-day AGTL treatment versus the 14-day untreated group ($p < 0.01$). Similarly, no significant increase was observed in the MMP-9 gene expression levels when compared to the 14-day AGTL treatment versus the 14-day untreated group ($p < 0.01$).

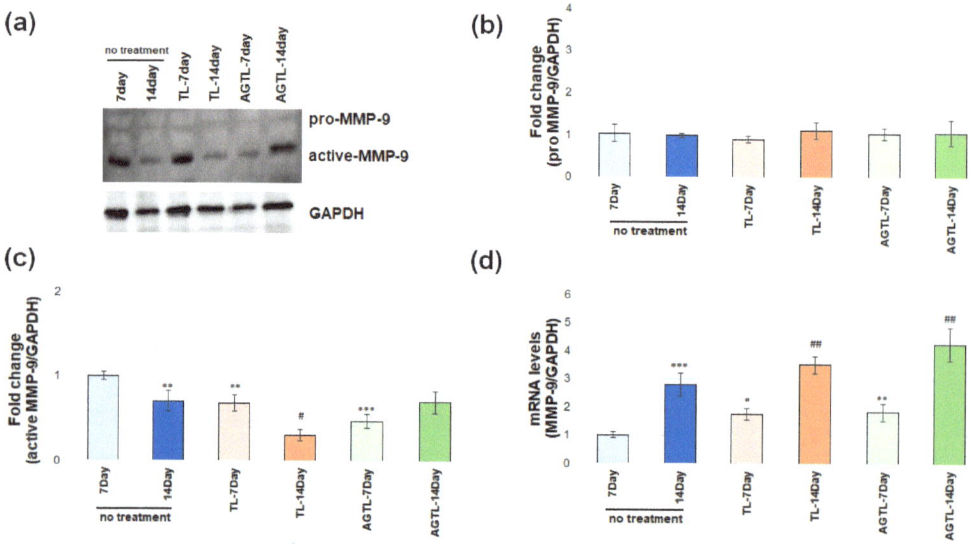

Figure 5. (**a**) MMP-9 western blot image of TL and AGTL treatment on day 7 and day 14, and TL and AGTL treatment on day 7 and day 14; (**b**) pro-MMP-9 band and (**c**) active MMP-9 band; (**d**) MMP-9 gene expression analyses of TL and AGTL treatment on day 7 and day 14 (* $p < 0.05$, ** $p < 0.01$, *** $p < 0.001$ on day 7, # $p < 0.05$, ## $p < 0.01$ on day 14 compared with no treatment group).

3.5. TL and AGTL Treatment Effects on COL1A1 and COL3A1 Gene Expression

As illustrated in Figure 6a, the data demonstrate a notable decrease in COL1A1 gene expression levels on the 14th day in the untreated group of the burn wound compared to the 7th day of the untreated group ($p < 0.001$). In contrast, a significant increase in COL1A1 gene expression levels was observed in the 7-day TL treatment group compared to the 7-day untreated group ($p < 0.05$). However, no significant change was observed when the 14-day TL treatment was compared to the 14-day untreated group. A significant increase in COL1A1 gene expression levels was also observed in the 7-day AGTL treatment group compared to the 7-day untreated group ($p < 0.001$). However, no significant increase was observed in COL1A1 gene expression levels when the 14-day AGTL treatment was compared to the 14-day untreated group ($p < 0.05$).

As illustrated in Figure 6b, the data demonstrate a notable increase in COL3A1 gene expression levels in the untreated group of the burn wound on the 14th day compared to the 7th day of the untreated group ($p < 0.001$). A significant change was observed when the 7-day TL treatment was compared to the 7-day untreated group ($p < 0.05$). A significant increase was observed when the 14-day TL treatment was compared to the 14-day untreated group ($p < 0.001$). No significant change was observed when the 7-day AGTL treatment was compared to the 7-day untreated group. However, a significant increase in COL3A1 gene expression levels was observed when the 14-day AGTL treatment was compared to the 14-day untreated group ($p < 0.001$).

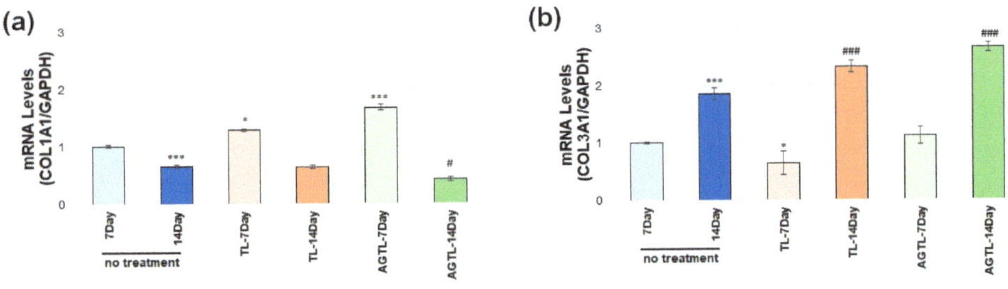

Figure 6. (**a**) COL1A1 and (**b**) COL3A1 gene expression of TL and AGTL treatment on day 7 and day 14 (* $p < 0.05$, *** $p < 0.001$ on day 7, # $p < 0.05$, ### $p < 0.001$ on day 14 compared with no treatment group).

3.6. TL and AGTL Treatment Effects on Skin Tissue Epithelization and Collagen Fibers

In histological analyses performed on tissues, an evaluation was made in terms of skin epithelialization and hair follicles in HE staining (Figure 7a). On the seventh day, the absence of treatment resulted in an increase in inflammatory cells and the initiation of epithelization in the skin tissue. On the seventh day, the TL-treated groups exhibited progenitor cell formation, hair follicles, and an augmented epithelium layer. On the seventh day, the AGTL-treated groups demonstrated adipocytic tissue that was clearly discernible and dense epithelization. On the seventh day, the groups that were not treated exhibited light epithelization and hair follicles. On the fourteenth day, the groups that were treated exhibited dense cell formations and increased epithelization. On the fourteenth day, the groups that were treated exhibited dense hair follicles, aggregated cells, epithelium layer, and regeneration of the epidermis. Wound epithelization data (Figure 7b) demonstrated a notable increase in the untreated group of the burn wound on the 14th day compared to the 7th day ($p < 0.001$). No significant change was observed when the 7-day TL treatment was compared to the 7-day untreated group, but the 7-day AGTL group increased compared to the untreated group ($p < 0.001$). A significant increase was observed when the 14-day TL or AGTL treatment was compared to the 14-day untreated group ($p < 0.01$, $p < 0.001$).

Histological analyses were performed with Masson trichrome staining to evaluate the amount and distribution of connective tissue in the burn area. Blue staining of collagen fibers indicates the presence and distribution of connective tissue. The results showed that blue staining was scattered in the untreated groups, while in TL and AGTL treatment, blue staining was found to be more intense and healing in terms of providing connective tissue integration. Additionally, it was observed that the connective tissue of the skin increased, and fibrosis occurred, especially in 14-day treatments (Figure 7c). The collagen ratio (Figure 7d) increased in the untreated group of the burn wound on the 14th day compared to the 7th day ($p < 0.01$). A significant change was observed when the 7th day of the TL or AGTL treatment was compared to the 7th day of the untreated group ($p < 0.001$). A significant increase was observed when the 14th day of the TL or AGTL treatment was compared to the 14th day of the untreated group ($p < 0.001$).

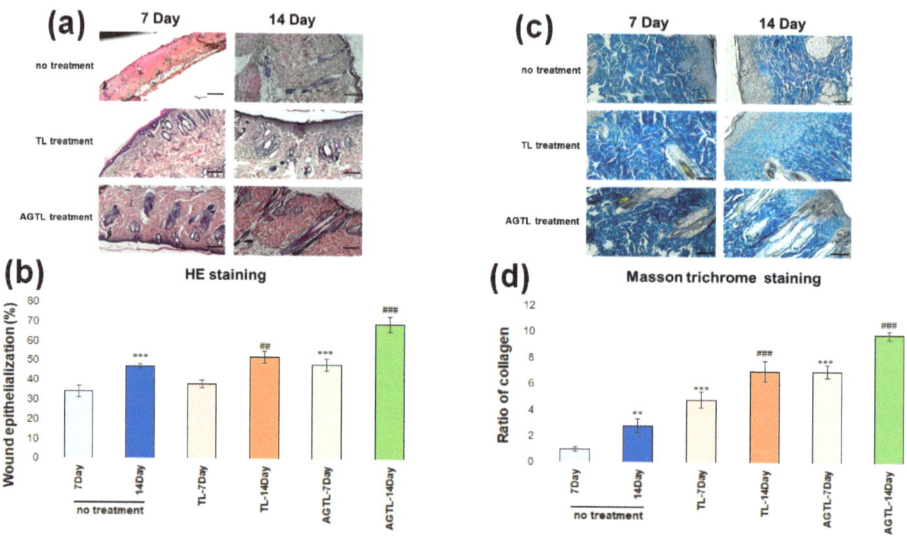

Figure 7. Microscopic images of (**a**) HE and (**b**) Masson trichrome staining (Scale bar: 100 μm); (**c**) wound epithelialization percentage; and (**d**) collagen ratio of TL and AGTL treatment on day 7 and day 14 (** $p < 0.01$, *** $p < 0.001$ on day 7, ## $p < 0.01$, ### $p < 0.001$ on day 14 compared to the no treatment group).

4. Discussion

Wound healing consists of the following three basic phases: inflammatory, proliferative, and maturation. The inflammatory phase is clot formation followed by inflammatory cell formation at the wound or injury site. This phase lasts 2–5 days and includes vasoconstriction, platelet aggregation, and clot formation followed by vasodilation and phagocytosis. The second phase, consisting of granulation, contraction, and epithelialization, lasts approximately 2–3 weeks. Then comes the maturation phase, where new tissue is formed, and this phase lasts for about 3 weeks. When a superficial wound in the skin is damaged, it quickly passes the threshold of hemostasis, which is the primary phase of the healing process. Excessive bleeding is prevented by the onset of coagulation in the blood vessels, which results in the accumulation of platelets around the endothelial region and the formation of a plug. This is how platelets form a blood clot. The formation of a clot not only stops bleeding, but it also serves as a temporary matrix for cell migration [31,32]. The primary events contributing to wound closure are epithelialization, wound contraction, and collagenous scar formation, which are the results of wound healing. MMPs are proteolytic enzymes that hydrolyze the tissue matrix. MMPs facilitate keratinocyte migration by destroying structures in the extracellular matrix that impede keratinocyte migration [33]. MMP-2 and MMP-9 facilitate the separation of keratinocytes in the basal layer from the basement membrane and their movement [34].

Research conducted in various laboratories has demonstrated that matrix metalloproteinases (MMPs) derived from epithelial cells regulate a series of events during the inflammatory phase of wound healing. These events include the transepithelial migration of leukocytes and the emergence of signaling proteins, such as chemokines. It has been confirmed that MMPs are prominently expressed in inflammatory cells and that stromal and epithelial cells in the wounded area also express multiple MMPs, including MMP 1, 2, 3, 7, 9, 10, and 28 [35]. It has been reported that MMP-2 can be secreted from macrophages during the wound healing process and plays a significant role in gene regulation in inflammatory regions [36]. With the exception of MMP-2 and MMP-9, which have been observed to have negative or inhibitory effects on cell proliferation, the vast majority of MMPs have been reported to positively influence the wound healing process [37]. During the initial phases of wound healing, there is an increase in collagen synthesis, followed by a subsequent decrease as the wound heals. MMP-2 and MMP-9, classified as gelatinase enzymes, are responsible for the degradation of collagen on the wound surface. This process involves the conversion of collagen fibrils, which are insoluble in nature, into soluble gelatin peptides [38]. During the normal wound healing process, the synthesis of type III collagen occurs initially, followed by type I collagen synthesis, which occurs later. As the wound heals, a process known as maturation occurs, which involves the reorganization of the wound healing process. This maturation results in a decrease in the ratio of type III collagen to type I collagen. Consequently, the amount of the extracellular matrix (ECM), water, and glycosaminoglycan also decreases [39,40].

In the present study, it was determined that there was a significant increase in hydroxyproline levels, which serves as an important marker for collagen levels, in the TL and AGTL groups. Furthermore, it was determined that epithelialization increased even more with AGTL, and that the humidity in the environment accelerated collagen formation. In terms of changes in collagen gene levels, it was observed that the 7-day AGTL application increased COL1A1 levels, which decreased in the 14-day application. This phenomenon can be attributed to the destruction of collagen scar tissue, facilitated by MMP-2 and MMP-9 activation on the 14th day, which enabled keratinocytes to approach each other. COL3A1, a pivotal regulator of fibroblasts, exhibited an increase in gene levels following 14-day TL and AGTL treatments. This increase indicates that fibroblasts are directed towards the surface in the proliferation phase and that tissue healing is in a permanent phase. Although tilapia fish skins have recently been used in clinical applications for wound treatment in many countries, they have not become widespread due to some limitations. However, the presence of collagen and peptides that contribute to collagen production in the structure of tilapia indicates its continued utilization in traditional burn treatment practices. The lack of experimental studies in the literature on this subject is limited in understanding how effective tilapia is in burn treatment. A smart dressing material made from tilapia fish skin containing allopurinol nanoparticles coated with silver nanoparticles has been shown to accelerate wound healing and suppress inflammation in rats [41]. In a separate study, silver nanoparticles (AgNPs) obtained from aloe vera via green synthesis were expressed as a suitable dressing material for burn wound healing because scaffolds decellularized with tilapia skin showed antimicrobial properties on the skin [25]. A study was conducted on the basis of the high collagen content of tilapia skin, and it was reported that collagen nanoparticles obtained in rats caused wound closure as of the 8th day, and the wound closed at the end of the 28th day [42]. It was ascertained that collagen, when present in a gel-like form and extracted from tilapia skin, promoted expedited wound closure and elevated relative gene expressions of transforming growth factor-beta 1 (TGF-β1), basic fibroblast growth factor (bFGF), and α-smooth muscle actin (αSMA) in rats with induced wounds over a period of 12 days [43]. Nanofiber scaffolds made with collagen purified from tilapia have been demonstrated to promote cell migration in wounds and to augment the expression of involucrin, filaggrin, and transglutaminase 1 (TGase1) in skin cells [44]. In the present study, we demonstrated that AGLT enhanced the healing of burn wounds in tilapia by increasing collagen levels, and it was effective in increasing COL3A1 levels in the wound in the 14-day post-burn period. The primary objective of the studies was to obtain collagen from the tilapia structure and subsequently develop formulations and products for utilization in wound healing. In our study, we demonstrated that the alginate hydrogel mixture we obtained by using the tilapia skin as a whole and containing both collagen and other molecules that support collagen production can be used in burn treatment.

It has been reported that alginate-based hydrogels are suitable for clinical applications in terms of holding and stabilizing protein molecules [45]. Alginate hydrogels have a high potential for use in burn treatments or wound dressings, especially because they are biodegradable [46,47]. A comparison of alginate hydrogels with traditional medical dressings reveals several notable advantages. Firstly, alginate hydrogels are non-toxic and highly water absorbent, qualities that enable them to form a hydrogel network on the wound surface, thereby maintaining a moist environment [48]. Secondly, alginate hydrogels do not adhere to wound tissues, and their removal does not cause secondary injury to the wound surface [49]. In this study, we hypothesized that protein-based molecules found in the tilapia structure that are effective in wound healing can accelerate the healing process in cells by providing both moisture and the stabilization of proteins with the help of alginate hydrogels. The treatment with tilapia skin resulted in less scar tissue formation. Thanks to the moisture provided by the hydrogel, collagen and the healing compounds in its structure contributed to the scar tissue formed during wound healing being more flexible and smoother. This may contribute to the scars remaining lighter by regular collagen production in the wound in skin grafts.

5. Conclusions

Our study demonstrated that the preparation of alginate-based tilapia fish skin forms and their subsequent storage in a humid environment over an extended period of time holds promise for facilitating wound healing. This study's limitations include its exclusive investigation of a single type of tilapia skin and its lack of experimentation with diverse hydrogel forms. Consequently, further research is necessary to investigate the effects of varying collagen amounts and hydrogel forms in different tilapia species. Such findings could contribute to the development of novel dressing materials for burn treatment. This finding is particularly significant in cases requiring prolonged use, such as for deep burns, where the ready availability of these products could significantly enhance the utilization of clinical treatments. Our study may contribute to researchers working in this field in the planning of dressing materials planned to be developed for burn wounds in the future and new biomaterials for the treatment of tilapia skin.

Author Contributions: Conceptualization, B.B. and O.C.; methodology, B.B., A.K., B.K., M.E. and B.I.A.; validation, B.B. and A.K.; formal analysis, B.B. and A.K.; investigation, B.B., A.K., B.K., M.E. and B.I.A.; resources, B.B. and O.C.; data curation, B.B. and B.I.A.; writing—original draft preparation, B.B. and O.C.; writing—review and editing, O.C.; visualization, B.B. and O.C.; supervision, O.C.; project administration, B.B. and O.C.; funding acquisition, O.C. All authors have read and agreed to the published version of the manuscript.

Funding: This research was funded by TUBITAK grant number 1919B012302802 and TUBA-GEBIP.

Institutional Review Board Statement: The animal study protocol was approved by the Experimental Animal Ethics Committee of Aydin Adnan Menderes University (2023/75 and 18 May 2023).

Informed Consent Statement: Not applicable.

Data Availability Statement: Data are contained within the article.

Acknowledgments: The authors would like to thank ODC Research and Development Inc. for providing access to facilities and supplying the reagents.

Conflicts of Interest: The authors declare no conflicts of interest.

References

1. Nişancı, M.; Öztürk, S. Yanık Fizyopatolojisi. *Turk. Klin. Plast. Surg.-Spec. Top.* **2010**, *2*, 8–14.
2. Hankins, C.L.; Tang, X.Q.; Phipps, A. Hot Beverage Burns: An 11-Year Experience of the Yorkshire Regional Burns Centre. *Burns* **2006**, *32*, 87–91. [CrossRef] [PubMed]
3. Elfawy, L.A.; Ng, C.Y.; Amirrah, I.N.; Mazlan, Z.; Wen, A.P.Y.; Fadilah, N.I.M.; Maarof, M.; Lokanathan, Y.; Fauzi, M.B. Sustainable Approach of Functional Biomaterials–Tissue Engineering for Skin Burn Treatment: A Comprehensive Review. *Pharmaceuticals* **2023**, *16*, 701. [CrossRef]
4. Davison-Kotler, E.; Marshall, W.S.; García-Gareta, E. Sources of Collagen for Biomaterials in Skin Wound Healing. *Bioengineering* **2019**, *6*, 56. [CrossRef]
5. Costa, B.A.; Júnior, E.M.L.; Filho, M.O.d.M.; Fechine, F.V.; de Moraes, M.E.A.; Júnior, F.R.S.; Soares, M.F.A.D.N.; Rocha, M.B.S. Use of Tilapia Skin as a Xenograft for Pediatric Burn Treatment: A Case Report. *J. Burn Care Res.* **2019**, *40*, 714–717. [CrossRef]
6. Mukherjee, S.; Bhattacherjee, S.; Keswani, K.; Nath, P.; Paul, S. Application of Tilapia Fish Skin in Treatment of Burn Patients. *Biocatal. Agric. Biotechnol.* **2024**, *59*, 103254. [CrossRef]
7. Fiakos, G.; Kuang, Z.; Lo, E. Improved Skin Regeneration with Acellular Fish Skin Grafts. *Eng. Regen.* **2020**, *1*, 95–101. [CrossRef]
8. Ibrahim, A.; Soliman, M.; Kotb, S.; Ali, M.M. Evaluation of Fish Skin as a Biological Dressing for Metacarpal Wounds in Donkeys. *BMC Vet. Res.* **2020**, *16*, 472. [CrossRef]
9. Hu, Z.; Yang, P.; Zhou, C.; Li, S.; Hong, P. Marine Collagen Peptides from the Skin of Nile Tilapia (*Oreochromis niloticus*): Characterization and Wound Healing Evaluation. *Mar. Drugs* **2017**, *15*, 102. [CrossRef]
10. Ge, B.; Wang, H.; Li, J.; Liu, H.; Yin, Y.; Zhang, N.; Qin, S. Comprehensive Assessment of Nile Tilapia Skin (*Oreochromis niloticus*) Collagen Hydrogels for Wound Dressings. *Mar. Drugs* **2020**, *18*, 178. [CrossRef]
11. Mei, F.; Liu, J.; Wu, J.; Duan, Z.; Chen, M.; Meng, K.; Chen, S.; Shen, X.; Xia, G.; Zhao, M. Collagen Peptides Isolated from Salmo salar and Tilapia nilotica Skin Accelerate Wound Healing by Altering Cutaneous Microbiome Colonization via Upregulated NOD2 and BD14. *J. Agric. Food Chem.* **2020**, *68*, 1621–1633. [CrossRef] [PubMed]
12. Zeng, S.K.; Zhang, C.H.; Lin, H.; Yang, P.; Hong, P.Z.; Jiang, Z. Isolation and Characterisation of Acid-Solubilised Collagen from the Skin of Nile Tilapia (*Oreochromis niloticus*). *Food Chem.* **2009**, *116*, 879–883. [CrossRef]
13. Shalaby, M.; Agwa, M.; Saeed, H.; Khedr, S.M.; Morsy, O.; El-Demellawy, M.A. Fish Scale Collagen Preparation, Characterization and Its Application in Wound Healing. *J. Polym. Environ.* **2020**, *28*, 166–178. [CrossRef]
14. Miranda, M.J.B.D.; Brandt, C.T. Nile Tilapia Skin Xenograft versus Silver-Based Hydrofiber Dressing in the Treatment of Second-Degree Burns in Adults. *Rev. Bras. Cir. Plástica* **2023**, *34*, 79–85. [CrossRef]
15. Guimarães, M.L.; da Silva, F.A.G.; de Souza, A.M.; da Costa, M.M.; de Oliveira, H.P. All-Green Wound Dressing Prototype Based on Nile Tilapia Skin Impregnated with Silver Nanoparticles Reduced by Essential Oil. *Appl. Nanosci.* **2022**, *12*, 129–138. [CrossRef]
16. Guimarães, M.L.; da Silva, F.A.G., Jr.; da Costa, M.M.; de Oliveira, H.P. Coating of Conducting Polymer-Silver Nanoparticles for Antibacterial Protection of Nile Tilapia Skin Xenografts. *Synth. Met.* **2022**, *287*, 117055. [CrossRef]
17. Elshahawy, A.M.; Mahmoud, G.A.E.; Mokhtar, D.M.; Ibrahim, A. The Optimal Concentration of Silver Nanoparticles in Sterilizing Fish Skin Grafts. *Sci. Rep.* **2022**, *12*, 19483. [CrossRef]
18. Hingorani, D.V.; Lippert, C.N.; Crisp, J.L.; Savariar, E.N.; Hasselmann, J.P.C.; Kuo, C.; Nguyen, Q.T.; Tsien, R.Y.; Whitney, M.A.; Ellies, L.G. Impact of MMP-2 and MMP-9 Enzyme Activity on Wound Healing, Tumor Growth and RACPP Cleavage. *PLoS ONE* **2018**, *13*, e0198464. [CrossRef]
19. Fanjul-Fernández, M.; Folgueras, A.R.; Cabrera, S.; López-Otín, C. Matrix Metalloproteinases: Evolution, Gene Regulation and Functional Analysis in Mouse Models. Biochim. Biophys. *Acta Mol. Cell Res.* **2010**, *1803*, 3–19. [CrossRef]
20. Feng, C.; Chen, X.; Yin, X.; Jiang, Y.; Zhao, C. Matrix Metalloproteinases on Skin Photoaging. *J. Cosmet. Dermatol.* **2024**, *23*, 3847–3862. [CrossRef]
21. Wu, M.; Sapin-Minet, A.; Stefan, L.; Perrin, J.; Raeth-Fries, I.; Gaucher, C. Heparinized Collagen-Based Hydrogels for Tissue Engineering: Physical, Mechanical and Biological Properties. *Int. J. Pharm.* **2025**, *670*, 125126. [CrossRef] [PubMed]
22. Gardeazabal, L.; Izeta, A. Elastin and Collagen Fibres in Cutaneous Wound Healing. *Exp. Dermatol.* **2024**, *33*, e15052. [CrossRef]
23. Ghosh, S.; Haldar, S.; Gupta, S.; Chauhan, S.; Mago, V.; Roy, P.; Lahiri, D. Single Unit Functionally Graded Bioresorbable Electrospun Scaffold for Scar-Free Full-Thickness Skin Wound Healing. *Biomater. Adv.* **2022**, *139*, 212980. [CrossRef] [PubMed]
24. Cao, L.; Zhang, Z.; Yuan, D.; Yu, M.; Min, J. Tissue Engineering Applications of Recombinant Human Collagen: A Review of Recent Progress. *Front. Bioeng. Biotechnol.* **2024**, *12*, 1358246. [CrossRef] [PubMed]
25. Adhikari, S.P.; Paudel, A.; Sharma, A.; Thapa, B.; Khanal, N.; Shastri, N.; Rai, S.; Adhikari, R. Development of Decellularized Fish Skin Scaffold Decorated with Biosynthesized Silver Nanoparticles for Accelerated Burn Wound Healing. *Int. J. Biomater.* **2023**, *2023*, 8541621. [CrossRef]
26. Council, N.R. *Guide for the Care and Use of Laboratory Animals*, 8th ed.; The National Academic Press: Washington, DC, USA, 2011.
27. Kilkenny, C.; Browne, W.J.; Cuthill, I.C.; Emerson, M.; Altman, D.G. Improving bioscience research reporting: The ARRIVE guidelines for reporting animal research. *PLoS Biol.* **2010**, *8*, e1000412. [CrossRef]

28. Herndon, D.N. *Total Burn Care*; Elsevier Health Sciences: Amsterdam, The Netherlands, 2007.
29. Kocabiyik, B.; Gumus, E.; Abas, B.I.; Anik, A.; Cevik, O. Human Wharton-Jelly Mesenchymal Stromal Cells Reversed Apoptosis and Prevented Multi-Organ Damage in a Newborn Model of Experimental Asphyxia. *J. Obstet. Gynaecol.* **2022**, *42*, 3568–3576. [CrossRef]
30. Oh, E.J.; Gangadaran, P.; Rajendran, R.L.; Kim, H.M.; Oh, J.M.; Choi, K.Y.; Chung, H.Y.; Ahn, B.-C. Extracellular Vesicles Derived from Fibroblasts Promote Wound Healing by Optimizing Fibroblast and Endothelial Cellular Functions. *Stem Cells* **2021**, *39*, 266–279. [CrossRef]
31. Toriseva, M.; Kähäri, V.-M. Proteinases in Cutaneous Wound Healing. *Cell. Mol. Life Sci.* **2009**, *66*, 203–224. [CrossRef]
32. Schultz, G.S.; Wysocki, A. Interactions Between Extracellular Matrix and Growth Factors in Wound Healing. *Wound Repair Regen.* **2009**, *17*, 153–162. [CrossRef]
33. Erfan, G. *Sıçanlarda ER: Yağ ile Oluşturulmuş Yarada Bitki Ekstrelerinin Karışımı Topikal Hemostatik Bir Ajanın Yara İyileşmesine Etkisi*; İstanbul İl Sağlık Müdürlüğü: Istanbul, Turkey, 2009.
34. Han, Y.-P.; Tuan, T.-L.; Hughes, M.; Wu, H.; Garner, W.L. Transforming Growth Factor-β- and Tumor Necrosis Factor-α-Mediated Induction and Proteolytic Activation of MMP-9 in Human Skin. *J. Biol. Chem.* **2001**, *276*, 22341–22350. [CrossRef] [PubMed]
35. Parks, W.C.; Wilson, C.L.; López-Boado, Y.S. Matrix Metalloproteinases as Modulators of Inflammation and Innate Immunity. *Nat. Rev. Immunol.* **2004**, *4*, 617–629. [CrossRef] [PubMed]
36. Eerola, L.M.; Alho, H.S.; Maasilta, P.K.; Inkinen, K.A.; Harjula, A.L.; Litmanen, S.H.; Salminen, U.-S. Matrix Metalloproteinase Induction in Post-Transplant Obliterative Bronchiolitis. *J. Heart Lung Transplant.* **2005**, *24*, 426–432. [CrossRef]
37. Bullard, K.M.; Lund, L.; Mudgett, J.S.; Mellin, T.N.; Hunt, T.K.; Murphy, B.B.; Ronan, J.B.; Werb, Z.; Banda, M.J. Impaired Wound Contraction in Stromelysin-1–Deficient Mice. *Ann. Surg.* **1999**, *230*, 260. [CrossRef]
38. Carrilho, M.R.O.; Geraldeli, S.; Tay, F.; de Goes, M.; Carvalho, R.; Tjäderhane, L.; Reis, A.; Hebling, J.; Mazzoni, A.; Breschi, L.; et al. In Vivo Preservation of the Hybrid Layer by Chlorhexidine. *J. Dent. Res.* **2007**, *86*, 529–533. [CrossRef]
39. Monaco, J.L.; Lawrence, W.T. Acute Wound Healing: An Overview. *Clin. Plast. Surg.* **2003**, *30*, 1–12. [CrossRef]
40. Niessen, F.B.; Spauwen, P.H.; Schalkwijk, J.; Kon, M. On the Nature of Hypertrophic Scars and Keloids: A Review. *Plast. Reconstr. Surg.* **1999**, *104*, 1435–1458. [CrossRef]
41. Javed, A.; Tariq, A.; Khan, M.F.A.; Mirza, R.; Usman, M.; Nadir, A.; Khan, A. Multifunctional Tilapia Skin Based Smart Dressing of Silver Capped Allopurinol Nanoparticles for Treatment of Infectious Burn Wounds. *J. Drug Deliv. Sci. Technol.* **2025**, *107*, 106804. [CrossRef]
42. Shalaby, M.; Hamouda, D.; Khedr, S.M.; Mostafa, H.M.; Saeed, H.; Ghareeb, A.Z. Nanoparticles Fabricated from the Bioactive Tilapia Scale Collagen for Wound Healing: Experimental Approach. *PLoS ONE* **2023**, *18*, e0282557. [CrossRef]
43. Elbialy, Z.I.; Atiba, A.; Abdelnaby, A.; Al-Hawary, I.I.; Elsheshtawy, A.; El-Serehy, H.A.; Abdel-Daim, M.M.; Fadl, S.E.; Assar, D.H. Collagen Extract Obtained from Nile Tilapia (*Oreochromis niloticus* L.) Skin Accelerates Wound Healing in Rat Model via Up Regulating VEGF, bFGF, and α-SMA Genes Expression. *BMC Vet. Res.* **2020**, *16*, 352. [CrossRef]
44. Zhou, T.; Wang, N.; Xue, Y.; Ding, T.; Liu, X.; Mo, X.; Sun, J. Electrospun Tilapia Collagen Nanofibers Accelerating Wound Healing via Inducing Keratinocytes Proliferation and Differentiation. *Colloids Surf. B* **2016**, *143*, 415–422. [CrossRef] [PubMed]
45. Shao, X.; Tian, M.; Yin, J.; Duan, H.; Tian, Y.; Wang, H.; Xia, C.; Wang, Z.; Zhu, Y.; Wang, Y.; et al. Biofunctionalized Dissolvable Hydrogel Microbeads Enable Efficient Characterization of Native Protein Complexes. *Nat. Commun.* **2024**, *15*, 8633. [CrossRef] [PubMed]
46. Chang, R.; Gruebele, M.; Leckband, D.E. Protein Folding Stability and Kinetics in Alginate Hydrogels. *Biomacromolecules* **2023**, *24*, 5245–5254. [CrossRef] [PubMed]
47. Somsesta, N.; Jinnapat, A.; Fakpiam, S.; Suksanguan, C.; Wongsan, V.; Ouneam, W.; Wattanaeabpun, S.; Hongrattanavichit, I. Antimicrobial and Biodegradable Hydrogel Based on Nanocellulose/Alginate Incorporated with Silver Nanoparticles as Active Packaging for Poultry Products. *Sci. Rep.* **2024**, *14*, 27135. [CrossRef]
48. Zhang, M.; Zhao, X. Alginate Hydrogel Dressings for Advanced Wound Management. *Int. J. Biol. Macromol.* **2020**, *162*, 1414–1428. [CrossRef]
49. Gumus, E.; Abas, B.I.; Cevik, E.; Kocabiyik, B.; Cenik, M.; Cevik, O. Alginate Encapsulation Induces Colony Formation with Umbilical Cord-Derived Mesenchymal Stem Cells. *Exp. Biomed. Res.* **2021**, *4*, 113–121. [CrossRef]

BioChem

MDPI

Article

Mammarenavirus Z Protein Myristoylation and Oligomerization Are Not Required for Its Dose-Dependent Inhibitory Effect on vRNP Activity

Haydar Witwit and Juan C. de la Torre *

Department of Immunology and Microbiology, The Scripps Research Institute, La Jolla, CA 92037, USA; hwitwit@scripps.edu
* Correspondence: juanct@scripps.edu

Abstract: Background/Objectives: N-Myristoyltransferase inhibitors (NMTi) represent a novel antiviral strategy against mammarenaviruses such as Lassa and Junin viruses. The Z matrix protein inhibits viral ribonucleoprotein (vRNP) activity in a dose-dependent manner. Here, we investigated whether Z-mediated vRNP inhibition depends on Z myristoylation or oligomerization. **Methods**: We used HEK293T cells transfected with wild-type (WT) or G2A-mutated Z constructs in LCMV minigenome (MG) assays. Cells were treated with the NMTi IMP-1088 and the proteasome inhibitor MG132. Z protein expression, vRNP activity, and VLP production were analyzed by immunofluorescence, western blotting, and colocalization analyses. **Results**: IMP-1088 treatment led to proteasome-mediated degradation of Z, reducing its inhibition of vRNP activity, which was restored by MG132. The non-myristoylated Z G2A mutant retained vRNP inhibitory activity but showed impaired oligomerization and budding capacity. These findings demonstrate that Z-mediated vRNP inhibition is independent of myristoylation and oligomerization. **Conclusions**: Z myristoylation and oligomerization are not required for its inhibitory vRNP activity. Targeting Z myristoylation with NMTi impairs virus assembly and budding without affecting Z-mediated inhibition of vRNP activity, supporting the development of NMTi as a promising broad-spectrum antiviral strategy against mammarenaviruses.

Keywords: mammarenavirus; multimerization; ribonucleoprotein inhibition; Z matrix protein; myristoylation; N-myristoyltransferases; antiviral; LCMV; LASV; proteasome; glycoprotein; signal peptide

1. Introduction

Several mammarenaviruses, chiefly Lassa virus (LASV) in Western Africa and Junin virus (JUNV) in the Argentine Pampas cause hemorrhagic fever (HF) diseases associated with high morbidity and mortality, posing important public health problems in their endemic regions. In addition, mounting evidence indicates that the globally distributed mammarenavirus lymphocytic choriomeningitis virus (LCMV) is a neglected human pathogen of clinical significance in pediatric and transplantation medicine [1–3]. Current anti-mammarenavirus therapy is limited to an off-label use of ribavirin, for which efficacy remains controversial [4]. Hence, the importance of developing novel therapeutics to combat human pathogenic mammarenaviruses.

Mammarenaviruses are enveloped viruses with a bi-segmented negative-stranded (NS) RNA genome [5]. Each genome segment, L (ca 7.3 kb) and S (ca 3.5 kb), uses an

 https://doi.org/10.3390/biochem5020010

ambisense coding strategy to direct the synthesis of two polypeptides in opposite orientations, separated by a non-coding intergenic region (IGR). The L segment encodes the viral RNA-dependent RNA polymerase (L) and Z matrix protein, whereas the S segment encodes the viral glycoprotein precursor (GPC) and the viral nucleoprotein (NP). GPC is co- and post-translationally processed to produce a stable signal peptide (SSP), and the mature GP1 and GP2 subunits that together with the SSP form the spikes that decorate the virus surface and mediate cell entry via receptor-mediated endocytosis [6–8]. GP1 mediates binding to the cellular receptor and GP2 mediates the pH-dependent fusion event in the late endosome required for the release of vRNP into the cell cytoplasm, where it directs replication and transcription of the viral genome [9,10].

Early studies have shown that N-myristoylation is required for the role of Z protein in assembly and budding [11,12] and for the role of the SSP in the GP2-mediate fusion event [9]. These findings were based on the use of 2-hydroxy-myristic acid (2-HMA) and 2-HMA analogs as inhibitors of N-myristoyltransferase 1 (NMT1) and 2 (NMT2) responsible for catalyzing N-myristoylation in mammalian cells [13]. However, recent studies have demonstrated that 2-HMA acts off-target and does not inhibit N-myristoylation within a concentration range consistent with activity on NMT [14]. Therefore, we revisited the contribution of N-myristoylation in mammarenavirus infection using the validated on-target specific pan-NMTi (DDD85646 and IMP-1088) [15]. We found that DDD85646 and IMP-1088 exhibit very potent antiviral activity against LCMV and LASV in cultured cells. Cell-based assays probing different steps of the LCMV life cycle revealed that NMTi exerted its anti-LCMV activity by interfering with Z budding activity and GP2-mediated fusion [16] (Figure 1). Our findings support the use of NMTi as a novel host-targeted antiviral strategy to combat LASV and other human pathogenic mammarenaviruses. Moreover, NMTi can be incorporated into combination therapies with direct-acting antivirals. By targeting a host cell factor, NMTi can impose a high genetic barrier against the selection of drug-resistant variants, a common problem with direct-acting antiviral drugs. While Z dose-dependent suppression of vRNP activity is well documented [17–19], there is limited knowledge about the underlying mechanism [20–22]. Z protein has been shown to exhibit different degrees of oligomerization [16,23,24], whose biological implications remain poorly understood. Here, we present evidence that Z oligomerization might be required for its budding activity, but not for its ability to inhibit vRNP activity.

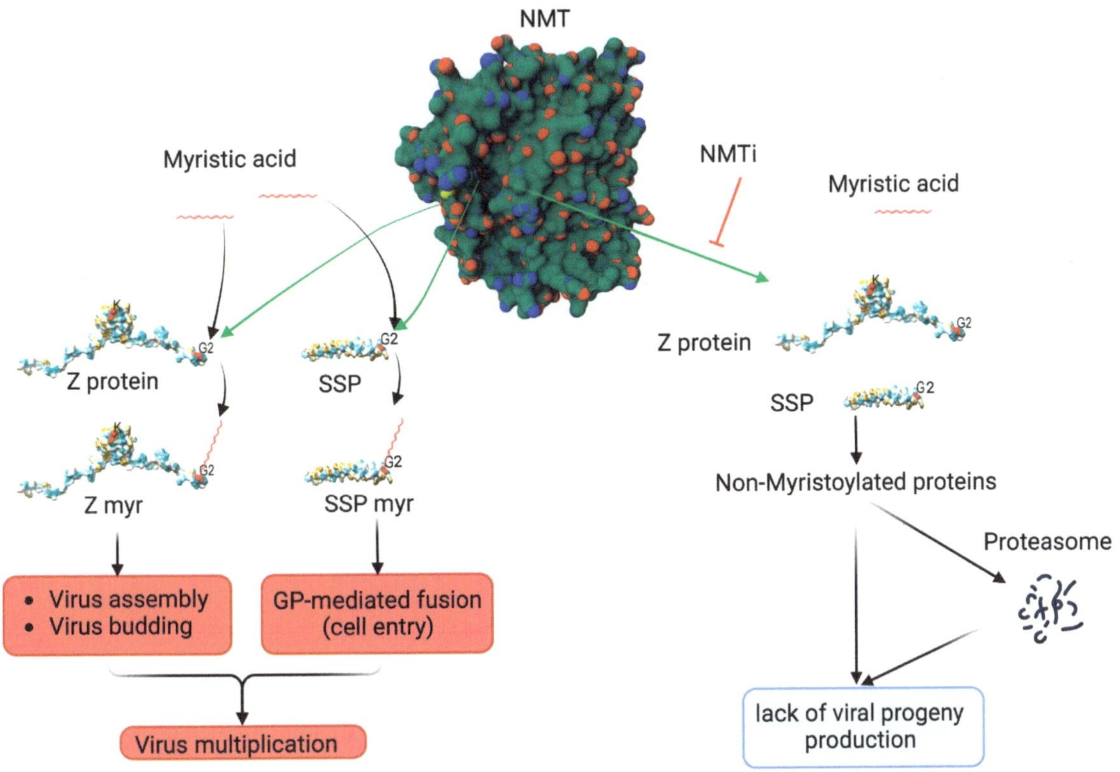

Figure 1. Proposed model of the effect of the NMT inhibitor on mammarenavirus cell entry and budding. NMT isozymes facilitate the addition of myristic acid to glycine (2) of SSP and Z protein, which protect them from proteasome-mediated degradation. Myristoylated SSP interacts with GP2 to facilitate the fusion event in the late endosome required to complete the virus cell entry process, whereas myristoylated Z directs the virus assembly and budding process. Inhibition of SSP and Z myristoylation by NMT inh results in proteasome-mediated degradation of SSP and Z, which results in the inhibition of virus multiplication. Z myr and SSP myr indicate myristoylated Z and SSP, respectively. NMT pdb, 3IWE [25], and LASV Z matrix protein pdb, 2M1S [26], were used to generate the 3D figure using ChimeraX [27].

2. Materials and Methods

2.1. Cell Line and Compounds

Human embryonic kidney cells stably expressing T antigen using the SV40 promoter (HEK293T) (ATCC CRL-3216) were maintained in Dulbecco's Modified Eagle's Medium (DMEM) supplemented with 10% fetal bovine serum, 1 mM penicillin and streptomycin, and 1 mM glutamine. Cells were incubated at 37C and supplemented with 5% CO_2. NMTi IMP-1088, (5-[3,4-difluoro-2-[2-(1,3,5-trimethyl-1H-pyrazol-4-yl)ethoxy]phenyl]-N,N,1-trimethyl-1H-indazole-3-methanamine, Cat No. 25366-1) was purchased from Cayman Chemical (Ann Arbor, MI, USA), dissolved in methyl acetate at 11 mM, and kept in aliquots at −20. MG132 (MFG No. M7449) was purchased from Sigma-Aldrich (St. Louis, MO, USA).

2.2. Plasmids

The ORF of the LCMV Z gene was cloned into pCAGGS and labeled with an HA-tag, referred to here as Z-WT-HA, as described previously [16]. T7 polymerase in the pCAGGS system and the S segment of LCMV, flanked by a T7 promoter at the amino terminus, was used to express GFP. Both LCMV L polymerase and NP are expressed in pCAGGS systems.

2.3. Minigenome Assay

HEK293 cells were seeded at 2 × 10^5 per well in poly-L-lysine-treated M12-well plates. The next day, cells were transfected with plasmids expressing LCMV L and NP proteins, the minimal transacting viral factors required for replication and transcription of the viral genome. Also included was a plasmid directing intracellular synthesis of an LCMV minigenome directing expression of GFP (MG-GFP), together with a plasmid expressing the T7 RNA polymerase to launch primary synthesis of the MG-GFP, and incremental amounts of a plasmid expressing LCMV-Z tagged with HA (Z-WT-HA). At 18 h post-transfection, cells were treated with the indicated compounds, and at 24 h post-treatment, cells were fixed with 4% paraformaldehyde (PFA), and stained with anti-HA, followed by a secondary fluorescence antibody to identify cells expressing the Z protein. The activity of the LCMV MG-GFP was assessed based on expression levels of GFP. Images were acquired at 4X magnification (Keyence Corporation, Itasca, IL, USA).

2.4. VLP Assay

Viral-like particle (VLP) production was performed as previously described [11]. Briefly, HEK293T cells were transfected with pC_E as the control, Z-WT or the G2A mutant, and cell culture supernatant (CCS) was collected at 48 or 72 h post-transfection (h pt). CCS were clarified by centrifugation (5 min, 5000 RPM, 4 °C). VLPs in clarified CCS were collected by ultracentrifugation (100,000× *g*, 2 h, 4 °C) through a 20% sucrose cushion in 50 mM Tris-HCl pH 7.5, 62.5 mM EDTA, 1% NP-40, 0.4% Na deoxycholoate. The pellet containing VLPs was resuspended in 60 µL of PBS and 20 µL of 4X Laemmli buffer was added.

2.5. Immunoblotting

Protein concentration in samples was determined by BCA assay (Pierce™ BCA Protein Assay Kits, Cat# 23227, ThermoFisher, Waltham, MA, USA). Samples (12 µg) were heated for 5 min at 95 °C and separated by SDS-PAGE using a stain-free gel (Bio-Rad, Hercules, CA, USA). Total protein (TP) was detected after the activation of the stain-free gel for one minute. Proteins were transferred to a low-fluorescence PVDF membrane (Bio-Rad, Hercules, CA, USA), and EveryBlot blocking buffer (Bio-Rad, Hercules, CA, USA) was used to block nonspecific antibody binding. The membrane was immunoreacted with an anti-HA antibody (Genscript, Piscataway, NJ, USA) overnight at 4 °C. After three washes, the membrane was immunoreacted with a horseradish peroxidase (HRP) conjugated with anti-mouse antibody. Primary and secondary antibodies were diluted in a OneBlock western-CL blocking buffer (Genesee Scientific, San Diego, CA, USA). Protein bands were visualized using a chemiluminescent substrate (ThermoFisher Scientific).

2.6. Immunofluorescence Imaging

Images were acquired at 4X magnification using a BZ-X710 Keyence fluorescence microscope. The quantification and depiction of Z and GFP were carried out using the BIOP JaCoP plugin, a colocalization quantification software, in FIJI imagej software (1.54f). The Otsu threshold setting was selected to quantify the degree of colocalization. The BIOP JaCoP generates, based on a threshold setting, an automated figure that on the left side displays unmasked (top) and masked (bottom) panels that represent the Pearson correlation coefficient (*X* axis) and probability (*Y* axis) plots, and on the right side displays a scatter plot representing the GFP (green, *Y* axis) and Z-HA (red, *X* axis) signals. The rationale to use the BIOP Jacob plug is to quantify colocalization of GFP (a surrogate of the MG activity) and the AlexaFLuor 568 red signal (a surrogate of Z matrix protein expression).

3. Results

3.1. Effect of NMTi on Z-Mediated Inhibition of vRNP Activity

The Z protein has been shown to inhibit, in a dose-dependent manner, the activity of the mammarenavirus vRNP, which is responsible for directing the replication and transcription of the viral genome, in cell-based minigenome (MG) assays. We have documented that treatment with NMTi targets Z for degradation via the proteasome pathway [16]. We therefore predicted that the Z inhibitory effect on MG activity would be reduced in the presence of the pan-NMTi IMP-1088, and that treatment with the proteasome inhibitor MG132 would restore, in the presence of IMP-1088, the Z inhibitory effect on MG activity. Using a cell-based LCMV minigenome (MG) system we found that treatment with the NMTi IMP-1088 counteracts the Z-mediated inhibitory effect on MG activity (Figure 2). Treatment with the proteasome inhibitor MG132 restored expression levels of Z to those observed in the absence of IMP-1088, which resulted in the corresponding inhibitory effect on the activity of the LCMV MG (Figure 2).

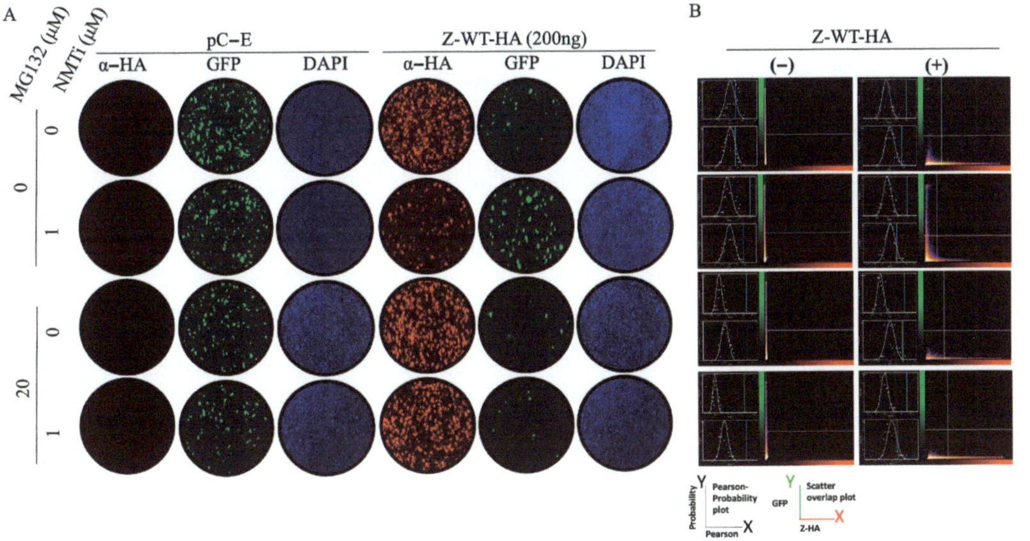

Figure 2. Effect of treatments with the NMTi (IMP-1088) and proteasome (MG132) inhibitors on the activity of the LCMV MG in the presence of LCMV WT Z protein. HEK293T cells were seeded at 2×10^5 cells per well in poly-L-lysine-treated M12-well plates. The next day, cells were transfected with pCAGGS plasmids expressing LCMV L and NP proteins, as well as T7RP, and a plasmid expressing the LCMV T7MG-GFP, together with 200 ng of a pCAGGS empty (pC-E) plasmid or expressing LCMV-Z -WT-HA. At 18 h post-transfection, cells were treated with the indicated compounds, and at 24 h post-treatment, cells were fixed with 4% PFA. Z-WT-HA expression was detected by IF using a rabitt polyclonal antibody to α-HA as primary antibody and an anti-mouse goat polyclonal antibody conjugated to Alexafluor 568 as secondary antibody. GFP was detected directly by epifluorescence. (**A**). The activity of LCMV MG-GFP was assessed based on expression levels of GFP. (**B**). The left side of each panel displays unmasked (**top**) and masked (**bottom**) panels that represent the Pearson correlation coefficient (*X* axis) and probability (*Y* axis) plots, and on the right side displays a scatter plot representing the GFP (green, *Y* axis) and Z-HA (red, *X* axis) signals. The BIOP Jacob plug quantified colocalization of GFP (a surrogate of the MG activity) and AlexaFLuor 568 red signal (a surrogate of Z matrix protein expression).

3.2. Role of N-Myristoylation on Z-Mediated Inhibition of vRNP Activity

Treatment with NMTi targeted Z protein for degradation, which prevented us from assessing whether myristoylation was required for Z-mediated inhibition of vRNP activity. To address this question, we used the Z (G2A) mutant that cannot undergo N-terminal myristoylation. As predicted, treatment with IMP-1088 did not significantly affect expression levels of Z (G2A) (Figure 3). Notably, Z (G2A) inhibited MG-directed GFP expression with a similar efficiency to Z WT (Figure 3), indicating that myristoylation is not required for Z-mediated inhibition of vRNP activity.

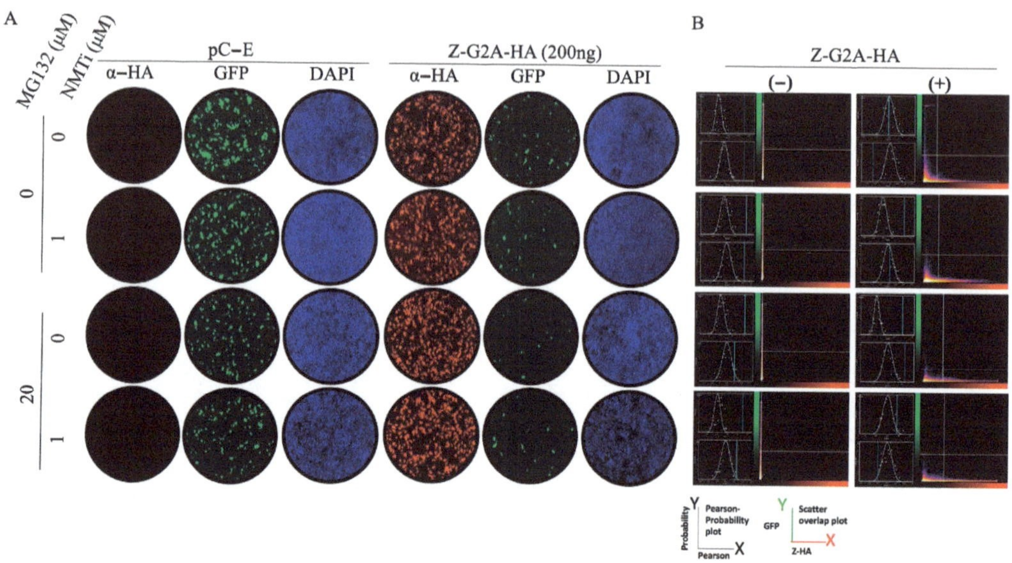

Figure 3. Effect of treatments with the NMT (IMP-1088) and proteasome (MG132) inhibitors on the activity of the LCMV MG in the presence of the LCMV Z G2A mutant. HEK293T cells were seeded at 2×10^5 cells per well in poly-L-lysine-treated M12 well plates. The next day, cells were transfected with pCAGGS plasmids expressing LCMV L and NP proteins, as well as T7RP, and a plasmid expressing the LCMV T7 MG-GFP, together with 200 ng of a pCAGGS empty (pC-E) plasmid or a plasmid expressing LCMV-Z-G2A-HA. At 18 h post-transfection, cells were treated with the indicated compounds, and at 24 h post-treatment, cells were fixed with 4% PFA. Z-G2A-HA expression was detected by IF using a rabitt polyclonal antibody to α-HA as a primary antibody and an anti-mouse goat polyclonal antibody conjugated to Alexafluor 568 as a secondary antibody. GFP was detected directly by epifluorescence. (**A**). The activity of LCMV MG-GFP was assessed based on the expression levels of GFP. (**B**). The left side of each panel displays unmasked (**top**) and masked (**bottom**) panels that represent the Pearson correlation coefficient (*X* axis) and probability (*Y* axis) plots, and the right side displays a scatter plot representing the GFP (green, *Y* axis) and Z-G2A-HA (red, *X* axis) signals. The BIOP Jacob plug quantified colocalization of GFP (a surrogate of the MG activity) and the AlexaFLuor 568 red signal (a surrogate of Z matrix protein expression).

3.3. Role of Z Oligomerization on Z-Mediated Inhibition of vRNP Activity and Z Budding Activity

We have documented the detection of different oligomer species of Z protein in lysates of Z transfected cells using Western blotting [16]. We detected these Z oligomeric species in VLPs produced in cells expressing WT, but not in cells expressing mutant G2A or Z protein (Figure 4) mutant. In addition, treatment with the proteasome inhibitor MG132 rescued high expression levels of dimer and oligomers of G2A mutant form of Z in WCL, but these Z(G2A) species lacked detectable levels of budding activity as determined by their absence in the corresponding VLP samples (Figure 4B). These findings are consistent with published results showing the lack of Z oligomeric species in lysates and Z-containing VLP in cell culture supernatants of cells transfected with Z-G2A-HA [11,16]. We observed four Z species in VLP produced in cells transfected with Z-WT-HA, with monomeric Z being the most abundant Z species. Inhibition of Z myristoylation using NMTi or mutation G2A results in the degradation of Z oligomeric species [16], which complicates discerning the relative contributions of myristoylation and oligomerization to Z budding activity.

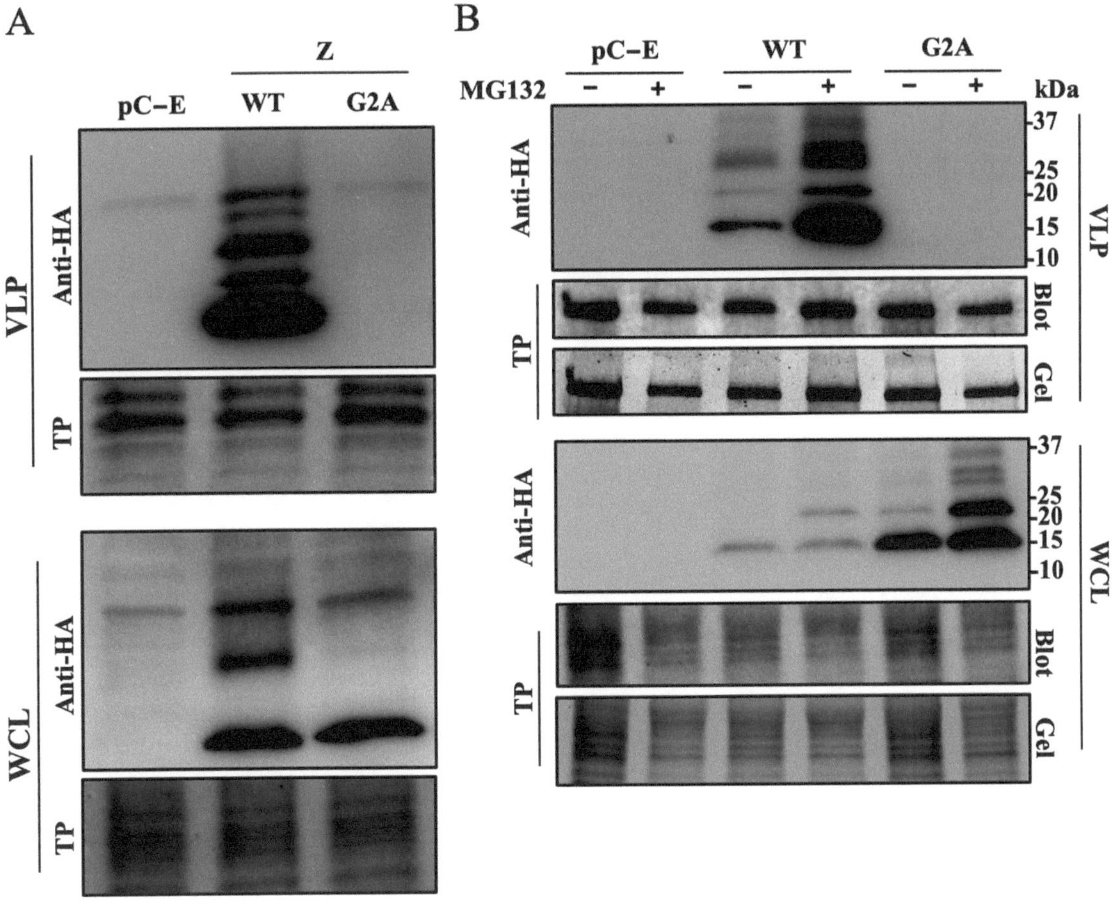

Figure 4. Oligomerization and cell egress of Z protein. (**A**). HEK293T cells were seeded at 1×10^6 cells per well in poly-L-lysine-treated M6-well plates. The next day, cells were transfected with plasmids expressing LCMV Z WT or G2A mutant proteins tagged with an HA tag or empty pCAGGS (pC-E). At 72 h post-transfection, supernatant and protein cell lysate were collected to extract VLP or whole-cell lysate (WCL) protein, respectively. (**B**). HEK293T cells were seeded at 1×10^6 cells per well in poly-L-lysine-treated M6-well plates. The next day, cells were transfected with plasmids expressing LCMV Z WT or G2A mutant proteins tagged with an HA tag or pC-E. The next day, media were aspirated and replaced with media containing MG132 (10 μg/mL) or a vehicle control. At 48 h post-transfection, supernatants were collected and whole cell lysate (WCL) was prepared. VLPs were collected by ultracentrifugation. Levels of Z in VLP and WCL samples were determined by Western blotting using an antibody to HA. Membranes were imaged for total protein (TP) prior to probing with an anti-HA antibody.

4. Discussion

The mammarenavirus Z matrix protein plays critical roles in the assembly and budding of matured infectious particles, processes that require Z myristoylation and oligomerization [11,24,28]. In addition, as with the matrix protein of other negative strand RNA viruses [29], Z protein has been shown to exhibit a dose-dependent inhibitory effect on mammarenavirus vRNP activity [18,19,30]. Biochemical and structural studies have indicated that Z can lock a polymerase-promoter complex [31] and induce conformational changes in L catalytic domains [20], which may account for Z-mediated inhibition of vRNP-directed synthesis of viral RNA. These studies, however, did not address the question of whether myristoylation and oligomerization are required for Z-mediated inhibition of vRNP activity in infected cells.

Here, we have documented that NMTi potent antiviral activity against LCMV and other mammarenaviruses correlated with proteasome-mediated degradation of non-myristoylated Z protein, which disrupted virus particle assembly and budding, resulting in reduced production of infectious progeny and restricted virus propagation [16] without inhibition of vRNP activity. Treatment with the proteasome inhibitor MG132 in the presence of NMTi restored Z-mediated inhibition of vRNP activity (Figure 2), and the Z G2A mutant inhibited the activity of the LCMV MG with similar efficiency to Z WT (Figure 3). These findings indicate that Z myristoylation is not required for its inhibitory effect on vRNP activity. We also observed that Z G2A exhibited significantly reduced levels of oligomerization compared to Z WT (Figure 4), suggesting that oligomerization is not required for the Z inhibitory effect on vRNP activity.

We have shown that dimers are the most abundant oligomer forms of Z in lysates from LCMV-infected or Z-transfected cells, and that these Z homodimers are efficiently targeted for degradation in the presence of NMTi, questioning the conclusion that Z homo-oligomerization is required for Z accumulation at the plasma membrane [24]. Our results also question whether G2 residue is required for oligomerization [24] because treatment with proteasome inhibitor MG132 resulted in similar expression levels of dimers of Z WT and Z G2A [16]. We propose that myristoylation at G2 prevents the targeting of Z for degradation, thus allowing the Z dose-dependent inhibitory effect on viral RNA synthesis mediated by the vRNP. Inhibition of Z myristoylation with NMTi results in proteasome-mediated degradation of Z, thus preventing the Z inhibitory effect on vRNP activity, which can be restored by treatment with the proteasome inhibitor MG132 (Figure 5). We observed a prominent Z species of ~33 kDa, likely a trimer of Z, in cells transfected with Z WT that was absent in cells transfected with the G2A Z mutant and in the presence of MG132. The biological role of this Z species remains to be determined [16]. We also showed that oligomers of the Z-G2A mutant lacked budding activity, compared to Z WT, and were not detected in VLPs, despite being expressed at high levels in WCL in the presence of MG132. These findings further support the notion that N-myristoylation is required for Z budding activity. It remains to be determined whether other protein lipidation modifications such as S-palmitoylation can substitute for N-myristoylation and facilitate Z association to membranes and budding.

The use of NMTi to target host-cell lipidation processes opens the door to a new class of broad-spectrum anti-mammarenavirus therapy. Moreover, NMTi could be also incorporated into combination therapy strategies with direct-acting antivirals, an approach expected to impose a significant genetic barrier to the emergence of drug-resistant viruses, and to facilitate drug formulations with reduced toxicity [32]. Notably, the small molecule NMTi PCLX-001 has been shown to be safe and is well tolerated by humans [33,34], supporting the interest in exploring the repurposing of NMTi to treat infections by human pathogenic mammarenaviruses. NMTi inhibitors have been shown to inhibit other viruses with myristoylated proteins, including picornaviruses [15] and vaccinia virus [32,35]. The use of specific pan-NMTi as a broad-spectrum antiviral strategy against viruses with myristoylated proteins warrants further investigation.

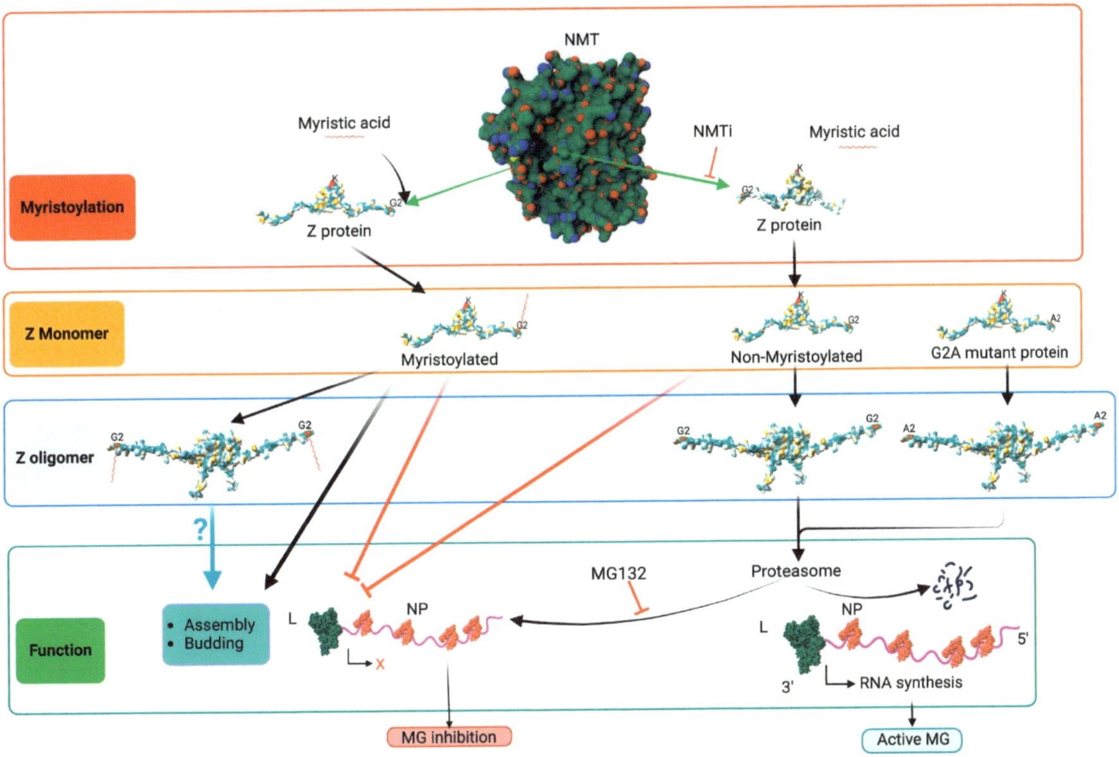

Figure 5. Proposed model of the effect of the NMT inh (IMP-1088) on the activity of the LCMV MG, budding, assembly and infectivity. NMT promotes Z myristoylation at G2, which prevents the targeting of Z for degradation by the proteasome machinery, thus allowing for the Z dose-dependent inhibitory effect on viral RNA synthesis mediated by vRNP and the consequent inhibition of MG-directed GFP expression. Inhibition of Z myristoylation by IMP-1088 results in the proteasome-mediated degradation of Z, which will relieve the Z inhibitory effect on the activity of the LCMV MG (right side of the graph). Treatment with the proteasome inhibitor MG132 prevents IMP-1088-induced Z protein degradation, thus resulting in LCMV MG inhibition. L polymerase (green), NP (red) and RNA (purple) are indicated. A red X indicates cessation activity, and a black angled arrow indicates RNA synthesis. Monomers of Z protein are sufficient to inhibit vRNP. A lack of Z myristoylation results in the degradation of Z oligomers, which complicates the assessment of the role of Z oligomers in virus assembly, budding and infectivity. NMT pdb, 3IWE [25], and LASV Z matrix protein pdb, 2M1S [26] were used to generate the 3D figure using ChimeraX [27].

Author Contributions: Conceptualization, H.W. and J.C.d.l.T.; methodology, H.W.; software, H.W.; validation, H.W., formal analysis, H.W. and J.C.d.l.T.; investigation, H.W.; resources, J.C.d.l.T.; writing—H.W.; writing—review and editing, H.W. and J.C.d.l.T.; visualization, H.W.; supervision, J.C.d.l.T.; project administration, H.W.; funding acquisition, J.C.d.l.T. All authors have read and agreed to the published version of the manuscript.

Funding: This research was supported by NIH/NIAID grants RO1 AI142985 and R21 AI128556 to J.C.d.l.T.

Data Availability Statement: Data is contained within the article.

Conflicts of Interest: The authors declare no conflicts of interest.

References

1. Yadav, K.; Mathur, G.; Ford, B.; Miller, R.; Group, C.W. A Case Cluster of Lymphocytic Choriomeningitis Virus Transmitted Via Organ Transplantation.: Abstract# D2381. *Transplantation* **2014**, *98*, 768.
2. Sayyad, L.E.; Smith, K.L.; Sadigh, K.S.; Cossaboom, C.M.; Choi, M.J.; Whitmer, S.; Cannon, D.; Krapiunaya, I.; Morales-Betoulle, M.; Annambhotla, P.; et al. Severe Non–Donor-Derived Lymphocytic Choriomeningitis Virus Infection in 2 Solid Organ Transplant Recipients. *Open Forum Infect. Dis.* **2025**, *12*, ofaf002. [CrossRef]
3. Schafer, I.J.; Miller, R.; Ströher, U.; Knust, B.; Nichol, S.T.; Rollin, P.E.; Centers for Disease Control and Prevention (CDC). Notes from the Field: A Cluster of Lymphocytic Choriomeningitis Virus Infections Transmitted Through Organ Transplantation—Iowa, 2013. *MMWR Morb. Mortal. Wkly. Rep.* **2014**, *63*, 249. [CrossRef]
4. Carrillo-Bustamante, P.; Nguyen, T.H.T.; Oestereich, L.; Günther, S.; Guedj, J.; Graw, F. Determining Ribavirin's Mechanism of Action against Lassa Virus Infection. *Sci. Rep.* **2017**, *7*, 11693. [CrossRef]
5. Radoshitzky, S.R.; Buchmeier, M.; de la Torre, J.C. Emerging Viruses: Arenaviridae. In *Fields Virology*, 7th ed.; Knipe, D., Howley, P., Whelan, S., Eds.; Wolters Kluwer: Philadelphia, PA, USA, 2021; Volume I, ISBN 978-1-975112-54-7.
6. Kunz, S.; Edelmann, K.H.; de la Torre, J.-C.; Gorney, R.; Oldstone, M.B.A. Mechanisms for Lymphocytic Choriomeningitis Virus Glycoprotein Cleavage, Transport, and Incorporation into Virions. *Virology* **2003**, *314*, 168–178. [CrossRef]
7. Rojek, J.M.; Lee, A.M.; Nguyen, N.; Spiropoulou, C.F.; Kunz, S. Site 1 Protease Is Required for Proteolytic Processing of the Glycoproteins of the South American Hemorrhagic Fever Viruses Junin, Machupo, and Guanarito. *J. Virol.* **2008**, *82*, 6045–6051. [CrossRef]
8. Fedeli, C.; Moreno, H.; Kunz, S. Novel Insights into Cell Entry of Emerging Human Pathogenic Arenaviruses. *J. Mol. Biol.* **2018**, *430*, 1839–1852. [CrossRef]
9. York, J.; Nunberg, J.H. Role of the Stable Signal Peptide of Junín Arenavirus Envelope Glycoprotein in pH-Dependent Membrane Fusion. *J. Virol.* **2006**, *80*, 7775–7780. [CrossRef]
10. York, J.; Nunberg, J.H. Myristoylation of the Arenavirus Envelope Glycoprotein Stable Signal Peptide Is Critical for Membrane Fusion but Dispensable for Virion Morphogenesis. *J. Virol.* **2016**, *90*, 8341–8350. [CrossRef]
11. Perez, M.; Greenwald, D.L.; de La Torre, J.C. Myristoylation of the RING Finger Z Protein Is Essential for Arenavirus Budding. *J. Virol.* **2004**, *78*, 11443–11448. [CrossRef]
12. Capul, A.A.; Perez, M.; Burke, E.; Kunz, S.; Buchmeier, M.J.; de la Torre, J.C. Arenavirus Z-Glycoprotein Association Requires Z Myristoylation but Not Functional RING or Late Domains. *J. Virol.* **2007**, *81*, 9451–9460. [CrossRef]
13. Cordo, S.M.; Candurra, N.A.; Damonte, E.B. Myristic Acid Analogs Are Inhibitors of Junin Virus Replication. *Microbes Infect.* **1999**, *1*, 609–614. [CrossRef]
14. Kallemeijn, W.W.; Lueg, G.A.; Faronato, M.; Hadavizadeh, K.; Grocin, A.G.; Song, O.-R.; Howell, M.; Calado, D.P.; Tate, E.W. Validation and Invalidation of Chemical Probes for the Human N-Myristoyltransferases. *Cell Chem. Biol.* **2019**, *26*, 892–900.e4. [CrossRef]
15. Ramljak, I.C.; Stanger, J.; Real-Hohn, A.; Dreier, D.; Wimmer, L.; Redlberger-Fritz, M.; Fischl, W.; Klingel, K.; Mihovilovic, M.D.; Blaas, D.; et al. Cellular N-Myristoyltransferases Play a Crucial Picornavirus Genus-Specific Role in Viral Assembly, Virion Maturation, and Infectivity. *PLoS Pathog.* **2018**, *14*, e1007203. [CrossRef]
16. Witwit, H.; Betancourt, C.A.; Cubitt, B.; Khafaji, R.; Kowalski, H.; Jackson, N.; Ye, C.; Martinez-Sobrido, L.; de la Torre, J.C. Cellular N-Myristoyl Transferases Are Required for Mammarenavirus Multiplication. *Viruses* **2024**, *16*, 1362. [CrossRef]
17. Cornu, T.I.; de la Torre, J.C. Characterization of the Arenavirus RING Finger Z Protein Regions Required for Z-Mediated Inhibition of Viral RNA Synthesis. *J. Virol.* **2002**, *76*, 6678–6688. [CrossRef]
18. Jácamo, R.; López, N.; Wilda, M.; Franze-Fernández, M.T. Tacaribe Virus Z Protein Interacts with the L Polymerase Protein To Inhibit Viral RNA Synthesis. *J. Virol.* **2003**, *77*, 10383–10393. [CrossRef]
19. López, N.; Jácamo, R.; Franze-Fernández, M.T. Transcription and RNA Replication of Tacaribe Virus Genome and Antigenome Analogs Require N and L Proteins: Z Protein Is an Inhibitor of These Processes. *J. Virol.* **2001**, *75*, 12241–12251. [CrossRef]
20. Liu, L.; Wang, P.; Liu, A.; Zhang, L.; Yan, L.; Guo, Y.; Xiao, G.; Rao, Z.; Lou, Z. Structure Basis for Allosteric Regulation of Lymphocytic Choriomeningitis Virus Polymerase Function by Z Matrix Protein. *Protein Cell* **2023**, *14*, 703–707. [CrossRef]
21. Cornu, T.I.; de la Torre, J.C. RING Finger Z Protein of Lymphocytic Choriomeningitis Virus (LCMV) Inhibits Transcription and RNA Replication of an LCMV S-Segment Minigenome. *J. Virol.* **2001**, *75*, 9415–9426. [CrossRef]
22. Iwasaki, M.; de la Torre, J.C. A Highly Conserved Leucine in Mammarenavirus Matrix Z Protein Is Required for Z Interaction with the Virus L Polymerase and Z Stability in Cells Harboring an Active Viral Ribonucleoprotein. *J. Virol.* **2018**, *92*, e02256-17. [CrossRef] [PubMed]

23. Hastie, K.M.; Zandonatti, M.; Liu, T.; Li, S.; Woods, V.L.; Saphire, E.O. Crystal Structure of the Oligomeric Form of Lassa Virus Matrix Protein Z. *J. Virol.* **2016**, *90*, 4556–4562. [CrossRef] [PubMed]
24. Loureiro, M.E.; Wilda, M.; Levingston Macleod, J.M.; D'Antuono, A.; Foscaldi, S.; Buslje, C.M.; Lopez, N. Molecular Determinants of Arenavirus Z Protein Homo-Oligomerization and L Polymerase Binding. *J. Virol.* **2011**, *85*, 12304–12314. [CrossRef] [PubMed]
25. Qiu, W.; Hutchinson, A.; Wernimont, A.; Lin, Y.-H.; Kania, A.; Ravichandran, M.; Kozieradzki, I.; Cossar, D.; Schapira, M.; Arrowsmith, C.H.; et al. Crystal Structure of Human Type-I N-Myristoyltransferase with Bound Myristoyl-CoA and Inhibitor DDD85646. 2010, *To be published*. [CrossRef]
26. Volpon, L.; Osborne, M.J.; Capul, A.A.; de la Torre, J.C.; Borden, K.L.B. Structural Characterization of the Z RING-eIF4E Complex Reveals a Distinct Mode of Control for eIF4E. *Proc. Natl. Acad. Sci. USA* **2010**, *107*, 5441–5446. [CrossRef]
27. Meng, E.C.; Goddard, T.D.; Pettersen, E.F.; Couch, G.S.; Pearson, Z.J.; Morris, J.H.; Ferrin, T.E. UCSF ChimeraX: Tools for Structure Building and Analysis. *Protein Sci.* **2023**, *32*, e4792. [CrossRef]
28. Perez, M.; Craven, R.C.; De La Torre, J.C. The Small RING Finger Protein Z Drives Arenavirus Budding: Implications for Antiviral Strategies. *Proc. Natl. Acad. Sci. USA* **2003**, *100*, 12978–12983. [CrossRef]
29. Baudin, F.; Petit, I.; Weissenhorn, W.; Ruigrok, R.W. In Vitro Dissection of the Membrane and RNP Binding Activities of Influenza Virus M1 Protein. *Virology* **2001**, *281*, 102–108. [CrossRef]
30. Cornu, T.I.; Feldmann, H.; de la Torre, J.C. Cells Expressing the RING Finger Z Protein Are Resistant to Arenavirus Infection. *J. Virol.* **2004**, *78*, 2979–2983. [CrossRef]
31. Kranzusch, P.J.; Whelan, S.P.J. Arenavirus Z Protein Controls Viral RNA Synthesis by Locking a Polymerase–Promoter Complex. *Proc. Natl. Acad. Sci. USA* **2011**, *108*, 19743–19748. [CrossRef]
32. Witwit, H.; Cubitt, B.; Khafaji, R.; Castro, E.M.; Goicoechea, M.; Lorenzo, M.M.; Blasco, R.; Martinez-Sobrido, L.; de la Torre, J.C. Repurposing Drugs for Synergistic Combination Therapies to Counteract Monkeypox Virus Tecovirimat Resistance. *Viruses* **2025**, *17*, 92. [CrossRef]
33. Sangha, R.; Davies, N.M.; Namdar, A.; Chu, M.; Spratlin, J.; Beauchamp, E.; Berthiaume, L.G.; Mackey, J.R. Novel, First-in-Human, Oral PCLX-001 Treatment in a Patient with Relapsed Diffuse Large B-Cell Lymphoma. *Curr. Oncol.* **2022**, *29*, 1939–1946. [CrossRef] [PubMed]
34. Sangha, R.S.; Jamal, R.; Spratlin, J.L.; Kuruvilla, J.; Sehn, L.H.; Weickert, M.; Berthiaume, L.G.; Mackey, J.R. A First-in-Human, Open-Label, Phase I Trial of Daily Oral PCLX-001, an NMT Inhibitor, in Patients with Relapsed/Refractory B-Cell Lymphomas and Advanced Solid Tumors. *JCO* **2023**, *41*, e15094. [CrossRef]
35. Priyamvada, L.; Kallemeijn, W.W.; Faronato, M.; Wilkins, K.; Goldsmith, C.S.; Cotter, C.A.; Ojeda, S.; Solari, R.; Moss, B.; Tate, E.W.; et al. Inhibition of Vaccinia Virus L1 N-Myristoylation by the Host N-Myristoyltransferase Inhibitor IMP-1088 Generates Non-Infectious Virions Defective in Cell Entry. *PLoS Pathog.* **2022**, *18*, e1010662. [CrossRef] [PubMed]

BioChem

Review

Oxidative Stress-Induced Gastrointestinal Diseases: Biology and Nanomedicines—A Review

Maryam Rezvani [1,2]

[1] Department of Life and Environmental Sciences, Drug Science Division, University of Cagliari, 09124 Cagliari, Italy; rezvani.mary@yahoo.com

[2] Advanced Nanobiotechnology and Nanomedicine Research Group (ANNRG), Iran University of Medical Sciences, Tehran 14496-4535, Iran

Abstract: Gastrointestinal diseases have been among the main concerns of medical and scientific societies for a long time. Several studies have emphasized the critical role of oxidative stress in the pathogenesis of the most common gastrointestinal diseases. To provide a comprehensive overview of gastrointestinal diseases caused by oxidative stress, their biological aspects, molecular mechanisms and specific pathways, the results of the most recent published articles from the online databases were studied considering both the upper and lower parts of the digestive tract. The results revealed that although the oxidative stress in each part of the digestive system manifests itself in a specific way, all these diseases arise from the imbalance between the generation of the reactive intermediates (especially reactive oxygen species) and the antioxidant defense system. Annual incidence and mortality statistics of gastrointestinal diseases worldwide emphasize the urgent need to find an effective and non-invasive treatment method to overcome these life-threatening problems. Therefore, in the next step, a variety of nanomedicurfines developed to treat these diseases and their effect mechanisms were investigated precisely. Furthermore, the most important nanomedicines responsive to endogenous and exogenous stimuli were evaluated in detail. This review could pave the way to open a new horizon in effectively treating gastrointestinal diseases.

Keywords: gastrointestinal diseases; oxidative stress; nanomedicine; stimuli-responsive nanoparticles; reactive oxygen species

1. Introduction

Oxidative stress (OS) in living organisms occurs arising from the excessive generation of reactive intermediates such as reactive oxygen species (ROS), reactive nitrogen species (RNS) and reactive sulfur species (RSS) that cannot be neutralized by the endogenous antioxidant system [1]. ROS at appropriate physiological concentrations plays a prominent role in intracellular homeostasis, gene expression, cell proliferation and differentiation, signal transduction and apoptosis [2]. Various external triggers including trans-fatty acids and acrylamide in processed foods, radiation, alcohol, drugs, organic solvents, heavy metals, smoking and pollutants can induce OS [3]. Plasma membrane, cytosol, lysosomes, peroxisomes, endoplasmic reticulum and mitochondria are intracellular sources of chemical reactive intermediates. Chemical reactions for ROS generation are catalyzed by a variety of enzymes including the enzymes of the mitochondrial electron transport respiratory chain (the most important ones) and nicotinamide adenine dinucleotide phosphate (NADPH) oxidase (with homologs of dual oxidase (DUOX) 1-2 and NOX1-5), cyclooxygenases (COXs), lipoxygenases (LOXs), xanthine oxidase (XO), phospholipase A2, glucose oxidase, myeloperoxidase (MPO) and uncoupled nitric oxide synthase (NOS) [4]. Hydroxyl radical ($^{\bullet}OH$), superoxide anion radical ($O_2^{\bullet-}$), hypochlorous acid (HOCl) and hydrogen peroxide (H_2O_2) are the major ROS, and nitric oxide (NO) and peroxynitrite radicals ($ONOO^-$) are the main RNS. Amongst the RSS, $GSSG^{\bullet-}$, reactive sulfur substances (SO_2, SO_3) and reactive sulfane species (RSR) are the most important ones [5,6]. Endogenous antioxidants can counteract and remove the reactive species by oxidizing themselves, delaying/inhibiting the oxidation of other compounds, chelating metal ions and blocking radical formation. Superoxide dismutase (SOD), catalase (CAT), glutathione peroxidase (GPx) and glutathione reductase (GSR) are the major endogenic enzymatic antioxidants [7]. Glutathione (GSH) and thioredoxin (Trx) are the prominent non-enzymatic endogenous antioxidants [8]. GSR converts oxidized glutathione (GSSG) to GSH by transferring electrons from NADPH to GSSG [9]. The thioredoxin system is composed of Trx and thioredoxin reductase (TrxR) [10]. Figure 1 demonstrates the generation of reactive intermediates and the function of the antioxidant defense system under normal physiological conditions. The imbalance between ROS generation and the neutralizing ability of the antioxidant system can cause various diseases resulting from damage to the vital molecules and cellular structures including DNA, proteins and lipids [11].

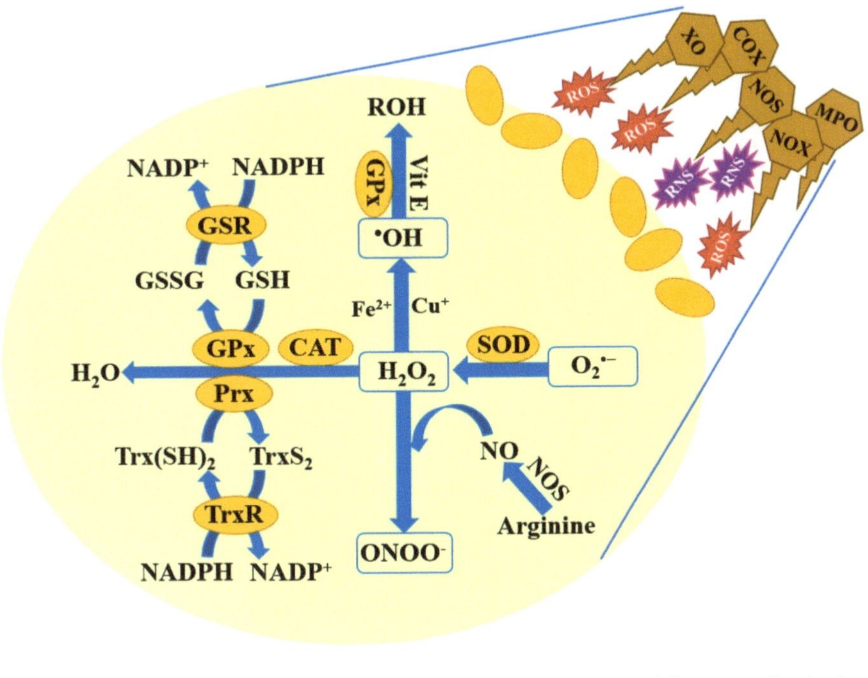

Figure 1. Generation of the reactive intermediates and the mechanisms of their neutralization by the antioxidant defense system. The orange ovals illustrate the antioxidant enzymes, the rectangles demonstrate the reactive intermediates and the brown hexagons refer to some of the enzymes generating reactive oxygen species (ROS) and reactive nitrogen species (RNS): NOX: nicotinamide adenine dinucleotide phosphate oxidase, NOS: nitric oxide synthase, COX: cyclooxygenase, XO: xanthine oxidase and MPO: myeloperoxidase. The generated $O_2^{\bullet-}$ can be converted to H_2O_2 by superoxide dismutase (SOD). H_2O_2 can be converted to $^{\bullet}OH$ through the Fenton reaction with Fe^{2+}, which can initiate lipid peroxidation. Glutathione peroxidase (GPx) and vitamin E (Vit E) can neutralize the free radicals and result in ROH formation as the product. Nitric oxide (NO) can be generated from arginine by NOS, which results in $ONOO^-$ production through its reaction with H_2O_2. Catalase (CAT), peroxiredoxin (Prx) and GPx can convert H_2O_2 to water. Glutathione reductase (GSR) and thioredoxin reductase (TrxR) keep glutathione (GSH) and thioredoxin (Trx) in their reduced forms by transferring the electrons from NADPH to their oxidized forms. GSSG is the oxidized form of GSH and $Trx(SH)_2$ and $TrxS_2$ are the reduced and oxidized states of Trx, respectively.

ROS can activate the nuclear factor kappa-light-chain-enhancer of activated B cells (NF-κB) pathway, which can contribute to regulating the generation of proinflammatory cytokines and the inflammatory response. ROS (especially H_2O_2) can stimulate the activation of NF-κB by promoting IkBα phosphorylation and its subsequent degradation, which causes the transrelocation of the dimerized NF-κB proteins into the nucleus to start the transcription of their target genes. It can promote the expression of proinflammatory cytokines including Interleukin-8 (IL-8), IL-1 and tumor necrosis factor-α (TNF-α), and lead to inflammation and carcinogenesis [12]. Another ROS-activated pathway is mitogen-activated protein kinase (MAPK). This pathway is activated by mitogen cascades including JNK (c-Jun N-terminal kinases), p38 (p38 kinase), ERK1/2 (offline cell-related kinases) and BMK1/ERK5 (significant MAP kinase 1 pathway). During the OS, H_2O_2 induces the oxidation of the cysteine residue of Trx, resulting in apoptosis signal-regulated kinase-1 (ASK-1) dissociation for activating JNK and p38 cascade-inducing apoptosis. Translocation of the activated MAPKs into the nucleus and phosphorylation of transcription factors can lead to the generation of pro-inflammatory mediators, enhancing the inflammatory response. Excessive ROS demonstrates a positive correlation with arresting the cell cycle, senescence and apoptotic condition through the signaling cascade of ASK1/JNK/p38 [13].

The gastrointestinal (GI) tract, as a passageway for delivering, swallowing, digesting and absorbing food and liquids and excreting waste from the body, consists of the oral cavity, esophagus, stomach, small and large intestines and rectum. The GI tract plays a prominent role in modulating energy homeostasis, providing vital nutrition, defending against infectious stimuli, secreting hormones and neurotransmitters and regulating body physiology [14]. All parts of the GI tract must be in normal condition to carry out their work. Otherwise, the food cannot be processed properly, and the health is adversely affected. OS-induced disease and inflammation in the GI tract can effectively influence the total homeostasis of the body, threatening life.

Conventional medical treatments including drug administration and clinical surgical solutions have possessed several shortcomings including the need to use high doses, poor bioavailability and the lack of effective targeted delivery of drugs, off-target toxicity and adverse side effects, drug resistance, low therapeutic efficacy and high recurrence rate. Nanomedicines with large surface-to-volume ratios, nanoscale dimensions and biocompatibilities have been developed to overcome the obstacles in treating OS-induced diseases. High drug loading capacity, providing synergistic effects of multi-loading, prolonging drug circulation time, specific targeting capability, improving bioavailability, promoting drug accumulation at diseased sites, controlled drug release and reducing drug resistance and systemic toxicity are some benefits of using nanomedicines in treating the GI diseases [15]. Furthermore, these nanomedicines can be designed and developed to respond to endogenous/exogenous stimuli for on-demand controlled release in the target tissues. Active targeting and passive targeting are two categories of targeting mechanisms. Active targeting provides selective drug delivery to the target area through affinity between nanomedicine ligands and the receptors of the target cells, which facilitates the internalization of nanomedicines and the selective release of the drugs into the target cells. Passive targeting exploits the differences between the tumor tissues and the normal ones. Blood vessels of the tumor tissue have large gaps between their endothelial cells, through which nanomedicines can leak and penetrate the tumor. Because of the poor lymphatic drainage, the accumulated nanomedicines are retained within the tumor microenvironment for a prolonged duration, providing sustained drug delivery. This is known as the enhanced permeability and retention (EPR) effect [16]. A wide range of nanomedicines has been developed to treat the GI disorders caused by OS, including nanovesicles, micelles, carbon nanotubes, polymeric nanoparticles (NPs), quantum dots, silica NPs, exosomes, dendrimers, lipid NPs and metallic (such as gold (Au), copper (Cu) and silver (Ag)) NPs. This study aimed to provide a comprehensive overview of how OS affects the biology and pathophysiology of each part of the GI tract from the mouth to the anus, involving specific molecular mechanisms and pathways and the nanomedicines developed to combat diseases. Some of the most common diseases caused by OS in the GI tract are illustrated in Figure 2, which will be discussed in detail in the following sections.

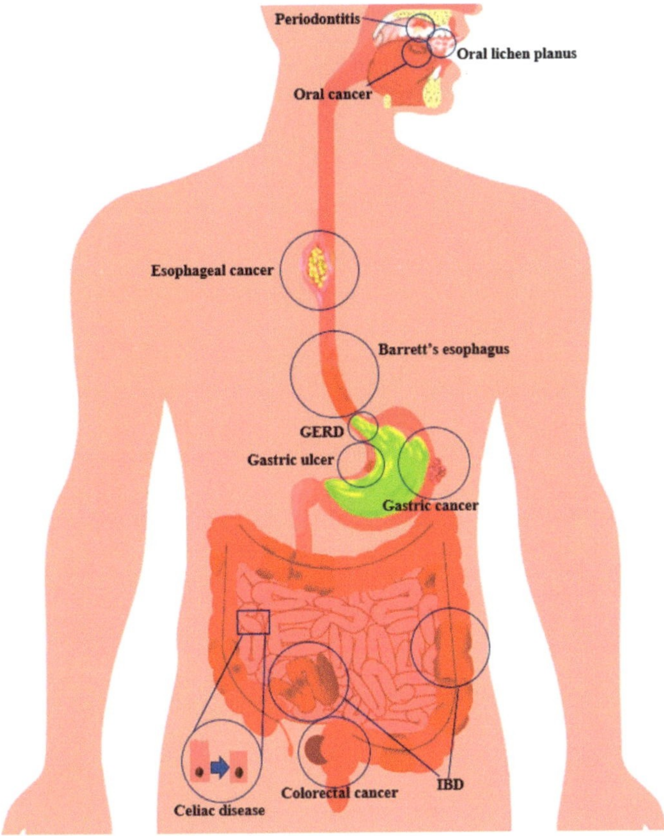

Figure 2. The most common oxidative stress-induced gastrointestinal diseases. GERD and IBD are the abbreviations of gastroesophageal reflux disease and inflammatory bowel disease, respectively.

2. Methodology

English language articles published from 2016 to 2024 were selected from online databases, particularly PubMed, MEDLINE, Google Scholar and Scopus. Publications on OS-induced GI disease, the involved mechanisms and nanomedicines developed for the treatment of GI diseases were separately searched for each of the related diseases including oral lichen planus, oral cancer, periodontitis, gastroesophageal reflux disease, Barrett's esophagus, esophageal cancer, *H. pylori*-induced gastric diseases, gastric ulcers, gastric cancer, celiac disease, inflammatory bowel disease and colorectal cancer. Each of the found articles was independently screened for quality, validity and eligibility, and finally, the best papers were precisely selected for the preparation of this review. The inclusion criteria were longitudinal studies, *in vitro* and *in vivo* studies, observational studies, book chapters, review articles, case reports and systematic reviews with special attention to more recent works published in high-impact international journals. The exclusion criteria included conferences, expert opinions, letters to the director, compendiums and symposia.

3. OS-Induced GI Diseases and Developed Nanomedicines

3.1. Oral Diseases

The mouth is the first part of the GI tract and is composed of the teeth (surrounded with the gums), tongue and salivary glands as the accessories. OS-induced oral diseases including oral cancer, lichen planus and periodontitis impose a significant economic burden on society and cause high mortality and morbidity worldwide [17,18]. In many cases, the colonization of bacteria in the oral cavity can stimulate the defense systems to combat pathogens by ROS generation. However, ROS targets both microorganisms and host cells [19]. In addition to salivary enzymes such as CAT, SOD and peroxidase, the non-enzymatic molecules including uric acid, glutathione, ascorbic acid and albumin play antioxidant roles. Among these molecules, most of the total antioxidant capacity is provided by uric acid as the main salivary antioxidant, which plays a crucial role in monitoring OS [20]. In a weakened antioxidant system, OS in the mouth results in various inflammatory diseases, which can affect mucosal, gingival and even underlying bone tissues. One of the relatively common chronic inflammatory diseases with unknown etiology is oral lichen planus (OLP), which can be developed by the activation of signaling pathways and autoimmune reactions caused by excessive OS products. Despite the unclear mechanism of OS in OLP incidence, various studies have revealed that ROS-induced OS can especially contribute to the inflammatory signaling and metabolic dysregulation of OLP [21]. The detection of a decreased level of total antioxidant activity and increased levels of salivary ROS, malondialdehyde (MDA) and NO in OLP patients compared to the healthy control group are the proofs for this claim [22,23]. Research demonstrated several possible pathomechanisms for ROS in OLP. The antigen-specific mechanism through lipid peroxidation and antigen-nonspecific mechanism via the Fas/FasL pathway, Bcl-2 family proteins, matrix metalloproteinase-9, TNF-α, granzyme B and p53 participate in OLP pathogenesis, leading to lymphocyte infiltration and keratinocyte apoptosis. On the other hand, ROS can cause tissue damage and the release of DAMPs (damp-associated molecular patterns), which can lead to inflammation and tissue dysfunction. OS in OLP exhibits a close correlation with the activation of NF-κB [21]. A current treatment strategy for OLPs is the local administration of corticosteroids, often for a long period and a subsequent increasing dose or systemic applications in recalcitrant cases, which cannot show necessary effectiveness many times and are followed by side effects [24]. Mucus as a strong barrier prevents the adherence and penetration of the topical corticosteroids prescribed in the forms of gel, lotion, cream and ointment to reach the epithelial surface and causes their rapid clearance during the first seconds of the application [25]. The use of nanoformulations has demonstrated promising results in this regard, possessing biodegradability and mucoadhesiveness. Sadeghian et al. revealed the effect of mucoadhesive nano-triamcinolone acetonide gel on the recovery acceleration of OLP, pain severity, lesion size and appearance compared to the conventional gel formulation. However, they reported statistically significant improvement just for ulcer appearance on the 6th and 14th days of treatment with the nanoformulation in comparison to the non-nano one [24]. Studies have shown the healing effects of nanoformulations containing curcumin on oral mucosal lesions. Curcumin exposes its antioxidant and anti-inflammatory activities by preventing free radical-induced damage and the downregulation of the pro-inflammatory cytokines (such as TNF-α, IL-8, IL-6 and IL-1), inflammatory transcription factors (like NF-κB) and some enzymes (such as LOX, COX-2 and COX-5) [26]. Kia et al. proposed oral nanomicellar soft gel capsules loaded with curcumin as an alternative strategy in treating OLP patients with contraindications on the use of corticosteroids and reoccurrence of the disease. They reported a reduction in lesion size, burning sensation and pain severity in OLP patients by oral administration of nano-curcumin with no significant differences with the prednisolone administration. [27]. Similar results were obtained by oral administration of the nanomicelle containing curcumin which significantly improved the clinical appearance of oral lesions and decreased pain intensity in OLP patients [28]. Combination therapy with 1% nanomicelle curcumin gel and 0.1% triamcinolone acetonide caused a higher reduction in the reticular-erosive-ulcerative clinical score and higher efficacy in reducing the extent of oral lesions than the triamcinolone acetonide alone in OLP patients [29]. Popovska et al. recommended the topical application of the nano-bio fusion gingival gel, a nanoemulsion form of propolis and vitamins E and C, as a substitute for the topical or systematic administration of steroids in the treatment of OLP. They reported significant clinical improvements during 5 weeks of treatment with stable achievements even after 3 months [30].

Dysbiosis in periodontal tissues results in periodontitis, destructing the supporting structures of teeth, chronic inflammation and tooth loss. OS plays a vital role in the pathogenesis of periodontitis. Gram-negative anaerobic bacteria on the dental plaque can activate neutrophils, which causes an excessive release of ROS through the NADPH oxidase pathway and subsequent damage of periodontal tissues, apoptosis of gingival fibroblasts, osteoclastogenesis and alveolar bone resorption [31]. ROS can lead to direct damage to periodontal tissues through damaging the nucleic acid (including base pair mutations or strand breaks), the denaturing of proteins, the deactivating of enzymes, the peroxidation of lipids, cell membrane destruction and mitochondrial damage. The analysis of the biological samples prepared from patients with chronic periodontitis showed higher levels of MDA and 8-isoprostane (as biomarkers of lipid peroxidation) and 7,8-dihydro-8-oxoguanine (as a biomarker of OS-mediated DNA damage) [32,33]. Furthermore, ROS can lead to periodontal pathogenesis by regulating the signaling pathways of the c-Jun N-terminal kinase (JNK), NOD-like receptor family, pyrin domain-containing protein 3 (NLRP3) inflammasomes and NF-κB. Activated JNK signaling can lead to the E-cadherin dissociation and disruption of the periodontal junctional epithelium. Furthermore, activation of this pathway can lead to the initiation of the apoptosis cascade through the caspase-3-dependent pathway. ROS-mediated NLRP3 inflammasome activation can result in IL-1β secretion and pyroptosis. Excessive levels of IL-1β can lead to periodontal destruction. ROS-induced activation of NF-κB can cause periodontal destruction by triggering inflammatory responses, the expression of pro-inflammatory cytokines and osteoclastic differentiation [34]. Various metal NPs have been developed to treat periodontitis, exhibiting anti-inflammatory, antibacterial and photothermal enhancement effects. ROS generated by metal NPs can damage the bacterial cell membranes and destroy their nucleic acids and proteins. Furthermore, these NPs can physically interact with bacterial cell walls, interrupting the electron transfer. The released positive ions from metal NPs adversely affect bacterial proteins and DNA [35]. Chlorhexidine/metronidazole-conjugated Ag NPs were evaluated for their effects on periodontal diseases. They showed an inhibitory effect on fungi and bacteria and also reduced the concentration of metalloproteinases MMP3 and MMP8 and the generation of the TNF-α, IL-1β, IL-6 and IL-8 [36]. Castangia et al. developed a mouthwash to counteract the OS-induced and bacterial damage in the oral cavity. This mouthwash formulation included quercetin and mint oil co-loaded liposomes improved with glycol and ethanol and demonstrated antioxidant activity against hydrogen peroxide-induced cell damage. The vesicles with 200 mg/mL mint oil in their formulations exerted an antibacterial effect on cariogenic bacteria (*Lactobacillus acidophilus* and *Streptococcus mutans*) [37]. Surfactin-loaded κ-carrageenan oligosaccharides linked cellulose nanofibers NPs diminished the generation of ROS and pro-inflammatory cytokines and exhibited an anti-inflammatory effect on lipopolysaccharide (LPS)-stimulated human gingival fibroblast cells. Furthermore, these NPs induced OS, reduced the viability of *Pseudomonas aeruginosa* and *Fusobacterium nucleatum* and prevented biofilm formation. However, these results were obtained from the *in vitro* studies and needed to be evaluated by the animal clinical ones [38]. Local injection of the curcumin-loaded NPs reduced the inflammatory infiltrate and osteoclast numbers and completely prevented bone resorption in the LPS-induced periodontal rat model. These NPs attenuated the activation of NF-κB and p38 mitogen-activated protein kinases in the gingival tissues [39].

Reduced antioxidants and increased lipid peroxidation have also been reported in oral cancers, indicating the OS contribution to this disease [32]. Furthermore, ROS-induced OS can damage DNA, proteins and lipids to develop oral cancer [40]. Oral squamous cell carcinoma (OSCC) is amongst the most common cancers worldwide, in which OS and the accumulation of ROS play key roles in carcinogenesis and progression. An overexpression of LOX, COX, NOX and NOS in oral cancer leads to an increase in the reactive intermediates (ROS and RNS) levels, which regulates the activity of NF-κB, p53 and STAT (signal transducer and activation of transcription), affecting cancer progression [41]. A variety of NPs have been developed to treat oral cancer. Shi et al. developed hyaluronic acid (HA)-based mesoporous silica NPs for the delivery of TH287 (protein MutT homolog 1 inhibitor) and P-glycoprotein 1 siRNA to treat OSCC. These NPs possessed a controlled release of the drug and uptake by CAL27 cells, an inhibitory effect on the function of P-glycoprotein 1 and a killing effect on oral cancer tissue. Treating the CAL27 xenograft model with NPs caused a significant reduction in the tumor burden in comparison to the free TH287 and control [42]. Arabic gum-encapsulated Au NPs demonstrated a reduction in cell viability, c-Myc antibody levels and the expression of hypoxia-regulating miRNAs (miR-21 and miR-210). They suppressed the expression of the hypoxia-inducible factor-1α protein, showed a dose-dependent inhibitory effect on hypoxia in CAL27 and caused death in tongue squamous carcinoma cells [43].

3.2. Esophageal Diseases

Gastroesophageal reflux disease (GERD), Barrett's esophagus (BE), eosinophilic esophagitis and esophageal cancer are the most common diseases induced by OS in the esophagus. Repeated backflow of the stomach acid and other gastric contents into the esophagus causes esophagitis. This problem is called GERD, which is characterized by regurgitation and heartburn [44]. Studies have revealed the involvement of OS in the development and progression of GERD, which in turn plays a pivotal role in the induction of other OS-related esophageal diseases [45]. As shown in Figure 3, GERD can lead to BE and replacement of the normal esophageal squamous epithelium with metaplastic columnar epithelium that may develop into esophageal adenocarcinoma, a rare cancer [46].

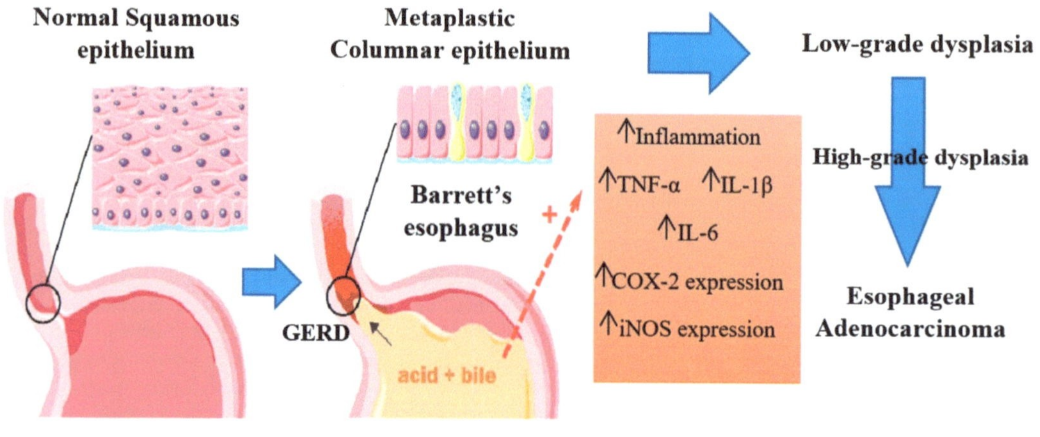

Figure 3. Gastroesophageal reflux disease (GERD) and its consequences as Barrett's esophagus and esophageal adenocarcinoma. TNF-α: tumor necrosis factor-α, IL-1β: interleukin-1β, IL-6: interleukin-6, COX-2: cyclooxygenase-2 and iNOS: inducible nitric oxide synthase. Reproduced from [46].

Unconjugated bile acids as potent COX-2 inducers and acidic pH cause ROS increase and the stabilization of HIF-2α (hypoxia-inducible factor 2 alpha) in BE. The inflammatory response to reflux injury is regulated by HIF-2α in association with NF-κB signaling. The activation of NF-κB in the distal esophagus causes persistent inflammation. From esophagitis to BE and esophageal adenocarcinoma, the levels of IL-1β, IL-8 and NF-κB increase, which results in epithelial cell proliferation, apoptosis inhibition and carcinogenesis. IL-6 is another pivotal cytokine in the BE pathogenesis. The generation of IL-6 in the metaplastic epithelium activates and translocates STAT3 to the nucleus, which leads to the synthesis of Bcl-xL and Mcl-1 as antiapoptotic proteins in cancer cells. The activation and interaction of NF-κB and IL-6/STAT3 pathways results in persistent inflammation and cancer progression [47,48]. In BE, Rho kinase ROCK2 is activated in a calcium-dependent manner and causes the overexpression of NOX5 and NOX5-S, which leads to the overproduction of $O_2^{\bullet-}$ and H_2O_2, respectively. SOD inactivation and lipid peroxidation have been reported in BE. The production of ROS is enhanced by esophageal epithelial cells under high gastric acid, which activates macrophages, platelet-activating factor and proteinase-activated receptor 2 and causes OS, inflammation and mucosal damage [49]. Preventing the incidence of GERD has been evaluated as an effective treatment strategy for other OS-induced diseases in the esophagus. Dysfunction of the lower esophageal sphincter (LES) contractility is one of the contributors to GERD pathogenesis. The use of nano-based delivery systems for the controlled release of therapeutic agents to regulate the LES tone or neutralize gastric acid is one of the strategies for effectively treating GERD [50]. Hydrogen sulfide (H_2S), a gastrotransmitter, has been reported as a probable key agent in the regulation of LES contractility [51]. The design and development of the NPs exploiting this regulatory role of H_2S can be a promising approach for treating GERD. The oral delivery of therapeutics to the esophagus faces several challenges including rapid clearance and short transit time [52]. Hammad et al. proposed an alternative to the oral administration of drugs in GERD treatment. Intranasal administration of the surface-modified nanostructured lipid carriers loaded with mosapride citrate showed great potential in enhancing the bioavailability of the drug for treating GERD (4.54-fold more than oral marketed tablets and 2.44-fold more than drug suspension). A strong relationship between the *in vivo* absorption in rabbits and the *in vitro* permeation of the nanostructures across sheep nasal mucosa was shown by a point-to-point correlation [53]. GERD in advanced stages can lead to esophageal cancer. The combined administration of Cu–cysteamine NPs and disulfiram in xenograft nude mice showed a noticeable inhibitory effect on the growth of esophageal tumors. This strategy led to ROS accumulation, cell apoptosis and blocking of the translocation of NF-κB into the nucleus in esophageal squamous cell carcinoma [54]. Zhuang et al. synthesized Cu NPs via an eco-friendly method using the extract of *Mentha piperita* as a reducing/stabilizing agent to treat esophageal cancer and attributed their anticancer effects to their high antioxidant activity [55]. Exploiting the overexpression of PI3K in esophageal squamous cell carcinoma, a combination of the disulfide cross-linked micelle loaded with docetaxel and the disulfide cross-linked micelle loaded with AZD8186 (PI3K inhibitor) was developed for tumor-specific targeting. Evaluating these formulations *in vitro* and in KYSE 70 xenograft mouse models showed potent synergistic anticancer activity of the nanoformulations with reduced hemato-toxicity [56].

3.3. Gastric Diseases

Consumption of alcohol and nonsteroidal anti-inflammatory drugs, smoking, stress and especially *Helicobacter pylori* (*H. pylori*) infection are among the OS-causing factors in the stomach, which can result in gastric inflammation, ulcers and cancer [57]. *H. pylori*, a microaerophilic and Gram-negative bacterium, possesses antibiotic resistance and has been reported as the major inducer of gastric cancer [58]. The lack of an effective treatment and unsuccessful attempts of the host to eradicate the infection of *H. pylori* lead to bacterial colonization and generation of ROS and RNS by epithelial and immune cells. This condition results in continued OS, chronic inflammation, DNA damage and gastric carcinogenesis [59]. Activated gastric epithelial cells produce ROS to combat the *H. pylori* infection. Reactive species such as $O_2^{\bullet-}$ and H_2O_2 are produced by NOX and spermine oxidase, respectively. DUOX2 and NOX1 play vital roles in gastric inflammation caused by *H. pylori*, which can lead to peptic ulcers and gastric cancer [60]. *H. pylori* causes OS in gastric epithelial cells and the activation of several signaling pathways including NF-κB, AMPK (AMP-activated protein kinase), ERK, NF-κB-mediated NLRP3 inflammasomes,

JAK/STAT3, PI3K/Akt/mTOR and PTEN/MAPK. NF-κB can result in angiogenesis and inflammation. AMPK and PI3K/Akt/mTOR are involved in cell survival and cell proliferation, respectively. The JAK/STAT3 pathway contributes to cell migration. ERK is involved in cell proliferation and gene expression. PTEN/MAPK can lead to inflammation and apoptosis [61].

Various nanomedicines have been developed to counteract this bacterial infection and its complications. Metal-based NPs have been studied as long-lasting antimicrobial agents to resolve the antibiotic resistance of *H. pylori*. Three main mechanisms have been proposed for this antibacterial activity of metal NPs including ROS generation, the release of metal ions and contribution to phototherapy [62]. Green synthesized zinc oxide NPs using *Quercus infectoria* gall extracts (as reducing/capping/stabilizing agents) demonstrated potent and dose-dependent antibacterial activity against *H. pylori* with a greater inhibitory effect than clarithromycin and amoxicillin (standard antibiotics). The combination of these nanoformulations with amoxicillin decreased MIC_{90}, demonstrating the synergistic effect between NPs and antibiotics against *H. pylori* [63]. Nanovesicles with an outer lipid layer of rhamnolipid and loaded with cholesterol–PEG, calcitriol and clarithromycin showed antibacterial efficacy against *H. pylori* in the classical infected mice model. These vesicles could reach the infected region by rapidly penetration through the mucosal layer. They demonstrated their anti-*H. pylori* effect by destroying the bacterial biofilms, exposing the interior bacteria, killing dispersed bacteria, preventing residual bacteria from re-adhesion and blocking the regeneration of biofilms. These nanovesicles activated downstream immune responses by releasing cholesterol–PEG from collapsed vesicles, repairing lipid rafts and reconstructing cytokine receptors. Furthermore, they killed intracellular bacteria by contributing to the regulation of H^+ and Ca^{2+} balance and restoring lysosomal acidification and degradation capabilities [64]. Activity of NOXs and especially infiltration of the neutrophils and leukocytes lead to the overproduction of $O_2^{\bullet-}$ and H_2O_2 in peptic ulcers. Damage in gastric mucosa is often followed by a decrease in H_2S levels. Therefore, one of the therapeutic strategies for gastric ulcers could be the administration of H_2S [65]. Studies have shown the antioxidative and anti-inflammatory effects of H_2S. Chitosan NPs loaded with tetrathiomolybdate and omeprazole released H_2S and molybdenum in a sustained manner and showed a healing effect on gastric ulcers in rat models, which was approved by the reduced gastric juice content, stimulated catalase activity and increased nitrite levels and histopathological evaluations [66]. The other member of the gasotransmitter family, carbon monoxide (CO), has also been proposed to treat gastric ulcers. Elsisi et al. reported the gastroprotective and healing effects of CO on gastric ulcers. CO released from its donor (CORM-2) activated the nuclear factor erythroid 2-related factor 2 (NRF2)/heme oxygenase-1 pathway and reduced OS, lipid peroxidation and COX-2 in indomethacin-induced gastric ulcer models. They highlighted that the PEGylated CORM-2 NPs could outperform CORM-2 in prophylaxis against gastric ulcers, indicating the promoted bioavailability of their payloads [67]. Chitosan–bilirubin conjugate NPs demonstrated a protective effect against OS and inflammatory-induced gastric injuries, an inhibitory effect on the secretion of TNF-α and IL-6 and improved cellular uptake. They accumulated in the stomach and provided long-lasting therapeutic effects and anti-ulcer activity in the ethanol-induced acute gastric ulcer model [68]. Metallic nanosystems such as Ag NPs exhibited a protective effect against gastric ulcers in rats, which could be attributed to the capability of NPs to scavenge free radicals and decrease lipid peroxidation and OS. These NPs were prepared using a combination of *Melissa* extract and Arabic gum and reduced ulcer index, MDA levels and Bax protein expression. An increase in the pH of stomach content, mucus secretion, CAT and SOD activities and HSP70 expression was observed upon using Ag NPs [69]. Studies revealed the effect of upregulated spermine oxidase and NOX1 activity in the oxidative damage of DNA and gastric oncogenesis induced by *H. pylori* infection [60,70]. *In vitro* and *in vivo* studies demonstrated the strong efficiency of glucosamine-decorated 7-ethyl-10-hydroxycamptothecin-poly lactic acid-loaded NPs in targeted gastric cancer therapy. Intravenous injection of these NPs into the MKN45 xenograft mouse model effectively prevented tumor growth. NPs possessed the capability of accumulation in the gastric tumor site through both EPR effect (passive targeting) and specific binding to the overexpressed glucose transporters on the tumor cells (active targeting) with no probable off-target effect and enhanced cytotoxicity against gastric cancer cells [71]. Poly (lactic-co-glycolic acid) (PLGA) NPs loaded with curcumin demonstrated strong anti-gastric cancer and anti-*H. pylori* activity with emphasis on the greater effectiveness of nanocurcumin than the native curcumin [72].

3.4. Intestinal and Rectal Diseases

OS in the lower part of the GI tract can lead to a variety of diseases such as colorectal cancer, inflammatory bowel disease (IBD) and celiac disease [73,74]. In celiac disease, gluten peptides can induce cytotoxic effects on the epithelial cells of the small intestine through OS induction, which causes total villous atrophy (a flat mucosa with no villi) [73,75]. An increase in the ROS level and inducible NOS (iNOS) and COX-2 expression and the activation of NF-κB signaling pathways and inflammatory cascade have been reported in celiac disease [75]. An injection of gliadin-encapsulated PLGA NPs into the celiac disease mouse models effectively caused immune tolerance to gliadin and decreased inflammatory markers and gluten-induced enteropathy [76]. Similar results were reported in another study where gliadin-encapsulated PLGA-antigen NPs inhibited gluten-induced immune activation in celiac patients [77]. These findings revealed that restoring T cell tolerance to gliadin could be used as a promising strategy to treat celiac disease. The administration of an NP-in-microsphere oral system encapsulating IL-15 and tissue transglutaminase-2 silencing siRNA sequences to the polyinosinic/polycytidylic acid-based celiac disease mouse model led to a decrease in the infiltration of neutrophils and expression of pro-inflammatory cytokines. These systems also improved the healing process and restored barrier function in the small intestine [78].

IBD is a complex and multifactorial disease manifesting in relapsing and chronic inflammation at different sites of the GI tract [79]. Despite the unknown exact etiopathology of IBD, the pivotal role of OS in the pathogenesis of IBD has been emphasized by both local and systemic detection of oxidative damages in the intestinal mucosa and peripheral blood leukocytes of the patients, respectively [80,81]. Based on the nature of the histological disorders and the location of the inflammation, IBD can be differentiated into two common types including ulcerative colitis (UC) and Crohn's disease (CD) [82]. In most cases, ulceration and inflammation in UC are limited to the mucosal and submucosal layers of the colon and rectum, while CD can extensively affect the GI tract and lead to inflammation in all wall layers from mouth to anus in a discontinuous fashion. However, the incidence of CD mostly occurs in the perianal site or terminal ileum [83]. The genesis and progression of UC are affected by NOS, ROS, pro-inflammatory cytokines, an increased concentration of MPO and a loss of mucosal antioxidant defense. The iNOS increases the synthesis of NO and stimulates the expression of TNF-α, activates the generation of the intercellular adhesion molecule (ICAM) and P-selectin and causes neutrophil infiltration, leading to cellular damage in the colon [84]. An association has been observed between the elevated activity of Mn-SOD, XO, iNOS and TNF-α in CD patients and decreased antioxidant levels. Memory T cells and major histocompatibility complex class II molecules (MHC) are increased in CD [85]. Despite the similar properties of both types of IBD, $O_2^{\bullet-}$ and $^{\bullet}OH$ plays a prominent role in CD occurrence, while HOCl and H_2O_2 have been reported as major players in the pathophysiology of UC [86]. The overproduction of ROS induced by mitochondrial dysfunction, a complicated balance between redox-sensitive pro-inflammatory pathways (NF-κB and NLRP3 inflammasome) and the adaptive upregulation of GPX2 and Mn-SOD determine the development of IBD [87]. Signaling pathways including extracellular signal-regulated kinase, RTKs, JNK and PKC (protein kinase-C) are activated by ROS, which can result in inflammation in IBD [88]. MAPK pathways have also been reported to be involved in initiating or developing IBD [89]. To overcome the obstacles of efficient IBD treatment, various nano-based strategies have been developed. Polymeric nanocarriers are amongst the most studied colonic delivery systems. Application of the PEGylated polyesterurethane NPs loaded with infliximab, an anti-TNF-α antibody, resulted in a higher cellular uptake and interaction in inflamed epithelial cells compared with polycaprolactone NPs and PLGA NPs. These NPs reduced the level of cytokines in inflamed monocytes, provided a rapid recovery of the epithelial barrier function and showed good potential to treat GI inflammation [90]. Tragacanth–whey liposomes loaded with gingerol possessed a high protective effect on intestinal cells against hydrogen peroxide-induced damage [91]. Biotechnological hyalonutriosomes fabricated by using the bioactive whey prepared by dark fermentation effectively internalized the colonic cells and showed great potential to counteract OS. The application of these nanoformulations proposed a strategy to promote intestinal health through their antioxidative effect on intestinal cells and proliferative effect on *Streptococcus salivarius*, a human commensal bacterium possessing an inhibitory effect on pathogen-induced inflammatory pathways [92]. Intrarectal administration of HA NPs loaded with budesonide, a second-generation glucocorticoid, diminished MPO activity and TNF-α concentration. These NPs targeted the inflamed tissues via interactions with overexpressed CD44 receptors, relieved inflammation, alleviated colitis and improved endoscopic appearance in an acetic acid-induced colitis rat model [93]. Oral delivery of tacrolimus by using cationic lipid-assisted NPs could alleviate lesions in dextran sodium sulfate (DSS)-induced colitis through the ability of NPs to accumulate in the colon, increase retention time and pass through the enterocytes [94]. Silencing and inhibiting the pro-inflammatory pathways is one of the strategies used to treat IBD and inhibit the development of colorectal cancer. Cationic liposomes loaded with NLRP3 siRNA efficiently delivered siNLRP3 into peritoneal macrophages and prevented activation of the NLRP3 inflammasome. They also exhibited an inhibitory effect on the secretion of IL-18 and IL-1β from macrophages and infiltration of the immune cells. Modulating the macrophage polarization, downregulating the $CD4^+$ T cell production, proper anti-inflammatory effects and alleviating intestinal injury in DSS-induced UC mouse models were achieved by using these nanocarriers [95]. Inflammatory mediator S100A9 correlates with the severity of UC and can be used as a target molecule to treat UC. An evaluation of the macrophage membrane-coated PLGA NPs developed for oral delivery of tasquinimod demonstrated drug accumulation in the inflamed tissue, decreased S100A9 and other cytokines, attenuated the symptoms of UC and reduced systemic toxicity in chemically induced UC mouse models [96]. Immunomodulators and corticosteroids have been employed for the treatment of IBD. However, various systemic side effects and immunosuppression limit their long-term use. 5-aminosalicylate (5-ASA), as a nonsteroidal drug, is commonly prescribed to treat mild to moderate IBD, especially UC [97]. However, a small amount of this therapeutic agent can reach the colon because of its rapid absorption by the small intestine [98]. To overcome this obstacle, various nanocarriers have been developed to deliver the 5-ASAs into the colon. For instance, 5-ASA-loaded silicon dioxide NPs with lower drug dosage showed therapeutic effects similar to high dosages of free drug, improved colonic histopathology scores and disease activity index and decreased the severity of mucosal injury in DSS-induced UC mouse models. Furthermore, lower levels of TNF-α and IL-6 in serum and expression of the related mRNA in colonic mucosa were reported with the administration of these NPs [99]. It has been reported that a combination of ZnO NPs with 5-ASA could improve the therapeutic efficacy of 5-ASA in a DSS-induced colitis mouse model. Metal oxide NPs such as SiO_2 and ZnO NPs have been used for treating IBD. ZnO NPs increased the length of the colon, decreased the histological lesion score and disease activity index and attenuated the colitis in the mouse model. Their therapeutic effect was attributed to their antioxidant and anti-inflammatory properties. They elevated the level of GSH, activated the NRF2 pathway and suppressed IL-1β, TNF-α, MPO, MDA and ROS generation. It was reported that released Zn^{2+} ions from the NPs likely possessed an attenuating effect on colonic injuries [100]. Another common OS-induced disease involving the colon and rectum is colorectal cancer. Activation of the NF-κB and MAPK pathways involves the expression of carcinogenesis-related genes, which increases the risk of colorectal cancer development [89]. 5-Fluorouracil is a standard chemotherapy drug for colorectal cancer. However, its treatment efficacy is hindered by its short half-life and rapid metabolism. The development of nanoformulations for loading this therapeutic agent has been proposed as a solution to overcome this limitation. Intragastric administration of bacteria bioinspired NPs composed of zinc gallogermanate mesoporous silica coated with *Lactobacillus reuteri* biofilm was carried out to target the colorectal tumor by 5-fluorouracil. Chemotherapy with these nanosystems diminished the number of tumors per mouse by one-half compared to the use of 5-fluorouracil alone. The targeted delivery of the drug to the colorectal site was supported by the ability of these NPs to withstand the gastric acid. These nanosystems localized to the colorectal tumor, prevented the tumor growth and systemic toxicity and increased the survival time of mice [101]. PEGylated solid lipid NPs loaded with 5-Fluorouracil caused high toxicity in HCT-116 cells. In subcutaneous xenograft mouse models, these NPs improved the pharmacokinetic parameters of the drug, prevented the growth of the tumor and downregulated the expression of human epidermal growth factor receptor 2, a key oncogenic driver in the progression of colorectal cancer [102].

3.5. Stimuli-Responsive Nanomedicines for Treatment of OS-Induced GI Diseases

In recent decades, a variety of nanomedicines have been developed and characterized for treating GI diseases, which can react to stimuli in a special predictable manner. This reaction can be in response to the exogenous (laser irradiation, ultrasound, magnetic field, changes in temperature) or endogenous (ROS, changes in pH values) stimulus or a combination of them. OS-induced diseases in the GI tract are characterized by an excessive level of ROS arising from an imbalance between ROS generation and the antioxidant system. Therefore, various ROS-responsive nanomedicines have been developed to treat GI diseases, especially for IBD. A prodrug as a pharmacologically inactive compound can be metabolized into a pharmacologically active drug by *in vivo* biotransformation through enzymatic or chemical cleavages [103]. NPs self-assembled from aromatized thioketal-linked budesonide and tempol prodrugs could be hydrolyzed via exposure to H_2O_2 in a time- and concentration-dependent manner. ROS-responsive NPs showed a simultaneous and almost complete release of drugs in inflammatory macrophages. These NPs accumulated in the inflamed colon and enhanced the maximum concentration of the drugs, preventing the expression of proinflammatory cytokines and oxidative mediators and relieving colitis in the IBD mouse model [104]. D-α-tocopherol polyethylene glycol succinate-b-poly(β-thioester) copolymer NPs loaded with luteolin alleviated the UC symptoms in the DSS-induced colitis murine model. They suppressed TNF-α, IL-6, IL-17A and interferon-γ and upregulated IL-4, IL-10 and GSH. A decrease in the numbers of T helper 1 (Th1) and Th17 cells and an increase in the numbers of Th2 and regulatory T cells were observed by the administration of these nanosystems, which regulated the inflammatory environment and accelerated intestinal wound healing. These NPs released the drug through size change in response to ROS due to the thioether bond in the polymer main chain [105].

Another endogenous stimulus for the targeted delivery of nanomedicines is changes in the pH value arising from a different condition in the specific tissues or cells or different parts of the GI tract. Administration of the clustered carbon dot NPs to dental biofilm rodent models reduced the viability of *Streptococcus mutans* without adverse effects on oral microbiota balance. Furthermore, these NPs showed an anti-biofilm activity that could be attributed to the interactions between NPs and bacterial membranes, the excessive production of ROS and the fragmentation of DNA [106]. Some pH-responsive nanodelivery systems have been developed to overcome the difficulties of drug delivery to the esophagus arising from short transit time and rapid clearance. Self-assembled chitosan–eggshell membrane NPs loaded with famotidine showed a controlled drug release up to 12 h at pH 1.2, indicating their great potential to treat GERD [107]. Curcumin and docetaxel co-loaded T7 peptide-modified targeted nanosystems exerted the pH-responsive drug release behavior. They promoted biodistribution to the esophageal tumor, increased the concentration of drugs in the tumor tissue and led to the synergistic therapeutic effect in the esophageal cancer cells of xenograft mice [108].

The harsh gastric environment leads to physiological obstacles for effectively treating *H. pylori* and the delivery of some therapeutic agents including antibiotics and proteins. The application of nanomotors, as a novel strategy for temporarily neutralizing the harsh acidic environment and adjusting the local physiological parameters, can inhibit decreasing the drug efficacy and avoid irreversible damage. Nanomotors can convert acid fuel into kinetic energy and propulsive force, and they can be an alternative to commonly used proton pump inhibitors. The application of a nanomotor composed of silica nanobottles loaded with Pt NPs, clarithromycin and nano-calcium peroxide in mouse models led to a reduction in *H. pylori* burden by 2.6 orders of magnitude compared with the negative control. The chemical reaction between CaO_2 and gastric fluid led to the rapid consumption of protons, temporarily neutralizing gastric acid. Pt NPs catalytically decomposed the product of this reaction (hydrogen peroxide) to a huge amount of oxygen. The resulting gas efflux through the narrow opening of the nanobottles caused them to push forward, providing maximum prodrug release and efficacy [109]. The pH-responsive NPs composed of Zn-based zeolitic imidazolate frameworks loaded with hydrogen-absorbed palladium were encapsulated in ascorbate palmitate hydrogel and used for treating *H. pylori* infection. The outer hydrogel could target the inflammatory site, was hydrolyzed by matrix metalloproteinase and released NPs. Hydrogen and Zn^{2+} resulted from the decomposition of these NPs by gastric acid, which effectively killed *H. pylori* by disrupting the cell membrane permeability of bacteria by hydrogen, accelerating the entrance of Zn^{2+} into the cells, promoting the cell leakage, interfering with the metabolism of bacterial cells and preventing urease activity. In this nanosystem, gastric acid invasion of the bacteria was accelerated by Zn^{2+}. Released hydrogen from NPs prevented hyperactive inflammatory response by regulating the secretion of inflammatory factors in macrophages, scavenging excessive oxygen free radicals and alleviating OS-induced damages to epithelial cells. It also restored the impaired gastric mucosa by upregulating the expression of mucosal repair protein. This anti-*H. pylori* treatment strategy had no adverse side effects on gut flora homeostasis [110].

Colon targeting via the oral route necessitates developing nanodelivery systems that can release therapeutic agents in a weak basic environment but not in acidic pHs. NPs coated with Eudragit L100, Eudragit S100, chitosan and other pH-responsive polymers can facilitate the delivery of the drug into the lower sites of the GI tract and increase the concentration of the drug in the colon with minimized side effects, which can be used in the effective treatment of IBD. One of the promising treatment strategies for IBD was loading the anti-miR-301a into an oral delivery system developed by using Eudragit S100, HA, PLGA and chitosan (Figure 4). These components ensured the targeted delivery of the payload without its degradation during its journey to the intestine. The epithelial cells and macrophages are the key factors in the progression of intestinal inflammation. Therefore, the CD44 receptor, which is overexpressed on the inflammatory epithelial cells and macrophages, was used as the target for this drug delivery. Eudragit S100 and HA endowed NPs with pH responsiveness and a high affinity to CD44 receptors. Coating the NPs with Eudragit S100 led to minimizing the release of the drug in the stomach. However, increasing the pH in the colon dissolved Eudragit S100 and led to the release of the payload. As shown in Figure 4, HA on the surface of the NPs promoted their cellular uptake through CD44 receptors at the inflamed site. NPs accumulated in the colon and improved the IBD symptoms in DSS-treated mice by relieving inflammation, restoring colon length, repairing the intestinal barrier and reducing the levels of IL-6, IL-1β, TNF-α and MPO [111].

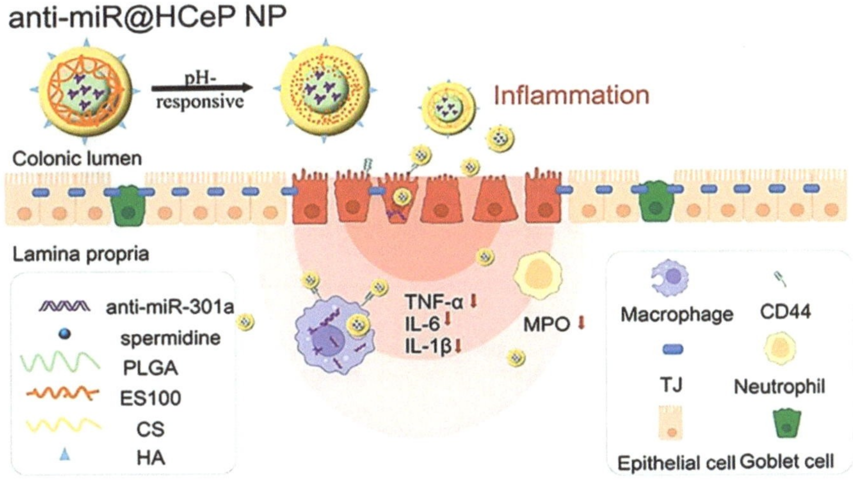

Figure 4. The pH-responsiveness and colon-targeting of anti-miR-301a-loaded nanoparticles with the high affinity of HA to the CD44 receptors at the inflamed site and ES100 dissolvability and releasing the payload in the colon. PLGA: poly (lactic-co-glycolic acid) copolymer, ES100: Eudragit S100, CS: chitosan, HA: hyaluronic acid. TJ: tight junction proteins. Reproduced from [111].

Eudragit–nutriosomes co-loaded with tocopherol and ascorbic acid demonstrated great potential for intestinal-targeted delivery, treating OS-induced diseases and healing intestinal wounds [3]. In another study, hydroxypropyl methylcellulose phthalate was used as a cross-linker in synthesizing chitosan NPs loaded with 5-ASA and berberine to treat UC and inhibit drug release in the upper parts of the GI tract. These NPs exhibited a reduction in the wet weight of the colon and an increase in the length of the colon. They improved the ulcer index and disease activity parameters and led to a proper drug release in the simulated intestinal fluid without release in simulated gastric fluid. They were effective in treating the acetic acid-induced UC in rats [112]. The pH-responsive nanocomposites composed of poly (acrylic acid) brushes anchored on the pore channels of mesoporous silica SBA-15 were designed for the colon-targeted delivery of doxorubicin. In this nanodelivery system, poly (acrylic acid) brushes acted as gatekeepers, which capped the pore outlets and kept the drug encapsulated in acidic pH (gastric environment). In colonic conditions, these brushes were swelled to allow for the release of doxorubicin through the open outlet [113].

The application of responsive nanomedicines to external stimuli such as electromagnetic field, ultrasound and laser irradiation in the treatment of the OS-induced GI disease has brought promising therapeutic results. Magnetic sphingomyelin-containing liposomes encapsulated with cisplatin used a combination of an external magnetic field and an endogenous disease marker (sphingomyelinase enzyme) to release the drug only at the tumor site and at the correct time. They prolonged the survival of the OSCC mouse model [114]. The application of the magnetic microbot was proposed as an effective strategy for treating esophageal cancer, which facilitated targeted locoregional drug delivery in response to the external magnetic field. These magnetic soft microbots were composed of mussel adhesive protein microparticles embedded with iron oxide magnetic NPs and doxorubicin and possessed a potent underwater adhesive capability and long-lasting retention in the hydrodynamic fluid of the esophagus. A sustainable release of doxorubicin and an effective anticancer activity were achieved by these microbots [115].

Sonodynamic therapy (SDT) by combining low-intensity ultrasound irradiation with sonosensitizers to produce ROS can be used as a noninvasive strategy to treat deep target tissues [116]. Verteporfin-preloaded lecithin bilayer-coated PLGA NPs neutralized the secreted virulence factor by *H. pylori*, vacuolating cytotoxin A. Localized ultrasound

(0.5 W/cm^2 for 10 min) exposure of the skin over the stomach of the infected mouse models decreased the *H. pylori* infection by these NPs as strong as the triple therapy (Figure 5).

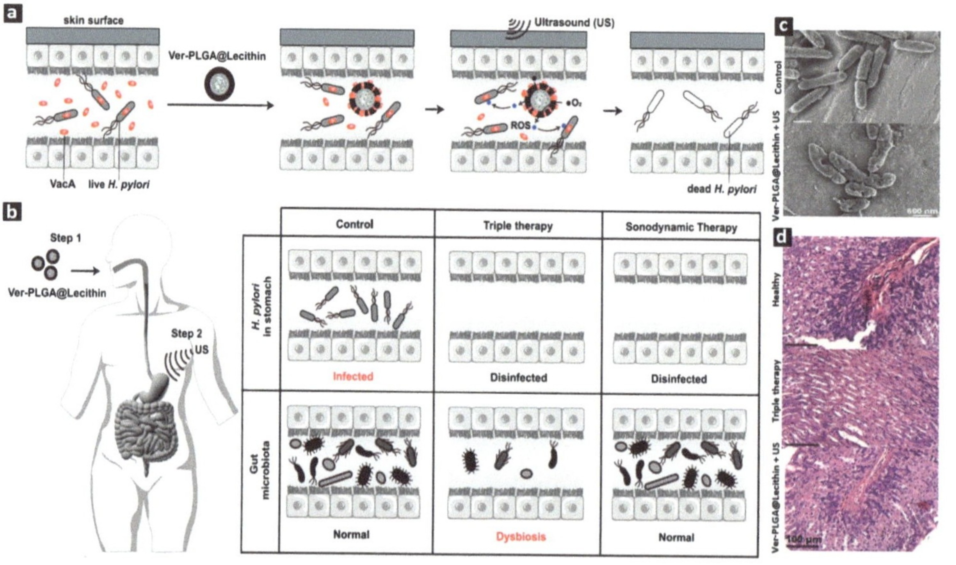

Figure 5. The effect of sonodynamic therapy on gastric *H. pylori* infection: the effect of the Ver-PLGA@Lecithin (verteporfin-preloaded lecithin bilayer-coated PLGA NPs) on VacA (vacuolating cytotoxin A) and *H. pylori* in the absence and presence of US (ultrasound) (**a**); the superior effect of sonodynamic therapy in treating the *H. pylori* infection compared to triple therapy (**b**); *H. pylori* cells after sonodynamic therapy in comparison to the control (**c**); stomach tissues of *H. pylori*-infected mouse models treated with triple therapy or the combination of NPs and US (**d**). Healthy tissues were used as the references. Reproduced from [117].

This effect was attributed to the generation of singlet oxygen ($^1O_2$) (neither $O_2^{\bullet-}$ nor $^{\bullet}OH$) as the ROS. This strategy revealed useful results in non-antibiotic-based therapy for the eradication of *H. pylori*, indicating not only no disruptive effect on the gut microbiota but also *Lactobacillus* upregulation. Stomach tissues in mouse models treated by this technique showed a similar appearance to the healthy ones. The results of the study revealed that cell apoptosis was increased due to the infection by *H. pylori*. The apoptotic condition became worse by triple therapy, while SDT by NPs decreased the relative ratio of TUNEL-positive cells in the *H. pylori*-infected stomach back to normal [117]. The combination effect of SDT and chemotherapy on the orthotopic colorectal cancer mouse model was evaluated by the oral administration of pH/ultrasonic dual-responsive enteric-coated granules produced via enwrapping the nanoprobes by carboxymethyl chitosan. Nanoprobes were fabricated by co-loading the chlorin e6 and doxorubicin hydrochloride into the mesoporous silicon-coated Au NPs and encapsulating the resulting structure in the folic acid-modified phospholipid. Carboxymethyl chitosan ensured the stability of granules in the gastric acidic environment and rapidly disintegrated in the colorectum, leading to the release of nanoprobes from the granules. Highly expressed folate receptors on the surface of the colon cancer cells facilitated tumor targeting by folic acid. Breaking down of the phospholipids could occur under ultrasound leading to the release of sonosensitizers and chemotherapy drug. Treatment guidance was provided by CT imaging using the Au NPs in mesoporous silicon, which showed the highest concentration of enteric-coated granules 7–9 h post-administration in the colorectum of the mouse model. The results revealed a better anticancer effect of combination therapy than the single treatment [118].

Phototherapy including photothermal therapy (PTT) and photodynamic therapy (PDT) is a light-induced therapeutic strategy using photothermal conversion agents and photosensitizers. In PTT, the energy of light is converted to heat and local hypothermia plays a pivotal role in the therapeutic effect [119]. PDT is used as an effective technique with minimal invasiveness for treating inflammatory and cancerous diseases, in which a photosensitizer is applied to a target site to be exposed to the light of a specific wavelength, preferentially in the red spectral region ($\lambda \geq 600$ nm), and conducts photochemical reactions leading to the generation of singlet oxygen and other ROS [120]. Phototherapy has been studied for the treatment of different GI diseases. Various limitations hinder the *in vivo* application of a single phototherapy method, particularly for deep-targeted cells. Dendritic Au Ag NPs coated with procyanidin–Fe networks photothermally eradicated the biofilm of periodontal pathogens under NIR irradiation and reduced OS and inflammation. They activated the phosphoinositide 3-kinase/protein kinase B signaling pathway, enhanced the polarization of M2 macrophages, upregulated NRF2, scavenged ROS and prevented the NF-κB signaling pathway. The combination of these nanosystems with NIR irradiation was proposed as an effective therapeutic strategy against periodontitis, which could repair periodontal damage by modulating tissue regeneration [121]. The application of doxorubicin–Ag NPs under laser irradiation resulted in the chemo-PTT dual effect on esophagus cancer cells, downregulated the expression of antiapoptotic genes and increased the level of caspase-3 mRNA in nanocomposite-treated cancer cells compared to the control cells [122]. Under NIR laser irradiation, pH-sensitive Au nanostars conjugated with *H. pylori*-antibodies nanoprobes could photothermally eradicate *H. pylori* isolated from clinical patients possessing antibiotic resistance and also in mice models. These nanoprobes did not disturb the intestinal microbiome within therapeutic doses and caused the restoration of gastric lesions to normal status one month after PTT [123]. Intraluminal PDT treatment using fiber-covered esophageal stents prolonged the survival of orthotopic esophageal cancer rabbit models. The electrospun fibers were embedded with albumin-Chlorin e6-manganese dioxide NPs, which were gradually released and diffused into the tumor. Chlorin e6 acted as a photosensitizer, and manganese dioxide significantly increased the generation of the $^1O_2$ by reacting with endogenous H^+ and H_2O_2 at the tumor site to produce O_2 and Mn^{2+}, promoting the PDT efficacy [124]. Combination therapy (photothermal-, photodynamic- and chemo-therapy) using photonic nanoporphyrin micelles loaded with 7-ethyl-10-hydroxycamptothecin (SN-38) exhibited higher antitumor efficacy than single treatments in HT-29 colon cancer xenograft mouse models. The enhancement of *in vitro* antitumor activity of this trimodal therapy was 350 and 78 times more than single phototherapy and single chemotherapy with SN-38, respectively [125]. The combination of PDT and ROS-responsiveness has been extensively studied for OS-induced diseases of the GI tract. Platinum nanozyme-loaded prodrug NPs consisting of thioketal bond-linked camptothecin and 2-(1-hexyloxyethyl)-2-devinyl pyropheophorbide-a were developed for chemo-PDT of colon cancer. In these nanosystems, NIR irradiation (660 nm) in the presence of 2-(1-hexyloxyethyl)-2-devinyl pyropheophorbide-a as a photosensitizer led to ROS generation, which not only possessed a PDT effect but also caused the camptothecin release, providing chemotherapeutic effect. The catalyzing activity of platinum nanozymes decomposed hydrogen peroxide and produced oxygen, promoting PDT and the on-demand release of camptothecin [126]. A similar therapeutic strategy was used for the targeted treatment of oral tongue squamous cell carcinoma. The synthesized PEGlated doxorubicin prodrug through thioketal linkage and cyclo-arginine-glycine-aspartic acid-d-phenylalanine-cysteine peptide modification was used to produce NPs for encapsulating hematoporphyrin (a photosensitizer). NPs internalized cancer cells, and under laser irradiation, they induced ROS production for PDT and the ROS-responsive release of doxorubicin. They could target the tumor and diminish its growth in the mouse model [127].

A summarized list of nanomedicines developed and evaluated for the treatment of GI diseases is illustrated in Table 1.

Table 1. Nanomedicines developed for treating oxidative stress-induced GI diseases.

GI Diseases	Nanomedicines	Stimulus	Study Design/Experimental Cells/Animal Model	Results	Ref.
Oral lichen planus	Gliadin/Ethylcellulose-based nanofibrous mats loaded with mycophenolate mofetil, ZnO NPs and aloe vera	-	*In vitro* study: Inflamed human gingival fibroblasts, *S. aureus* and *E. coli* Clinical study: RCT on patients suffering from OLP lesions	Nanofibrous mats showed antimicrobial activity and anti-inflammatory effects with a decrease in TNF-α, IL-6 and ROS of stimulated human gingival fibroblasts. The symptoms of patients were significantly improved with the mats compared to the treatment with commercial ointment.	[128]
	0.1% triamcinolone acetonide with nanoliposomal carriers in Orabase	-	Clinical study: RCT on patients suffering from OLP lesions with treatment duration of 1 month (3 times a day)	Nanoformulation effectively reduced the lesion size and pain intensity with 33.3% fully resolved lesions after 4 weeks.	[129]
Periodontitis	0.02% Ag NPs gel	-	Clinical study: RCT study on patients with chronic periodontitis for 3 months	The effectiveness of silver NPs gel with SRP was the same as sub-gingival delivery of tetracycline gel in chronic periodontitis patients, proposing a non-toxic therapeutic strategy with no increase in bacterial resistance, no side effects and no need for complicated armamentarium with good acceptance by the patients. Differences in microbiological (colony forming units) and clinical parameters (Plaque Index, Gingival Index, Probing Pocket Depth and Clinical Attachment Level were statistically significant from baseline (just before gel placement) to 3 months of gel application.	[130]
	Epigallocatechin gallate-modified Au NPs-loaded hydrogel	NIR (PTT)	*In vitro* study: *S. aureus*, *E. coli*, human umbilical vein endothelial cells and bone marrow mesenchymal stem cells *In vivo* study: Periodontitis rat model	Application of the NIR light spectrum effectively regulated the release of epigallocatechin gallate, increased the antibacterial effect, stimulated angiogenesis and boosted bone and periodontal tissue regeneration. This strategy caused a preventive effect on *E. coli* and *S. aureus* and its biofilm. The combination of NIR light irradiation and these nanomedicines inhibited the dental plaque biofilm and repaired the alveolar bone in the rat periodontitis model.	[131]

Table 1. Nanomedicines developed for treating oxidative stress-induced GI diseases.

GI Diseases	Nanomedicines	Stimulus	Study Design/Experimental Cells/Animal Model	Results	Ref.
	Methylene blue-loaded PLGA NPs	NIR (PDT)	*In vitro* study: Human dental plaque microorganisms (in planktonic and biofilm phase) *In vivo* study: Clinical pilot study on patients with chronic periodontitis	NP-mediated PDT showed a killing effect on the dental plaque bacteria and biofilms. The gingival bleeding index was more improved in the patients treated by NPs and ultrasonic scaling + SRP + PDT than in the cases just treated by ultrasonic scaling + SRP.	[132]
	Glutathione-stabilized Ag NPs	-	*In vitro* study: Human gingival fibroblast cells (HGF-1) and oral pathogens (*Porphyromonas gingivalis*, *Fusobacterium nucleatum*, and *Streptococcus mutans*	NPs showed an inhibitory effect on the growth of periodontal pathogens, increased the generation of cytokines and activated the inflammatory response in the oral epithelial model.	[133]
Oral squamous cell carcinoma	AE105-decorated dendritic mesoporous silica NPs encapsulating Rose Bengal and ultrasmall $Cu_{2x}S$ NPs	NIR(PTT)/US(SDT)/pH	*In vitro* study: OSC-19 cells *In vivo* study: OSCC xenografts mouse model	NPs targeted urokinase plasminogen activator receptors overexpressed in the tumor cell membrane and accumulated in the tumor site, and payloads were released by slowly biodegrading dendritic mesoporous silica NPs in the tumor acidic microenvironment and providing synergistic PTT/SDT nanotherapeutics and eradicated tumor cells/xenografts.	[134]
GERD and peptic ulcer	Montmorillonite famotidine/chitosan bio-nanocomposite hydrogels	pH	*In vitro* study: Sheep gastric mucosal tissue	The optimized bio-nanocomposite hydrogel formulation exerted proper mucoadhesion, long-lasting gastric retention time and sustainable release of the famotidine with potential therapeutic effect on GERD and peptic ulcers.	[135]
Esophageal cancer	DOX/Au-coated nanoturf esophageal stent	NIR (PTT)	*In vitro* study: Human Caucasian esophageal carcinoma cell *In vivo* study: Esophageal cancer mouse model	Nanoturfs provided a long-lasting DOX reservoir, sustained release and reproducible hyperthermia induced by localized surface plasmon resonances under NIR irradiation, promoting on-demand drug release (chemo-PTT). An *in vivo* application of thermo- and chemo-stents induced significant esophageal tumor apoptosis by the synergistic effect of the released drug and hyperthermia in response to NIR irradiation.	[136]

Table 1. Nanomedicines developed for treating oxidative stress-induced GI diseases.

GI Diseases	Nanomedicines	Stimulus	Study Design/Experimental Cells/Animal Model	Results	Ref.
	Trastuzumab-decorated cisplatin and fluoropyrimidine co-encapsulated lipid–polymer hybrid NPs	-	*In vitro* study: Human esophageal cancer cell line *In vivo* study: Esophageal adenocarcinoma-bearing xenograft mouse model	Trastuzumab-decorated NPs showed higher uptake by the esophageal cancer cell lines and more cytotoxicity than the non-decorated NPs. Cisplatin and fluoropyrimidine dual-loaded NPs prevented the growth of tumor cells more effectively than the single drug-loaded NPs *in vivo*. The cisplatin to fluoropyrimidine ratio (*w*/*w*) of 1/1 in the preparation of NPs led to the best synergistic effect. This nanosystem caused a sustained drug release with no systemic toxicity or side effects.	[137]
Gastric *H. pylori* infection	Superparamagnetic iron oxide NPs and amoxicillin co-loaded chitosan–poly (acrylic acid) particles	Magnetic field	*In vitro* study: Human gastric adenocarcinoma cell line, mammalian mouse fibroblast cell line, *H. pylori* strains (127–4, 125–54 and 125–57) *In vivo* study: Pathogen-free BALB/c mice, *H. pylori*-infected mouse model	The combination of these nanomedicines with the magnetic field caused an increase in mucopenetration and the residence time of the drug in the stomach for more effective *H. pylori* eradication therapy. Mucoadhesive properties, antibacterial and anti-biofilm activities and the sustained release of amoxicillin from the NPs were reported in this study.	[138]
Gastric cancer	GX1-modified nanostructured lipid carriers loaded with paclitaxel	-	*In vitro* study: Human umbilical vein endothelial cells (HUVEC cells), human gastric carcinoma cells (MKN45 cells), immortalized fetal gastric mucosal cells (GES-1) *In vivo* study: Tumor-bearing nude mouse model	GX1-modified nanocarriers exhibited higher uptake rates in CoHUVEC cells and a greater inhibitory effect on these cells compared to the unmodified nanocarriers and free paclitaxel. The minimum cytotoxicity effect of GX1-modified nanocarriers was observed in GES-1 cells. A potent antitumor effect with lower side effects in the mouse model was achieved by using GX1-modified nanocarriers.	[139]

Table 1. Nanomedicines developed for treating oxidative stress-induced GI diseases.

GI Diseases	Nanomedicines	Stimulus	Study Design/Experimental Cells/Animal Model	Results	Ref.
	AuNPs modified with the AS1411 aptamer and hairpin DNA loaded with DOX	pH/NIR	*In vitro* study: AGS cells, L929 cells	The nanosystem targeted AGS cells through the interaction of AS1411 with nucleolin on the AGS cytomembrane and more selectively entered the AGS cells than the L929 cells. These NPs showed dual responsiveness for laser irradiation and pH (pH 5), indicating more drug release after the internalization of NPs into the lysosomes or tumor intracellular environment. NIR laser irradiation enhanced the anticancer activity of the NPs on AGS cells.	[140]
IBD	Ginger-derived lipid NPs loaded with siRNA-CD98	-	*In vitro* study: Caco-2BBE cells, RAW 264.7 cells, colon-26 cells *In vivo* study: FVB mice	NPs possessed good biocompatibility and highly efficient cellular uptake, specifically targeting the colon and decreasing the expression of the colonic CD98 gene to treat ulcerative colitis.	[141]
	Glycyrrhizic acid-loaded Eudragit S100/PLGA NPs	pH	*In vivo* study: Dextran sodium sulfate-induced colitis mouse model	These NPs possessed anti-inflammatory and antioxidant activity, diminished the colitis progression, improved the retention time of the drug and its accumulation in the inflamed area and alleviated the symptoms of the disease.	[142]
Colorectal cancer	PEGylated hyaluronic acid–DOX NPs	pH	*In vitro* study: CT26 cells *In vivo* study: CT26 tumor-bearing mouse model	NPs increased the circulation time of DOX by 12.5 times and showed strong antitumor activity in the tumor-bearing mice model. They targeted CD44-positive cancer cells and accumulated in the tumor. The PEG shell of the NPs was dissolved in the acidic environment of the tumor and promoted the cellular endocytosis of NPs. The interaction of released DOX with the nucleus inhibited the growth and proliferation of CT26 cells.	[143]
	DOX-loaded magnetic mesoporous silica core-shell nanocarrier modified with gold gatekeepers, PEG and EpCAM aptamer	pH	*In vitro* study: Human colon cancer cell line (HT-29), Chinese hamster ovary (CHO) cell line *In vivo* study: C57BL/6 mice bearing HT-29 tumors	NPs specifically targeted cancer cells and released the drug at the acidic pH as a result of the functionality of aptamer and gold gatekeepers on their surfaces. EpCAM-positive HT-29 cells showed greater uptake of the NPs than EpCAM-negative CHO cells. These nanomedicines prevented tumor growth with a decrease in off-target toxicity in mouse models.	[144]

RCT: randomized clinical trial, SRP: scaling and root planing, US: ultrasound, NIR: near-infrared, PLGA: poly(lactic-co-glycolic acid), PDT: photodynamic therapy, PTT: photothermal therapy, SDT: sonodynamic therapy, GERD: gastroesophageal reflux disease, DOX: doxorubicin, PEG: polyethylene glycol, EpCAM: epithelial cell adhesion molecule.

4. Safety Issues of Nanomedicines

As mentioned previously, traditional pharmaceutical treatments require frequent administration and more drug dosages due to their non-selectivity, which can lead to severe side effects as well. The low solubility, poor biodistribution and disability to penetrate through different biological barriers adversely affect the efficacy of the traditional drug molecules. Targeted delivery, controlled release, reduced toxicity and the promoted solubility, lipophilicity, penetration proficiency, half-life and bioavailability of drugs achieved by the use of nanomedicines have overcome the obstacles of the traditional treatments and enhanced the efficacy of nanomedicine-based treatments [145]. For instance, in the treatment of tumor tissues where traditional pharmaceutical molecules face penetration difficulties, nanomedicines can provide effective penetration, prolonged circulation time and targeted accumulation in the tumor tissue with reduced off-target toxicity [146]. Despite these potent beneficial effects, our knowledge regarding the effects of long-term administration of nanomedicines, their probable toxicity and negative consequences is restricted. It is necessary to pay more attention to the pharmacodynamics and pharmacokinetics aspects of nanodrugs and nanomaterials used in the construction of nanomedicines. Prediction and inhibition of their potential toxicity may not be completely feasible because of their unknown interaction mechanisms with biological tissues, limited detection techniques and the difficulties of monitoring the nanomaterials' metabolism and deposition in the body [147]. Long-term oral administration of nanomedicines even prepared with safe substances approved by the FDA (Food and Drug Administration) may adversely influence gut microbiome homeostasis [148]. Some studies have reported genotoxicity, immunogenicity and irreversible changes and malfunction of the intracellular organelles due to the administration of nanomedicines [149]. Some of the nanomaterials may damage DNA, break its strands, fragment chromosomes, cause point mutations and increase the risk of cancer. Nanomedicines may induce this genotoxicity directly through their interaction with chromosomes or DNA, inhibiting its transcription or replication. They can also act indirectly via releasing toxic ions or ROS generation, which interfere with involved proteins in the genetic functions or oxidize nitrogenous bases and cause their mispairing and mutation [150]. Nanomedicines may cause immunogenicity, regional inflammation or allergic-like responses. The immunogenicity potential of nanomedicines depends on their type, dose and exposure time. The biocompatibility of nanomaterials cannot eliminate the risk of immunotoxicity [151]. Homeostasis of intracellular organelles such as the mitochondria, endoplasmic reticulum and plasma membrane may be negatively affected by nanomedicines. Especially, negatively charged NPs can induce lysosome membrane permeabilization, cytosolic acidification, cellular component disintegration and apoptosis. Positively charged NPs may damage the intracellular organelles and cause cell death [152]. Extensive *in vivo* studies and the use of artificial intelligence can help evaluate and predict the pharmacotoxicological, pharmacokinetic and toxicological characteristics of nanoparticles [153]. In this regard, a multidisciplinary collaboration of scientists, regulatory bodies and clinicians along with efforts for the cost-effective and safe clinical transformation of nanomedicines are required as well. Despite the current activities of WHO (World Health Organization), FDA, EMA (European Medicines Agency) and ICH (International Council for Harmonization of Technical Requirements for Pharmaceuticals for Human Use), it seems more regulation, standards and supervision are needed to ensure the production of nanomedicines with long-term safety.

5. Concluding Remarks

Despite the critical role of ROS in various physiological processes in the GI tract, excessive generation of ROS that cannot be neutralized by the antioxidant defense system leads to OS and the incidence of different GI diseases. In this regard, the role of pro-oxidants, reactive intermediates and antioxidants in each part of the alimentary canal was elucidated in the current review paper. Mitochondrial dysfunction, lipid peroxidation and DNA and tissue damage are the consequences of OS in the GI tract. This review revealed the great strides made in the treatment of GI diseases by application of nanomedicines including targeted and stimuli-responsive NPs. The results of the studies showed that nanomedicines can treat GI diseases by inducing antioxidant activity, anti-inflammatory effects, antibacterial and anti-biofilm activity and tissue repair, and promoting immune tolerance according to the specific conditions of each disease and the related GI environment. However, there is still a long way to go to find a potent strategy for the comprehensive management of GI diseases caused by OS. Further studies are needed to gain more insight into the oxidative pathway involved in the pathogenesis of each GI disease, the therapeutic mechanisms of nanomedicines and their potential risks. Future perspectives for more effective treatment of GI diseases may include the design and development of bioinspired and biomimetic smart multifunctional nanomedicines including nanomotors and nanobots by exploiting the specific environment of different GI organs.

Funding: This research received no external funding.

Data Availability Statement: No new data were created in this study.

Conflicts of Interest: The author declares no conflict of interest.

References

1. Medrano-Macías, J.; Flores-Gallegos, A.C.; Nava-Reyna, E.; Morales, I.; Tortella, G.; Solís-Gaona, S.; Benavides-Mendoza, A. Reactive Oxygen, Nitrogen, and Sulfur Species (RONSS) as a Metabolic Cluster for Signaling and Biostimulation of Plants: An Overview. *Plants* **2022**, *11*, 3203. [CrossRef] [PubMed]
2. Zhang, B.; Pan, C.; Feng, C.; Yan, C.; Yu, Y.; Chen, Z.; Guo, C.; Wang, X. Role of mitochondrial reactive oxygen species in homeostasis regulation. *Redox Rep.* **2022**, *27*, 45–52. [CrossRef] [PubMed]
3. Rezvani, M.; Manca, M.L.; Caddeo, C.; Escribano-Ferrer, E.; Carbone, C.; Peris, J.E.; Usach, I.; Diez-Sales, O.; Fadda, A.M.; Manconi, M. Co-Loading of Ascorbic Acid and Tocopherol in Eudragit-Nutriosomes to Counteract Intestinal Oxidative Stress. *Pharmaceutics* **2019**, *11*, 13. [CrossRef] [PubMed]
4. Juan, C.A.; de la Lastra, J.M.P.; Plou, F.J.; Pérez-Lebeña, E. The Chemistry of Reactive Oxygen Species (ROS) Revisited: Outlining Their Role in Biological Macromolecules (DNA, Lipids and Proteins) and Induced Pathologies. *Int. J. Mol. Sci.* **2021**, *22*, 4642. [CrossRef] [PubMed]
5. Ali, S.S.; Ahsan, H.; Zia, M.K.; Siddiqui, T.; Khan, F.H. Understanding oxidants and antioxidants: Classical team with new players. *J. Food Biochem.* **2020**, *44*, e13145. [CrossRef] [PubMed]
6. Giles, G.I.; Nasim, M.J.; Ali, W.; Jacob, C. The Reactive Sulfur Species Concept: 15 Years On. *Antioxidants* **2017**, *6*, 38. [CrossRef] [PubMed]
7. Al-Sheikh, Y.A.; Ghneim, H.K.; Alharbi, A.F.; Alshebly, M.M.; Aljaser, F.S.; Aboul-Soud, M.A. Molecular and biochemical investigations of key antioxidant/oxidant molecules in Saudi patients with recurrent miscarriage. *Exp. Ther. Med.* **2019**, *18*, 4450–4460. [CrossRef] [PubMed]
8. Ren, X.; Zou, L.; Zhang, X.; Branco, V.; Wang, J.; Carvalho, C.; Holmgren, A.; Lu, J. Redox Signaling Mediated by Thioredoxin and Glutathione Systems in the Central Nervous System. *Antioxid. Redox Signal.* **2017**, *27*, 989–1010. [CrossRef] [PubMed]
9. Dobreva, A.; Camacho, E.T.; Miranda, M. Mathematical model for glutathione dynamics in the retina. *Sci. Rep.* **2023**, *13*, 10996. [CrossRef] [PubMed]
10. Hasan, A.A.; Kalinina, E.; Tatarskiy, V.; Shtil, A. The Thioredoxin System of Mammalian Cells and Its Modulators. *Biomedicines* **2022**, *10*, 1757. [CrossRef]
11. Lee, D.; Jo, M.G.; Kim, S.Y.; Chung, C.G.; Lee, S.B. Dietary Antioxidants and the Mitochondrial Quality Control: Their Potential Roles in Parkinson's Disease Treatment. *Antioxidants* **2020**, *9*, 1056. [CrossRef] [PubMed]
12. Sarmiento-Salinas, F.L.; Perez-Gonzalez, A.; Acosta-Casique, A.; Ix-Ballote, A.; Diaz, A.; Treviño, S.; Rosas-Murrieta, N.H.; Millán-Perez-Peña, L.; Maycotte, P. Reactive oxygen species: Role in carcinogenesis, cancer cell signaling and tumor progression. *Life Sci.* **2021**, *284*, 119942. [CrossRef] [PubMed]
13. Kirtonia, A.; Sethi, G.; Garg, M. The multifaceted role of reactive oxygen species in tumorigenesis. *Cell. Mol. Life Sci.* **2020**, *77*, 4459–4483. [CrossRef] [PubMed]
14. Monteiro, M.P.; Batterham, R.L. The Importance of the Gastrointestinal Tract in Controlling Food Intake and Regulating Energy Balance. *Gastroenterology* **2017**, *152*, 1707–1717.e2. [CrossRef] [PubMed]
15. Liu, Q.; Zou, J.; Chen, Z.; He, W.; Wu, W. Current research trends of nanomedicines. *Acta Pharm. Sin. B* **2023**, *13*, 4391–4416. [CrossRef] [PubMed]
16. Ashique, S.; Garg, A.; Hussain, A.; Farid, A.; Kumar, P.; Taghizadeh-Hesary, F. Nanodelivery systems: An efficient and target-specific approach for drug-resistant cancers. *Cancer Med.* **2023**, *12*, 18797–18825. [CrossRef] [PubMed]
17. Peres, M.A.; Macpherson, L.M.; Weyant, R.J.; Daly, B.; Venturelli, R.; Mathur, M.R.; Listl, S.; Celeste, R.K.; Guarnizo-Herreño, C.C.; Kearns, C.; et al. Oral diseases: A global public health challenge. *Lancet* **2019**, *394*, 249–260. [CrossRef] [PubMed]

18. Subramanian, A.K.; Narayan, V.; Navaneethan, R. Oxidative Stress and Oral Diseases. In *Book Role of Oxidative Stress in Pathophysiology of Diseases*; Springer: Singapore, 2020; pp. 1–12. [CrossRef]
19. Sardaro, N.; DELLA Vella, F.; Incalza, M.A.; DI Stasio, D.; Lucchese, A.; Contaldo, M.; Laudadio, C.; Petruzzi, M. Oxidative Stress and Oral Mucosal Diseases: An Overview. *In Vivo* **2019**, *33*, 289–296. [CrossRef] [PubMed]
20. Čižmárová, B.; Tomečková, V.; Hubková, B.; Hurajtová, A.; Ohlasová, J.; Birková, A. Salivary Redox Homeostasis in Human Health and Disease. *Int. J. Mol. Sci.* **2022**, *23*, 10076. [CrossRef] [PubMed]
21. Li, X.; Wang, Z. Mechanisms of reactive oxygen species in oral lichen planus: A literature review. *Eur. J. Inflamm.* **2022**, *20*, 1721727X221104389. [CrossRef]
22. Dalirsani, Z.; Shirzaiy, M.; Salehian, M. Salivary antioxidants levels in patients with oral lichen planus. *Indian J. Dermatol.* **2022**, *67*, 651–656. [CrossRef]
23. Tvarijonaviciute, A.; Aznar-Cayuela, C.; Rubio, C.P.; Tecles, F.; Ceron, J.J.; López-Jornet, P. Salivary Antioxidant Status in Patients with Oral Lichen Planus: Correlation with Clinical Signs and Evolution during Treatment with *Chamaemelum nobile*. *BioMed Res. Int.* **2018**, *2018*, 5187549. [CrossRef] [PubMed]
24. Rohani, B.; Sadeghian, R.; Golestannejad, Z.; Sadeghian, S.; Mirzaee, S. Comparison of therapeutic effect of mucoadhesive nano-triamcinolone gel and conventional triamcinolone gel on oral lichen planus. *Dent. Res. J.* **2019**, *16*, 277–282. [CrossRef]
25. Jurczyszyn, K.; Trzeciakowski, W.; Kozakiewicz, M.; Kida, D.; Malec, K.; Karolewicz, B.; Konopka, T.; Zborowski, J. Fractal Dimension and Texture Analysis of Lesion Autofluorescence in the Evaluation of Oral Lichen Planus Treatment Effectiveness. *Materials* **2021**, *14*, 5448. [CrossRef] [PubMed]
26. Da Anunciação, T.A.; Garcez, L.S.; Pereira, E.M.; Queiroz, V.A.d.O.; Costa, P.R.d.F.; de Oliveira, L.P.M. Curcumin supplementation in the treatment of patients with cancer: A systematic review. *Braz. J. Pharm. Sci.* **2021**, *57*, e18008. [CrossRef]
27. Kia, S.J.; Basirat, M.; Mortezaie, T.; Moosavi, M.-S. Comparison of oral Nano-Curcumin with oral prednisolone on oral lichen planus: A randomized double-blinded clinical trial. *BMC Complement. Med. Ther.* **2020**, *20*, 328. [CrossRef]
28. Kia, S.J.; Basirat, M.; Estakhr, L. The effect of oral curcumin on pain and clinical appearance of oral lichen planus. *J. Dentomaxillofac.* **2017**, *6*, 1–7.
29. Bakhshi, M.; Gholami, S.; Mahboubi, A.; Jaafari, M.R.; Namdari, M. Combination Therapy with 1% Nanocurcumin Gel and 0.1% Triamcinolone Acetonide Mouth Rinse for Oral Lichen Planus: A Randomized Double-Blind Placebo Controlled Clinical Trial. *Dermatol. Res. Pract.* **2020**, *2020*, 4298193. [CrossRef] [PubMed]
30. Popovska, M.; Fidovski, J.; Mindova, S.; Dirjanska, K.; Ristoska, S.; Stefanovska, E.; Radojkova-Nikolovska, V.; Mitic, K.; Rusevska, B. The Effects of NBF Gingival Gel Application in the Treatment of the Erosive Lichen Planus: Case Report. *Open Access Maced. J. Med. Sci.* **2016**, *4*, 158–163. [CrossRef] [PubMed]
31. Shang, J.; Liu, H.; Zheng, Y.; Zhang, Z. Role of oxidative stress in the relationship between periodontitis and systemic diseases. *Front. Physiol.* **2023**, *14*, 1210449. [CrossRef] [PubMed]
32. Kumar, J.; Teoh, S.L.; Das, S.; Mahakknaukrauh, P. Oxidative Stress in Oral Diseases: Understanding Its Relation with Other Systemic Diseases. *Front. Physiol.* **2017**, *8*, 693. [CrossRef] [PubMed]
33. Mohideen, K.; Chandrasekar, K.; Ramsridhar, S.; Rajkumar, C.; Ghosh, S.; Dhungel, S. Assessment of Oxidative Stress by the Estimation of Lipid Peroxidation Marker Malondialdehyde (MDA) in Patients with Chronic Periodontitis: A Systematic Review and Meta-Analysis. *Int. J. Dent.* **2023**, *2023*, 6014706. [CrossRef]
34. Liu, C.; Mo, L.; Niu, Y.; Li, X.; Zhou, X.; Xu, X. The Role of Reactive Oxygen Species and Autophagy in Periodontitis and Their Potential Linkage. *Front. Physiol.* **2017**, *8*, 439. [CrossRef] [PubMed]
35. Nasiri, K.; Masoumi, S.M.; Amini, S.; Goudarzi, M.; Tafreshi, S.M.; Bagheri, A.; Yasamineh, S.; Alwan, M.; Arellano, M.T.C.; Gholizadeh, O. Recent advances in metal nanoparticles to treat periodontitis. *J. Nanobiotechnol.* **2023**, *21*, 283. [CrossRef] [PubMed]
36. Steckiewicz, K.P.; Cieciórski, P.; Barcińska, E.; Jaśkiewicz, M.; Narajczyk, M.; Bauer, M.; Kamysz, W.; Megiel, E.; Inkielewicz-Stepniak, I. Silver Nanoparticles as Chlorhexidine and Metronidazole Drug Delivery Platforms: Their Potential Use in Treating Periodontitis. *Int. J. Nanomed.* **2022**, *17*, 495–517. [CrossRef] [PubMed]
37. Castangia, I.; Manconi, M.; Allaw, M.; Perra, M.; Orrù, G.; Fais, S.; Scano, A.; Escribano-Ferrer, E.; Ghavam, M.; Rezvani, M.; et al. Mouthwash Formulation Co-Delivering Quercetin and Mint Oil in Liposomes Improved with Glycol and Ethanol and Tailored for Protecting and Tackling Oral Cavity. *Antioxidants* **2022**, *11*, 367. [CrossRef] [PubMed]
38. Johnson, A.; Kong, F.; Miao, S.; Thomas, S.; Ansar, S.; Kong, Z.-L. In-Vitro Antibacterial and Anti-Inflammatory Effects of Surfactin-Loaded Nanoparticles for Periodontitis Treatment. *Nanomaterials* **2021**, *11*, 356. [CrossRef] [PubMed]
39. Zambrano, L.M.G.; Brandao, D.A.; Rocha, F.R.G.; Marsiglio, R.P.; Longo, I.B.; Primo, F.L.; Tedesco, A.C.; Guimaraes-Stabili, M.R.; Junior, C.R. Local administration of curcumin-loaded nanoparticles effectively inhibits inflammation and bone resorption associated with experimental periodontal disease. *Sci. Rep.* **2018**, *8*, 6652. [CrossRef] [PubMed]
40. Iqbal, M.J.; Kabeer, A.; Abbas, Z.; Siddiqui, H.A.; Calina, D.; Sharifi-Rad, J.; Cho, W.C. Interplay of oxidative stress, cellular communication and signaling pathways in cancer. *Cell Commun. Signal.* **2024**, *22*, 7. [CrossRef]
41. Ionescu, C.; Kamal, F.Z.; Ciobica, A.; Halitchi, G.; Burlui, V.; Petroaie, A.D. Oxidative Stress in the Pathogenesis of Oral Cancer. *Biomedicines* **2024**, *12*, 1150. [CrossRef]
42. Shi, X.-L.; Li, Y.; Zhao, L.-M.; Su, L.-W.; Ding, G. Delivery of MTH1 inhibitor (TH287) and MDR1 siRNA via hyaluronic acid-based mesoporous silica nanoparticles for oral cancers treatment. *Colloids Surf. B Biointerfaces* **2019**, *173*, 599–606. [CrossRef] [PubMed]

43. Gamal-Eldeen, A.M.; Baghdadi, H.M.; Afifi, N.S.; Ismail, E.M.; Alsanie, W.F.; Althobaiti, F.; Raafat, B.M. Gum arabic-encapsulated gold nanoparticles modulate hypoxamiRs expression in tongue squamous cell carcinoma. *Mol. Cell. Toxicol.* **2021**, *17*, 111–121. [CrossRef]
44. Chapelle, N.; Ben Ghezala, I.; Barkun, A.; Bardou, M. The pharmacotherapeutic management of gastroesophageal reflux disease (GERD). *Expert Opin. Pharmacother.* **2021**, *22*, 219–227. [CrossRef] [PubMed]
45. Franceschelli, S.; Gatta, D.M.P.; Pesce, M.; Ferrone, A.; Di Martino, G.; Di Nicola, M.; De Lutiis, M.A.; Vitacolonna, E.; Patruno, A.; Grilli, A.; et al. Modulation of the oxidative plasmatic state in gastroesophageal reflux disease with the addition of rich water molecular hydrogen: A new biological vision. *J. Cell. Mol. Med.* **2018**, *22*, 2750–2759. [CrossRef] [PubMed]
46. Bilski, J.; Pinkas, M.; Wojcik-Grzybek, D.; Magierowski, M.; Korbut, E.; Mazur-Bialy, A.; Krzysiek-Maczka, G.; Kwiecien, S.; Magierowska, K.; Brzozowski, T. Role of Obesity, Physical Exercise, Adipose Tissue-Skeletal Muscle Crosstalk and Molecular Advances in Barrett's Esophagus and Esophageal Adenocarcinoma. *Int. J. Mol. Sci.* **2022**, *23*, 3942. [CrossRef] [PubMed]
47. Hu, Y.; Ye, X.; Wang, R.; Poon, K. Current research progress in the role of reactive oxygen species in esophageal adenocarcinoma. *Transl. Cancer Res.* **2021**, *10*, 1568–1577. [CrossRef] [PubMed]
48. Maslenkina, K.; Mikhaleva, L.; Naumenko, M.; Vandysheva, R.; Gushchin, M.; Atiakshin, D.; Buchwalow, I.; Tiemann, M. Signaling Pathways in the Pathogenesis of Barrett's Esophagus and Esophageal Adenocarcinoma. *Int. J. Mol. Sci.* **2023**, *24*, 9304. [CrossRef] [PubMed]
49. Imro'Ati, T.A.; Sugihartono, T.; Widodo, B.; Nefertiti, E.P.; Rovian, I.; Wibawa, I.G.N. The Relationship between Serum Total Oxidant Status, Total Antioxidant Status, and Oxidative Stress Index with Severity Levels of Gastroesophageal Reflux Disease: A Literature Review. *Open Access Maced. J. Med. Sci.* **2021**, *9*, 584–589. [CrossRef]
50. Herdiana, Y. Chitosan Nanoparticles for Gastroesophageal Reflux Disease Treatment. *Polymers* **2023**, *15*, 3485. [CrossRef] [PubMed]
51. Bai, X.; Ihara, E.; Hirano, K.; Tanaka, Y.; Nakano, K.; Kita, S.; Iwamoto, T.; Ogino, H.; Hirano, M.; Oda, Y.; et al. Endogenous Hydrogen Sulfide Contributes to Tone Generation in Porcine Lower Esophageal Sphincter Via Na^+/Ca^{2+} Exchanger. *Cell. Mol. Gastroenterol. Hepatol.* **2018**, *5*, 209–221. [CrossRef] [PubMed]
52. Krause, J.; Brokmann, F.; Rosenbaum, C.; Weitschies, W. The challenges of drug delivery to the esophagus and how to overcome them. *Expert Opin. Drug Deliv.* **2022**, *19*, 119–131. [CrossRef] [PubMed]
53. Hammad, R.W.; Sanad, R.A.B.; Abdelmalk, N.S.; Aziz, R.L.; Torad, F.A. Intranasal Surface-Modified Mosapride Citrate-Loaded Nanostructured Lipid Carriers (MOS-SMNLCs) for Treatment of Reflux Diseases: *In vitro* Optimization, Pharmacodynamics, and Pharmacokinetic Studies. *AAPS PharmSciTech* **2018**, *19*, 3791–3808. [CrossRef] [PubMed]
54. Chang, Y.; Wu, F.; Pandey, N.K.; Chudal, L.; Xing, M.; Zhang, X.; Nguyen, L.; Liu, X.; Liu, J.P.; Chen, W.; et al. Combination of Disulfiram and Copper–Cysteamine Nanoparticles for an Enhanced Antitumor Effect on Esophageal Cancer. *ACS Appl. Bio Mater.* **2020**, *3*, 7147–7157. [CrossRef]
55. Zhuang, X.; Kang, Y.; Zhao, L.; Guo, S. Design and synthesis of copper nanoparticles for the treatment of human esophageal cancer: Introducing a novel chemotherapeutic supplement. *J. Exp. Nanosci.* **2022**, *17*, 274–284. [CrossRef]
56. Wang, X.-S.; Ding, X.-Z.; Li, X.-C.; He, Y.; Kong, D.-J.; Zhang, L.; Hu, X.-C.; Yang, J.-Q.; Zhao, M.-Q.; Gao, S.-G.; et al. A highly integrated precision nanomedicine strategy to target esophageal squamous cell cancer molecularly and physically. *Nanomed. Nanotechnol. Biol. Med.* **2018**, *14*, 2103–2114. [CrossRef] [PubMed]
57. Lee, J.; Kim, M.-H.; Kim, H. Anti-Oxidant and Anti-Inflammatory Effects of Astaxanthin on Gastrointestinal Diseases. *Int. J. Mol. Sci.* **2022**, *23*, 15471. [CrossRef] [PubMed]
58. Reyes, V.E. *Helicobacter pylori* and Its Role in Gastric Cancer. *Microorganisms* **2023**, *11*, 1312. [CrossRef] [PubMed]
59. Salvatori, S.; Marafini, I.; Laudisi, F.; Monteleone, G.; Stolfi, C. *Helicobacter pylori* and Gastric Cancer: Pathogenetic Mechanisms. *Int. J. Mol. Sci.* **2023**, *24*, 2895. [CrossRef] [PubMed]
60. Gobert, A.P.; Wilson, K.T. Polyamine- and NADPH-dependent generation of ROS during Helicobacter pylori infection: A blessing in disguise. *Free Radic. Biol. Med.* **2017**, *105*, 16–27. [CrossRef] [PubMed]
61. Sah, D.K.; Arjunan, A.; Lee, B.; Jung, Y.D. Reactive Oxygen Species and *H. pylori* Infection: A Comprehensive Review of Their Roles in Gastric Cancer Development. *Antioxidants* **2023**, *12*, 1712. [CrossRef] [PubMed]
62. Franco, D.; Calabrese, G.; Guglielmino, S.P.P.; Conoci, S. Metal-Based Nanoparticles: Antibacterial Mechanisms and Biomedical Application. *Microorganisms* **2022**, *10*, 1778. [CrossRef] [PubMed]
63. Attia, H.G.; Albarqi, H.A.; Said, I.G.; Alqahtani, O.; Raey, M.A.E. Synergistic Effect between Amoxicillin and Zinc Oxide Nanoparticles Reduced by Oak Gall Extract against *Helicobacter pylori*. *Molecules* **2022**, *27*, 4559. [CrossRef] [PubMed]
64. Chen, X.; Zou, Y.; Zhang, S.; Fang, P.; Li, S.; Li, P.; Sun, Y.; Yuan, G.; Hu, H. Multi-functional vesicles improve Helicobacter pylori eradication by a comprehensive strategy based on complex pathological microenvironment. *Acta Pharm. Sin. B* **2022**, *12*, 3498–3512. [CrossRef] [PubMed]
65. Shen, M.-F.; Wang, Z.; Shen, F.; Zhao, C.-S.; Chen, G. The role of hydrogen sulfide in gastric mucosal damage. *Med. Gas Res.* **2019**, *9*, 88–92. [CrossRef] [PubMed]
66. Manivannan, S.; Sivaraman, H.; Murugesan, R.; Narayan, S. Omeprazole and H_2S releasing agents encapsulated in chitosan nanoparticles to enhance healing process against indomethacin-induced gastric ulcer model. *J. Mater. Res.* **2023**, *38*, 3089–3109. [CrossRef]

67. Elsisi, A.E.; Mekky, E.F.; Abu-Risha, S.E. Potential effects of carbon monoxide donor and its nanoparticles on experimentally induced gastric ulcer in rats. *Inflammopharmacology* **2023**, *31*, 1495–1510. [CrossRef] [PubMed]
68. Huang, Z.; Shi, Y.; Wang, H.; Chun, C.; Chen, L.; Wang, K.; Lu, Z.; Zhao, Y.; Li, X. Protective Effects of Chitosan-Bilirubin Nanoparticles Against Ethanol-Induced Gastric Ulcers. *Int. J. Nanomed.* **2021**, *16*, 8235–8250. [CrossRef] [PubMed]
69. Ibrahim, I.A.A.; Hussein, A.I.; Muter, M.S.; Mohammed, A.T.; Al-Medhtiy, M.H.; Shareef, S.H.; Aziz, P.Y.; Agha, N.F.S.; Abdulla, M.A. Effect of nano silver on gastroprotective activity against ethanol-induced stomach ulcer in rats. *Biomed. Pharmacother.* **2022**, *154*, 113550. [CrossRef]
70. Butcher, L.D.; Hartog, G.D.; Ernst, P.B.; Crowe, S.E. Oxidative Stress Resulting from Helicobacter pylori Infection Contributes to Gastric Carcinogenesis. *Cell. Mol. Gastroenterol. Hepatol.* **2017**, *3*, 316–322. [CrossRef] [PubMed]
71. Ding, N.; Xu, S.; Zheng, S.; Ye, Q.; Xu, L.; Ling, S.; Xie, S.; Chen, W.; Zhang, Z.; Xue, M.; et al. "Sweet tooth"-oriented SN38 prodrug delivery nanoplatform for targeted gastric cancer therapy. *J. Mater. Chem. B* **2021**, *9*, 2816–2830. [CrossRef] [PubMed]
72. Alam, J.; Dilnawaz, F.; Sahoo, S.K.; Singh, D.V.; Mukhopadhyay, A.K.; Hussain, T.; Pati, S. Curcumin Encapsulated into Biocompatible Co-Polymer PLGA Nanoparticle Enhanced Anti-Gastric Cancer and Anti-Helicobacter Pylori Effect. *Asian Pac. J. Cancer Prev.* **2022**, *23*, 61–70. [CrossRef] [PubMed]
73. Veres-Székely, A.; Bernáth, M.; Pap, D.; Rokonay, R.; Szebeni, B.; Takács, I.M.; Lippai, R.; Cseh, Á.; Szabó, A.J.; Vannay, Á. PARK7 diminishes oxidative stress-induced mucosal damage in celiac disease. *Oxid. Med. Cell. Longev.* **2020**, *2020*, 4787202. [CrossRef]
74. Wang, Y.; Chen, Y.; Zhang, X.; Lu, Y.; Chen, H. New insights in intestinal oxidative stress damage and the health intervention effects of nutrients: A review. *J. Funct. Foods* **2020**, *75*, 104248. [CrossRef]
75. Yu, T.; Xie, Y.; Yuan, J.; Gao, J.; Xiao, Z.; Wu, Y.; Chen, H. The Nutritional Intervention of Resveratrol Can Effectively Alleviate the Intestinal Inflammation Associated with Celiac Disease Induced by Wheat Gluten. *Front. Immunol.* **2022**, *13*, 878186. [CrossRef] [PubMed]
76. Freitag, T.L.; Podojil, J.R.; Pearson, R.M.; Fokta, F.J.; Sahl, C.; Messing, M.; Andersson, L.C.; Leskinen, K.; Saavalainen, P.; Hoover, L.I.; et al. Gliadin Nanoparticles Induce Immune Tolerance to Gliadin in Mouse Models of Celiac Disease. *Gastroenterology* **2020**, *158*, 1667–1681.e12. [CrossRef] [PubMed]
77. Kelly, C.P.; Murray, J.A.; Leffler, D.A.; Getts, D.R.; Bledsoe, A.C.; Smithson, G.; First, M.R.; Morris, A.; Boyne, M.; Elhofy, A.; et al. TAK-101 Nanoparticles Induce Gluten-Specific Tolerance in Celiac Disease: A Randomized, Double-Blind, Placebo-Controlled Study. *Gastroenterology* **2021**, *161*, 66–80.e8. [CrossRef] [PubMed]
78. Attarwala, H.Z.; Suri, K.; Amiji, M.M. Co-Silencing of Tissue Transglutaminase-2 and Interleukin-15 Genes in a Celiac Disease Mimetic Mouse Model Using a Nanoparticle-in-Microsphere Oral System. *Mol. Pharm.* **2021**, *18*, 3099–3107. [CrossRef]
79. Elhag, D.A.; Kumar, M.; Saadaoui, M.; Akobeng, A.K.; Al-Mudahka, F.; Elawad, M.; Al Khodor, S. Inflammatory Bowel Disease Treatments and Predictive Biomarkers of Therapeutic Response. *Int. J. Mol. Sci.* **2022**, *23*, 6966. [CrossRef] [PubMed]
80. Alemany-Cosme, E.; Sáez-González, E.; Moret, I.; Mateos, B.; Iborra, M.; Nos, P.; Sandoval, J.; Beltrán, B. Oxidative Stress in the Pathogenesis of Crohn's Disease and the Interconnection with Immunological Response, Microbiota, External Environmental Factors, and Epigenetics. *Antioxidants* **2021**, *10*, 64. [CrossRef] [PubMed]
81. Tian, T.; Wang, Z.; Zhang, J. Pathomechanisms of Oxidative Stress in Inflammatory Bowel Disease and Potential Antioxidant Therapies. *Oxidative Med. Cell. Longev.* **2017**, *2017*, 4535194. [CrossRef] [PubMed]
82. Sahoo, D.K.; Heilmann, R.M.; Paital, B.; Patel, A.; Yadav, V.K.; Wong, D.; Jergens, A.E. Oxidative stress, hormones, and effects of natural antioxidants on intestinal inflammation in inflammatory bowel disease. *Front. Endocrinol.* **2023**, *14*, 1217165. [CrossRef] [PubMed]
83. Kellermann, L.; Riis, L.B. A close view on histopathological changes in inflammatory bowel disease, a narrative review. *Dig. Med. Res.* **2021**, *4*, 3. [CrossRef]
84. Ashique, S.; Mishra, N.; Garg, A.; Sibuh, B.Z.; Taneja, P.; Rai, G.; Djearamane, S.; Wong, L.S.; Al-Dayan, N.; Roychoudhury, S.; et al. Recent updates on correlation between reactive oxygen species and synbiotics for effective management of ulcerative colitis. *Front. Nutr.* **2023**, *10*, 1126579. [CrossRef] [PubMed]
85. Guan, G.; Lan, S. Implications of Antioxidant Systems in Inflammatory Bowel Disease. *BioMed Res. Int.* **2018**, *2018*, 1290179. [CrossRef] [PubMed]
86. Miranda, M.R.; Vestuto, V.; Moltedo, O.; Manfra, M.; Campiglia, P.; Pepe, G. The ion channels involved in oxidative stress-related gastrointestinal diseases. *Oxygen* **2023**, *3*, 336–365. [CrossRef]
87. Pérez, S.; Taléns-Visconti, R.; Rius-Pérez, S.; Finamor, I.; Sastre, J. Redox signaling in the gastrointestinal tract. *Free. Radic. Biol. Med.* **2017**, *104*, 75–103. [CrossRef] [PubMed]
88. Rauf, A.; Khalil, A.A.; Awadallah, S.; Khan, S.A.; Abu-Izneid, T.; Kamran, M.; Hemeg, H.A.; Mubarak, M.S.; Khalid, A.; Wilairatana, P. Reactive oxygen species in biological systems: Pathways, associated diseases, and potential inhibitors—A review. *Food Sci. Nutr.* **2024**, *12*, 675–693. [CrossRef] [PubMed]
89. Lin, S.; Li, Y.; Zamyatnin, A.A., Jr.; Werner, J.; Bazhin, A.V. Reactive oxygen species and colorectal cancer. *J. Cell. Physiol.* **2018**, *233*, 5119–5132. [CrossRef] [PubMed]
90. Pabari, R.M.; Mattu, C.; Partheeban, S.; Almarhoon, A.; Boffito, M.; Ciardelli, G.; Ramtoola, Z. Novel polyurethane-based nanoparticles of infliximab to reduce inflammation in an in-vitro intestinal epithelial barrier model. *Int. J. Pharm.* **2019**, *565*, 533–542. [CrossRef] [PubMed]

91. Rezvani, M.; Manca, M.L.; Muntoni, A.; De Gioannis, G.; Pedraz, J.L.; Gutierrez, G.; Matos, M.; Fadda, A.M.; Manconi, M. From process effluents to intestinal health promotion: Developing biopolymer-whey liposomes loaded with gingerol to heal intestinal wounds and neutralize oxidative stress. *Int. J. Pharm.* **2022**, *613*, 121389. [CrossRef] [PubMed]
92. Manconi, M.; Rezvani, M.; Manca, M.L.; Escribano-Ferrer, E.; Fais, S.; Orrù, G.; Lammers, T.; Asunis, F.; Muntoni, A.; Spiga, D.; et al. Bridging biotechnology and nanomedicine to produce biogreen whey-nanovesicles for intestinal health promotion. *Int. J. Pharm.* **2023**, *633*, 122631. [CrossRef]
93. Vafaei, S.Y.; Abdolghaffari, A.H.; Mahjub, R.; Eslami, S.M.; Esmaeili, M.; Abdollahi, M.; Atyabi, F.; Dinarvand, R. Budesonide-Loaded Hyaluronic Acid Nanoparticles for Targeted Delivery to the Inflamed Intestinal Mucosa in a Rodent Model of Colitis. *BioMed Res. Int.* **2022**, *2022*, 7776092. [CrossRef] [PubMed]
94. Wang, J.-L.; Gan, Y.-J.; Iqbal, S.; Jiang, W.; Yuan, Y.-Y.; Wang, J. Delivery of tacrolimus with cationic lipid-assisted nanoparticles for ulcerative colitis therapy. *Biomater. Sci.* **2018**, *6*, 1916–1922. [CrossRef] [PubMed]
95. Huang, J.; Dai, M.; He, M.; Bu, W.; Cao, L.; Jing, J.; Cao, R.; Zhang, H.; Men, K. Treatment of Ulcerative Colitis by Cationic Liposome Delivered NLRP3 siRNA. *Int. J. Nanomed.* **2023**, *18*, 4647–4662. [CrossRef]
96. Li, Z.; Zhang, X.; Liu, C.; Peng, Q.; Wu, Y.; Wen, Y.; Zheng, R.; Yan, Q.; Ma, J. Macrophage-Biomimetic Nanoparticles Ameliorate Ulcerative Colitis through Reducing Inflammatory Factors Expression. *J. Innate Immun.* **2022**, *14*, 380–392. [CrossRef] [PubMed]
97. Zhang, S.; Fu, J.; Dogan, B.; Scherl, E.J.; Simpson, K.W. 5-Aminosalicylic acid downregulates the growth and virulence of Escherichia coli associated with IBD and colorectal cancer, and upregulates host anti-inflammatory activity. *J. Antibiot.* **2018**, *71*, 950–961. [CrossRef] [PubMed]
98. Yuri, T.; Kono, Y.; Fujita, T. Transport characteristics of 5-aminosalicylic acid into colonic epithelium: Involvement of sodium-coupled monocarboxylate transporter SMCT1-mediated transport system. *Biochem. Biophys. Res. Commun.* **2020**, *524*, 561–566. [CrossRef] [PubMed]
99. Tang, H.; Xiang, D.; Wang, F.; Mao, J.; Tan, X.; Wang, Y. 5-ASA-loaded SiO_2 nanoparticles-a novel drug delivery system targeting therapy on ulcerative colitis in mice. *Mol. Med. Rep.* **2017**, *15*, 1117–1122. [CrossRef] [PubMed]
100. Li, J.; Chen, H.; Wang, B.; Cai, C.; Yang, X.; Chai, Z.; Feng, W. ZnO nanoparticles act as supportive therapy in DSS-induced ulcerative colitis in mice by maintaining gut homeostasis and activating Nrf2 signaling. *Sci. Rep.* **2017**, *7*, 43126. [CrossRef] [PubMed]
101. Wang, Z.-H.; Liu, J.-M.; Li, C.-Y.; Wang, D.; Lv, H.; Lv, S.-W.; Zhao, N.; Ma, H.; Wang, S. Bacterial Biofilm Bioinspired Persistent Luminescence Nanoparticles with Gut-Oriented Drug Delivery for Colorectal Cancer Imaging and Chemotherapy. *ACS Appl. Mater. Interfaces* **2019**, *11*, 36409–36419. [CrossRef] [PubMed]
102. Smith, T.; Affram, K.; Nottingham, E.L.; Han, B.; Amissah, F.; Krishnan, S.; Trevino, J.; Agyare, E. Application of smart solid lipid nanoparticles to enhance the efficacy of 5-fluorouracil in the treatment of colorectal cancer. *Sci. Rep.* **2020**, *10*, 16989. [CrossRef] [PubMed]
103. Cho, S.; Yoon, Y.-R. Understanding the pharmacokinetics of prodrug and metabolite. *Transl. Clin. Pharmacol.* **2018**, *26*, 1–5. [CrossRef] [PubMed]
104. Li, S.; Xie, A.; Li, H.; Zou, X.; Zhang, Q. A self-assembled, ROS-responsive Janus-prodrug for targeted therapy of inflammatory bowel disease. *J. Control. Release* **2019**, *316*, 66–78. [CrossRef]
105. Tan, C.; Fan, H.; Ding, J.; Han, C.; Guan, Y.; Zhu, F.; Wu, H.; Liu, Y.; Zhang, W.; Hou, X.; et al. ROS-responsive nanoparticles for oral delivery of luteolin and targeted therapy of ulcerative colitis by regulating pathological microenvironment. *Mater. Today Bio* **2022**, *14*, 100246. [CrossRef]
106. Ostadhossein, F.; Moitra, P.; Altun, E.; Dutta, D.; Sar, D.; Tripathi, I.; Hsiao, S.-H.; Kravchuk, V.; Nie, S.; Pan, D. Function-adaptive clustered nanoparticles reverse *Streptococcus mutans* dental biofilm and maintain microbiota balance. *Commun. Biol.* **2021**, *4*, 846. [CrossRef] [PubMed]
107. Enwereuzo, O.O.; Akakuru, O.C.; Uwaoma, R.C.; Elemike, E.E.; Akakuru, O.U. Self-assembled membrane-polymer nanoparticles of top-notch tissue tolerance for the treatment of gastroesophageal reflux disease. *J. Nanostruct. Chem.* **2021**, *11*, 707–719. [CrossRef]
108. Deng, L.; Zhu, X.; Yu, Z.; Li, Y.; Qin, L.; Liu, Z.; Feng, L.; Guo, R.; Zheng, Y. Novel T7-Modified pH-Responsive Targeted Nanosystem for Co-Delivery of Docetaxel and Curcumin in the Treatment of Esophageal Cancer. *Int. J. Nanomed.* **2020**, *15*, 7745–7762. [CrossRef] [PubMed]
109. Wu, Y.; Song, Z.; Deng, G.; Jiang, K.; Wang, H.; Zhang, X.; Han, H. Gastric Acid Powered Nanomotors Release Antibiotics for *In Vivo* Treatment of *Helicobacter pylori* Infection. *Small* **2021**, *17*, 2006877. [CrossRef]
110. Zhang, W.; Zhou, Y.; Fan, Y.; Cao, R.; Xu, Y.; Weng, Z.; Ye, J.; He, C.; Zhu, Y.; Wang, X. Metal–Organic-Framework-Based Hydrogen-Release Platform for Multieffective *Helicobacter Pylori* Targeting Therapy and Intestinal Flora Protective Capabilities. *Adv. Mater.* **2022**, *34*, 2105738. [CrossRef] [PubMed]
111. Wang, J.; Yao, M.; Zou, J.; Ding, W.; Sun, M.; Zhuge, Y.; Gao, F. pH-Sensitive Nanoparticles for Colonic Delivery Anti-miR-301a in Mouse Models of Inflammatory Bowel Diseases. *Nanomaterials* **2023**, *13*, 2797. [CrossRef] [PubMed]
112. Mahami, S.; Salehi, M.; Mehrabi, M.; Vahedi, H.; Hassani, M.S.; Bitaraf, F.S.; Omri, A. pH-sensitive HPMCP-chitosan nanoparticles containing 5-aminosalicylic acid and berberine for oral colon delivery in a rat model of ulcerative colitis. *Int. J. Biol. Macromol.* **2023**, *244*, 125332. [CrossRef] [PubMed]

113. Tian, B.; Liu, S.; Wu, S.; Lu, W.; Wang, D.; Jin, L.; Hu, B.; Li, K.; Wang, Z.; Quan, Z.A. pH-responsive poly (acrylic acid)-gated mesoporous silica and its application in oral colon targeted drug delivery for doxorubicin. *Colloids Surf. B Biointerfaces.* **2017**, *154*, 287–296. [CrossRef] [PubMed]
114. Medina, T.P.; Gerle, M.; Humbert, J.; Chu, H.; Köpnick, A.-L.; Barkmann, R.; Garamus, V.M.; Sanz, B.; Purcz, N.; Will, O.; et al. Lipid-Iron Nanoparticle with a Cell Stress Release Mechanism Combined with a Local Alternating Magnetic Field Enables Site-Activated Drug Release. *Cancers* **2020**, *12*, 3767. [CrossRef] [PubMed]
115. Choi, H.S.; Jo, Y.K.; Ahn, G.; Joo, K.I.; Kim, D.; Cha, H.J. Magnetically Guidable Proteinaceous Adhesive Microbots for Targeted Locoregional Therapeutics Delivery in the Highly Dynamic Environment of the Esophagus. *Adv. Funct. Mater.* **2021**, *31*, 2104602. [CrossRef]
116. Jiang, Z.; Xiao, W.; Fu, Q. Stimuli responsive nanosonosensitizers for sonodynamic therapy. *J. Control. Release* **2023**, *361*, 547–567. [CrossRef] [PubMed]
117. Liu, T.; Chai, S.; Li, M.; Chen, X.; Xie, Y.; Zhao, Z.; Xie, J.; Yu, Y.; Gao, F.; Zhu, F.; et al. A nanoparticle-based sonodynamic therapy reduces Helicobacter pylori infection in mouse without disrupting gut microbiota. *Nat. Commun.* **2024**, *15*, 844. [CrossRef] [PubMed]
118. Zhang, R.-Y.; Cheng, K.; Xuan, Y.; Yang, X.-Q.; An, J.; Hu, Y.-G.; Liu, B.; Zhao, Y.-D. A pH/ultrasonic dual-response step-targeting enterosoluble granule for combined sonodynamic-chemotherapy guided via gastrointestinal tract imaging in orthotopic colorectal cancer. *Nanoscale* **2021**, *13*, 4278–4294. [CrossRef] [PubMed]
119. Ren, Y.; Yan, Y.; Qi, H. Photothermal conversion and transfer in photothermal therapy: From macroscale to nanoscale. *Adv. Colloid Interface Sci.* **2022**, *308*, 102753. [CrossRef]
120. Correia, J.H.; Rodrigues, J.A.; Pimenta, S.; Dong, T.; Yang, Z. Photodynamic Therapy Review: Principles, Photosensitizers, Applications, and Future Directions. *Pharmaceutics* **2021**, *13*, 1332. [CrossRef]
121. Wang, H.; Wang, D.; Huangfu, H.; Lv, H.; Qin, Q.; Ren, S.; Zhang, Y.; Wang, L.; Zhou, Y. Branched AuAg nanoparticles coated by metal–phenolic networks for treating bacteria-induced periodontitis via photothermal antibacterial and immunotherapy. *Mater. Des.* **2022**, *224*, 111401. [CrossRef]
122. Moawad, M.; Youssef, A.M.; Elsherbeni, S.A.E.; Fahmy, A.M.; El-Ghannam, G. Silver Nanoparticles Enhanced Doxorubicin treatment for Improving their Efficacy against Esophageal Cancer Cells. *Egypt. J. Chem.* **2024**, *67*, 505–512. [CrossRef]
123. Zhi, X.; Liu, Y.; Lin, L.; Yang, M.; Zhang, L.; Zhang, L.; Liu, Y.; Alfranca, G.; Ma, L.; Zhang, Q.; et al. Oral pH sensitive GNS@ab nanoprobes for targeted therapy of Helicobacter pylori without disturbance gut microbiome. *Nanomed. Nanotechnol. Biol. Med.* **2019**, *20*, 102019. [CrossRef] [PubMed]
124. Xiao, J.; Cheng, L.; Fang, T.; Zhang, Y.; Zhou, J.; Cheng, R.; Tang, W.; Zhong, X.; Lu, Y.; Deng, L.; et al. Nanoparticle-Embedded Electrospun Fiber–Covered Stent to Assist Intraluminal Photodynamic Treatment of Oesophageal Cancer. *Small* **2019**, *15*, e1904979. [CrossRef] [PubMed]
125. Yang, X.; Xue, X.; Luo, Y.; Lin, T.-Y.; Zhang, H.; Lac, D.; Xiao, K.; He, Y.; Jia, B.; Lam, K.S.; et al. Sub-100 nm, long tumor retention SN-38-loaded photonic micelles for tri-modal cancer therapy. *J. Control. Release* **2017**, *261*, 297–306. [CrossRef] [PubMed]
126. Hao, Y.; Chen, Y.; He, X.; Yu, Y.; Han, R.; Li, Y.; Yang, C.; Hu, D.; Qian, Z. Polymeric Nanoparticles with ROS-Responsive Prodrug and Platinum Nanozyme for Enhanced Chemophotodynamic Therapy of Colon Cancer. *Adv. Sci.* **2020**, *7*, 2001853. [CrossRef] [PubMed]
127. Shi, S.; Zhang, L.; Zhu, M.; Wan, G.; Li, C.; Zhang, J.; Wang, Y.; Wang, Y. Reactive Oxygen Species-Responsive Nanoparticles Based on PEGlated Prodrug for Targeted Treatment of Oral Tongue Squamous Cell Carcinoma by Combining Photodynamic Therapy and Chemotherapy. *ACS Appl. Mater. Interfaces* **2018**, *10*, 29260–29272. [CrossRef] [PubMed]
128. Alipour, M.; Habibivand, E.; Sekhavati, S.; Aghazadeh, Z.; Ranjkesh, M.; Ramezani, S.; Aghazadeh, M.; Ghorbani, M. Evaluation of therapeutic effects of nanofibrous mat containing mycophenolate mofetil on oral lichen planus: *In vitro* and clinical trial study. *Biomater. Investig. Dent.* **2023**, *10*, 2283177. [CrossRef] [PubMed]
129. Azizi, A.; Dadras, O.G.; Jafari, M.; Ghadim, N.M.; Lawaf, S.; Sadri, D. Efficacy of 0.1% triamcinolone with nanoliposomal carrier formulation in orabase for oral lichen planus patients: A clinical trial. *Eur. J. Integr. Med.* **2016**, *8*, 275–280. [CrossRef]
130. Kadam, P.; Mahale, S.; Sonar, P.; Chaudhari, D.; Shimpi, S.; Kathurwar, A. Efficacy of silver nanoparticles in chronic periodontitis patients: A clinico-microbiological study. *Iberoam. J. Med.* **2020**, *2*, 142–147. [CrossRef]
131. Dong, Z.; Lin, Y.; Xu, S.; Chang, L.; Zhao, X.; Mei, X.; Gao, X. NIR-triggered tea polyphenol-modified gold nanoparticles-loaded hydrogel treats periodontitis by inhibiting bacteria and inducing bone regeneration. *Mater. Des.* **2023**, *225*, 111487. [CrossRef]
132. De Freitas, L.M.; Calixto, G.M.F.; Chorilli, M.; Giusti, J.S.M.; Bagnato, V.S.; Soukos, N.S.; Amiji, M.M.; Fontana, C.R. Polymeric Nanoparticle-Based Photodynamic Therapy for Chronic Periodontitis *In Vivo*. *Int. J. Mol. Sci.* **2016**, *17*, 769. [CrossRef] [PubMed]
133. Zorraquín-Peña, I.; Cueva, C.; de Llano, D.G.; Bartolomé, B.; Moreno-Arribas, M.V. Glutathione-Stabilized Silver Nanoparticles: Antibacterial Activity against Periodontal Bacteria, and Cytotoxicity and Inflammatory Response in Oral Cells. *Biomedicines* **2020**, *8*, 375. [CrossRef] [PubMed]
134. Zuo, J.; Huo, M.; Wang, L.; Li, J.; Chen, Y.; Xiong, P. Photonic hyperthermal and sonodynamic nanotherapy targeting oral squamous cell carcinoma. *J. Mater. Chem. B* **2020**, *8*, 9084–9093. [CrossRef] [PubMed]
135. Farhadnejad, H.; Mortazavi, S.A.; Jamshidfar, S.; Rakhshani, A.; Motasadizadeh, H.; Fatahi, Y.; Mahdieh, A.; Darbasizadeh, B. Montmorillonite-Famotidine/Chitosan Bio-nanocomposite Hydrogels as a Mucoadhesive/Gastroretentive Drug Delivery System. *Iran. J. Pharm. Res.* **2022**, *21*, e127035. [CrossRef] [PubMed]

136. Lee, S.; Hwang, G.; Kim, T.H.; Kwon, S.J.; Kim, J.U.; Koh, K.; Park, B.; Hong, H.; Yu, K.J.; Chae, H.; et al. On-Demand Drug Release from Gold Nanoturf for a Thermo- and Chemotherapeutic Esophageal Stent. *ACS Nano* **2018**, *12*, 6756–6766. [CrossRef] [PubMed]
137. Fu, Q.; Wang, J.; Liu, H. Chemo-immune synergetic therapy of esophageal carcinoma: Trastuzumab modified, cisplatin and fluorouracil co-delivered lipid–polymer hybrid nanoparticles. *Drug Deliv.* **2020**, *27*, 1535–1543. [CrossRef] [PubMed]
138. Yang, S.-J.; Huang, C.-H.; Yang, J.-C.; Wang, C.-H.; Shieh, M.-J. Residence Time-Extended Nanoparticles by Magnetic Field Improve the Eradication Efficiency of *Helicobacter pylori*. *ACS Appl. Mater. Interfaces* **2020**, *12*, 54316–54327. [CrossRef] [PubMed]
139. Jian, Y.; Zhao, M.; Cao, J.; Fan, T.; Bu, W.; Yang, Y.; Li, W.; Zhang, W.; Qiao, Y.; Wang, J.; et al. A Gastric Cancer Peptide GXI-Modified Nano-Lipid Carriers Encapsulating Paclitaxel: Design and Evaluation of Anti-Tumor Activity. *Drug Des. Dev. Ther.* **2020**, *14*, 2355–2370. [CrossRef] [PubMed]
140. Zhang, Y.; Tan, J.; Zhou, L.; Shan, X.; Liu, J.; Ma, Y. Synthesis and Application of AS1411-Functionalized Gold Nanoparticles for Targeted Therapy of Gastric Cancer. *ACS Omega* **2020**, *5*, 31227–31233. [CrossRef] [PubMed]
141. Zhang, M.; Wang, X.; Han, M.K.; Collins, J.F.; Merlin, D. Oral administration of ginger-derived nanolipids loaded with siRNA as a novel approach for efficient siRNA drug delivery to treat ulcerative colitis. *Nanomedicine* **2017**, *12*, 1927–1943. [CrossRef]
142. Zeeshan, M.; Ali, H.; Khan, S.; Mukhtar, M.; Khan, M.I.; Arshad, M. Glycyrrhizic acid-loaded pH-sensitive Poly-(Lactic-Co-Glycolic Acid) Nanoparticles for the Amelioration of Inflammatory Bowel Disease. *Nanomedicine* **2019**, *14*, 1945–1969. [CrossRef] [PubMed]
143. Zhang, X.; Zhao, M.; Cao, N.; Qin, W.; Zhao, M.; Wu, J.; Lin, D. Construction of a tumor microenvironment pH-responsive cleavable PEGylated hyaluronic acid nano-drug delivery system for colorectal cancer treatment. *Biomater. Sci.* **2020**, *8*, 1885–1896. [CrossRef] [PubMed]
144. Iranpour, S.; Bahrami, A.R.; Nekooei, S.; Saljooghi, A.S.; Matin, M.M. Improving anti-cancer drug delivery performance of magnetic mesoporous silica nanocarriers for more efficient colorectal cancer therapy. *J. Nanobiotechnol.* **2021**, *19*, 314. [CrossRef] [PubMed]
145. Li, Y.; Lu, H.; Liang, S.; Xu, S. Dual Stable Nanomedicines Prepared by Cisplatin-Crosslinked Camptothecin Prodrug Micelles for Effective Drug Delivery. *ACS Appl. Mater. Interfaces* **2019**, *11*, 20649–20659. [CrossRef] [PubMed]
146. Krasteva, N.; Georgieva, M. Promising Therapeutic Strategies for Colorectal Cancer Treatment Based on Nanomaterials. *Pharmaceutics* **2022**, *14*, 1213. [CrossRef] [PubMed]
147. Lama, S.; Merlin-Zhang, O.; Yang, C. *In Vitro* and *In Vivo* Models for Evaluating the Oral Toxicity of Nanomedicines. *Nanomaterials* **2020**, *10*, 2177. [CrossRef] [PubMed]
148. Ghebretatios, M.; Schaly, S.; Prakash, S. Nanoparticles in the Food Industry and Their Impact on Human Gut Microbiome and Diseases. *Int. J. Mol. Sci.* **2021**, *22*, 1942. [CrossRef] [PubMed]
149. Hashem, M.M.; Abo-El-Sooud, K.; Abd-Elhakim, Y.M.; Badr, Y.A.-H.; El-Metwally, A.E.; Bahy-El-Dien, A. The long-term oral exposure to titanium dioxide impaired immune functions and triggered cytotoxic and genotoxic impacts in rats. *J. Trace Elements Med. Biol.* **2020**, *60*, 126473. [CrossRef]
150. Yao, Y.; Zhang, T.; Tang, M. The DNA damage potential of quantum dots: Toxicity, mechanism and challenge. *Environ. Pollut.* **2023**, *317*, 120676. [CrossRef] [PubMed]
151. Lee, Y.; Jeong, M.; Park, J.; Jung, H.; Lee, H. Immunogenicity of lipid nanoparticles and its impact on the efficacy of mRNA vaccines and therapeutics. *Exp. Mol. Med.* **2023**, *55*, 2085–2096. [CrossRef] [PubMed]
152. Liu, N.; Tang, M. Toxic effects and involved molecular pathways of nanoparticles on cells and subcellular organelles. *J. Appl. Toxicol.* **2020**, *40*, 16–36. [CrossRef] [PubMed]
153. Nasnodkar, S.; Cinar, B.; Ness, S. Artificial Intelligence in Toxicology and Pharmacology. *J. Eng. Res. Rep.* **2023**, *25*, 192–206. [CrossRef]

BioChem

MDPI

Review

Biotechnological Advances in Vanillin Production: From Natural Vanilla to Metabolic Engineering Platforms

Arnold William Tazon, Fatima Awwad , Fatma Meddeb-Mouelhi and Isabel Desgagné-Penix *

Department of Chemistry, Biochemistry and Physics, Université du Québec à Trois-Rivières, Trois-Rivières, QC G8Z 4M3, Canada; arnold.william.tazon@uqtr.ca (A.W.T.); fatima.awwad@usherbrooke.ca (F.A.); fatma.meddeb@uqtr.ca (F.M.-M.)
* Correspondence: isabel.desgagne-penix@uqtr.ca

Abstract: Vanillin, an aromatic aldehyde, is one of the most popular flavors worldwide, extensively used in the food, cosmetics, pharmaceutical, and agrochemical industries. Despite its widespread use, less than 1% of the total vanillin production is natural, with the majority being synthesized chemically. While chemical synthesis can help to meet the growing demand for vanillin, a strong market trend has rapidly developed for products created from natural ingredients, including natural vanillin. Given the labor-intensive process of extracting vanillin from vanilla pods, there is a critical need for new metabolic engineering platforms to support the biotechnological production of nature-identical vanillin. This review highlights the significance of vanillin in various markets, its diverse applications, and the current state of bio-engineered production using both prokaryotic and eukaryotic biological systems. Although recent advancements have demonstrated successful vanillin production through biocatalytic approaches, our focus was to provide a current and innovative overview of vanillin bioengineering across various host systems with special consideration placed on microalgae, which are emerging as promising platforms for vanillin production through metabolic engineering. The use of these systems to support the biotechnological production of vanillin, while leveraging the photosynthetic capabilities of microalgae to capture CO_2 and convert it into biomass, can significantly reduce the overall carbon footprint.

Keywords: microalgae; vanilla; synthetic biology; biotechnological production; sustainability; photosynthetic microorganisms

1. Introduction

In recent decades, the demand for flavor and fragrance compounds has surged, driven by the consumer preference for enhanced sensory experiences in food and other products [1]. Vanillin, in particular, has emerged as one of the most widely used flavoring agents worldwide, valued for its distinctive aroma and versatility [2,3]. Consequently, the aroma and fragrance industry has invested heavily in innovative production methods for vanillin to meet this growing demand sustainably and efficiently. Vanillin is the major organoleptic compound of vanilla aroma and is an aromatic aldehyde crystallized in the form of a white powder with a sweet taste and intense aroma [4]. The production of vanillin is possible through natural plant, biotechnological, and chemical processes [5]. However, only natural plant-extracted vanilla and vanillin produced using biotechnological platforms have been approved by the European Food Law (European Directive 88/388/EEC, OJ No. L184, 22 June 1988) and United States of America Food Law [6] as a "natural" food-grade spice [7]. Lately, a strong market trend has developed rapidly for products designed from natural ingredients [8], thus creating a strong demand for nature-identical vanillin, mainly for applications in beverages and foods with therapeutic properties. Currently, the global vanillin market is expected to grow from USD 12.73 billion in 2022 to USD 19.11 billion by 2030, with a Compound Annual Growth Rate (CAGR) of 5.2% over the forecast period between 2022 and 2030 according to the market research company SNS Insider Strategy and Stats [9] (Figure 1).

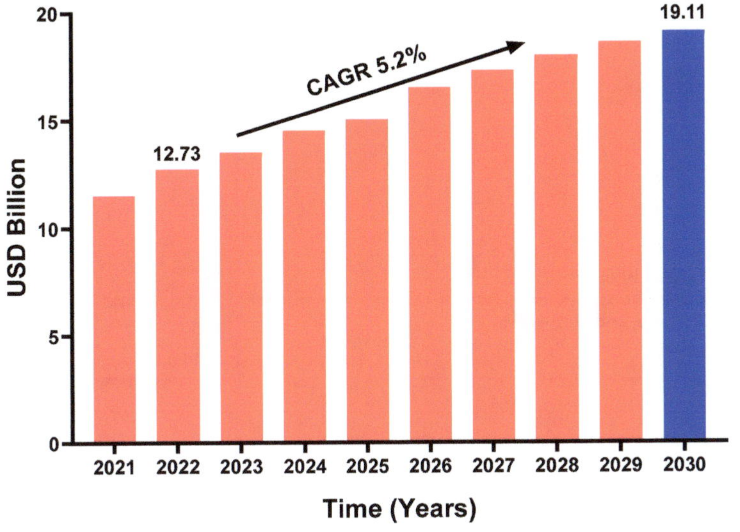

Figure 1. Vanillin market size reported by SNS Insider in 2022 [9].

Natural vanilla and vanillin, although often used interchangeably in everyday contexts, differ significantly in their source, composition, and production methods. Natural vanilla is derived from the beans of the vanilla orchid, involving a labor-intensive process of cultivation, hand-pollination, harvesting, and curing, resulting in a complex mixture of several hundred compounds that contribute to its rich and nuanced flavor profile. In contrast, vanillin is a single chemical compound, 4-hydroxy-3-methoxybenzaldehyde, which can be produced synthetically from precursors like guaiacol, or through biotechnological methods involving microbial fermentation. This simpler composition and more straightforward production process make vanillin more affordable and widely available, albeit with a less complex flavor compared to natural vanilla. Vanillin was isolated by Theodore Nicolas Gobley in 1858 [10]. Today, it is widely coveted by industries mainly for its applications in the areas of agri-food/beverages, perfumery/cosmetology, and medicine. Approximately 60% of industrial vanillin is used in the food/beverage industry, 33% in the perfumery/cosmetics industry, and 7% in the pharmaceutical industry [11].

The production of natural vanilla from vanilla pods is a costly and laborious process; to produce a single kilogram of vanilla, it requires approximately 500 kg of vanilla pods and a manual pollination of about 40,000 flowers [12]. This generates a high cost of natural vanilla on the market, between USD 1200 and 1400 per kg [13], compared to synthetic vanillin that costs less than USD 15 per kg [6,14]. For this reason, the biotechnological or bioengineering production of vanillin is of substantial interest for higher yields and more sustainable alternatives for spice production. As vanillin is derived from aromatic amino acid metabolism, screening *Vanilla planifolia* plants and other natural producers of vanillin such as *Glechoma hederacea*, *Oryza sativa*, and *Bruguiera gymnorhiza* [4,15,16] for the genes encoding the enzymes that are responsible for vanillin biosynthesis enables the heterologous production of vanillin in other organisms and its optimization into a sustainable high-yield system (Figure 2).

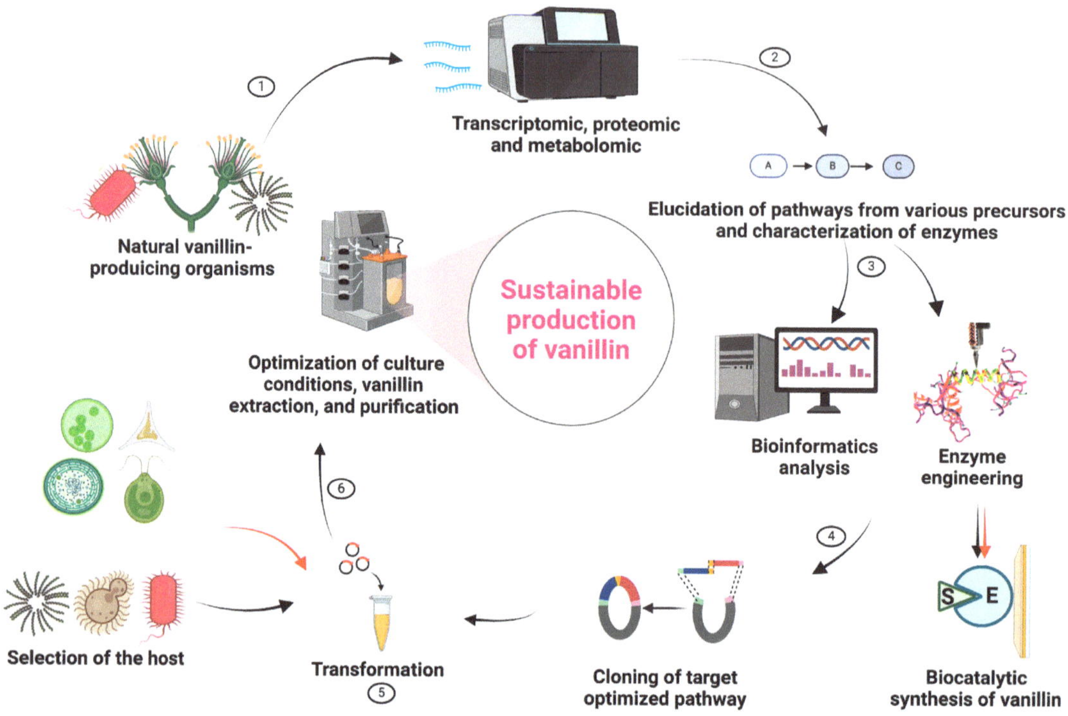

Figure 2. Current and future approaches for biologically engineered hosts for the production of vanillin. The black arrows indicate current approaches, while red arrows represent future approaches.

The biotechnological production of vanillin is carried out by the bioconversion of various natural substrates, such as ferulic acid, eugenol, isoeugenol, lignin, and glucose, using metabolically efficient biofactories, such as fungi, yeasts, or bacteria [17]. However, these current production systems often face several challenges, including inefficient metabolism in some cases, as well as cytotoxicity of the reaction products and/or intermediates in the culture media. Additionally, the formation of by-products due to the overexpression of endogenous vanillin dehydrogenase enzymes significantly impacts yield and quality, necessitating further purification steps. This is often viewed as a challenge for manufacturers and a bottleneck for environmental sustainability [7]. However, with the recent advances in genetic engineering, CRISPR-Cas9 technology, and purification techniques, the proportion of vanillin produced from these expression systems is increasingly expanding [3,18–20]. In order to support the biotechnological production of high-quality, sustainable, food-grade vanillin, photosynthetic organisms such as microalgae represent an ideal biomanufacturing system. Indeed, photosynthetic microorganisms can grow in the absence of a source of organic carbon. In addition, they can sequester carbon, allowing them to limit the industrial carbon dioxide (CO_2) emissions while reducing their impact in terms of greenhouse gases [21]. Several molecules of interest, including recombinant proteins such as pleiotropic antigens, and plant-specialized metabolites such as zeaxanthin, terpenoids, and growth hormones, have already been produced by this type of expression system [22–24]. Moreover, with the rapid advancement of metabolic engineering tools and techniques, there has been a surge in the algal production of high-value compounds. This review summarizes the importance of vanillin across various markets (food, perfumery/cosmetics, and pharmaceuticals/agrochemicals) and examines the current biological engineering production systems, including those based on bacteria, yeast, and plants. Special consideration is afforded to microalgae, which are emerging as promising platforms for supporting the biotechnological production of vanillin through metabolic engineering. The manuscript also addresses the challenges associated with implementing microalgal-based vanillin production systems. This work aims to promote the use of these systems to enhance the biotechnological production of vanillin, leveraging the photosynthetic capabilities of microalgae to capture CO_2 and convert it into biomass, thereby reducing the overall carbon footprint.

2. Industrial Applications of Vanillin

2.1. Food and Beverage Industry Applications

Flavor is a sensation caused jointly by the taste and smell of a food. This is an organoleptic criterion that is regularly monitored in the agri-food industry in order to meet consumer demands [25]. The iconic and unmistakable flavor of vanillin, often described as sweet, is widely used in cold or baked confectionery, chocolate, pastries, and in the malting industry for its application to therapeutic drinks [11]. More interestingly, vanillin is often used in athletic beverage preparation to mask the bitter taste usually provided by some amino acids, such as arginine [25]. In addition to the human food applications, vanillin is applied as an additive to improve the feeding of chickens while protecting them against gastrointestinal infections [26]. In addition, the benefits of vanillin could be directly applied to the science of climate change. Ruminants (goats and dairy cows) have microorganisms (protozoa, fungi, and bacteria) in their rumen that are capable of fermenting complex carbohydrates to produce simple sugars and some atmospheric gases such as methane [27] which is known to be a potent greenhouse gas that is responsible for climate change [28]. Thus, the addition of vanillin as an ingredient in ruminant feed could improve its taste while inhibiting the growth of microorganisms, thereby limiting the release of methane into the atmosphere.

2.2. Applications in Perfume/Cosmetics

The tasty and pleasant scent of vanillin has applications in perfumery and cosmetology. Vanilla is used as a fragrance ingredient in perfumery formulations and preparations; it masks unpleasant odors and provides a very pleasant aroma [14]. Vanillin is used as a perfume fixative to preserve the appreciated fragrance in time and space. Due to its biological structure, vanillin stimulates the expression of the transcriptional factors (pOct-4, Oct-4, and Nanog) involved in the maintenance of pluripotent cells, the regeneration of epithelial stem cells in humans, and the expression of epithelial adhesive protein (E-cadherin), which helps to maintain the freshness of the body and increases its resistance against skin infections [29–31]. These regenerative and antimicrobial properties make vanillin a potential ingredient used in cosmetology and various applications in addition to its pleasant organoleptic properties.

2.3. Bioactivities/Pharmaceutical and Agrochemical Applications

Vanillin is used in the pharmaceutical industry as an excipient for several therapeutic drugs, such as antidepressants and antimicrobials [32,33]. This is the case for the antihypertensive drug methyl-dopamine (mixture consisting of 0.01 M of dopamine, vanillin, and N-methylacetamide, respectively) [34] used to treat hypertension, L-dopamine used to treat Parkinson's disease, and a few anti-infectious agents such as trimethoprim [35]. In addition, numerous studies have demonstrated the pharmacological properties of vanillin, such as anticancer activity by mediating the antimutagenic potential and DNA damage [36], neuroprotective activity by attenuating biochemical impairments, learning and memory, and locomotor and motor coordination [37,38], antifungal activity through mitochondrial dysfunction [39], and, finally, antiviral and anti-inflammatory activities by neuraminidase inhibition and the modulation of the intracellular signaling pathways in THP-1 cells [40,41]. Moreover, it has recently been demonstrated that vanillin induces osteoblast-mediated bone-forming phenotypes through cell migration and F-actin polymerization [42]. Although vanillin is widely used in the food, pharmaceutical, and cosmetics industries, its application in agro-chemistry for the production of green pesticides has recently attracted significant attention from the scientific community. Indeed, the chemical structure of vanillin, which contains a hydroxyl group and a highly modifiable aldehyde function, enables the addition of a series of bioactive motifs via an ether bridge, reductive amination, and thioacetalization to generate vanillin-derived pesticides [43]. Lou, et al. [44] recently provided a comprehensive summary of the advancements in vanillin-derived pesticides, highlighting how vanillin's unique properties are being harnessed to create more sustainable and environmentally friendly antiviral pest control solutions. These innovations contribute to the growing interest in green chemistry and sustainable agriculture, offering alternatives to the conventional synthetic pesticides that are often harmful to the environment and human health.

Due to the wide range of vanillin's applications, farmers, industrialists, and scientists are increasingly interested in the production systems of this highly valuable molecule to ensure its timely availability in large quantities.

3. Vanillin Biosynthesis in Vanilla Plants

Vanillin, or 3-methoxy-4-hydroxybenzaldehyde ($C_8H_8O_3$), is an aromatic aldehyde found in vanilla pods. Natural vanilla is a spice derived from orchids of the genus *Vanilla*, mainly obtained from the pods of *V. planifolia*, *V. pompona*, and *V. tahitensis* [7,45]. The de novo biosynthesis of vanillin through the phenylpropanoid pathway in *V. planifolia* was investigated by Gallage, Hansen, Kannangara, Olsen, Motawia, Jørgensen, Holme, Hebelstrup, Grisoni and Møller [4]. Vanillin biosynthesis involved several steps and enzymes, such as phenylalanine ammonia lyase (PAL; EC 4.3.1.25), cinnamate-4-hydroxylase (C4H; EC 1.14.13.11), 4-hydroxycinnamoyl-CoA ligase (4CL; EC 6.2.1.12), hydroxycinnamoyl transferase (HCT; EC 2.3.1.133), and 4-coumaroyl ester 3′-hydroxylase (C3′H; EC 1.14.13.36) [46,47]. The intermediates obtained from the catalytic reaction of these enzymes are coenzyme A derivatives as well as shikimate- and quinate esters of coumaric acid and caffeic acid [48]. Caffeic acid is a direct precursor of ferulic acid. It is methylated to ferulic acid by the enzyme caffeate *O*-methyltransferase (COMT; EC 2.1.1.68). Using coupled transcription/translation assays and transient expression in tobacco and yeast, a gene *VpVAN* encoding an enzyme called vanillin synthase (*Vp*VAN; EC 4.1.2.41), which catalyzes the C2 side chain shortening of ferulic acid and its glucoside to vanillin or vanillin-b-D-glucoside, was identified in the chloroplasts and re-differentiated chloroplasts of *V. planifolia*. The discovery of this enzyme enabled Gallage, Hansen, Kannangara, Olsen, Motawia, Jørgensen, Holme, Hebelstrup, Grisoni and Møller [4] to elucidate the complete pathway of vanillin biosynthesis (Figure 3). A non-β-oxidative pathway for vanillin biosynthesis was also reported by Podstolski, et al. [49]. In this proposed pathway, 4-coumaric acid is first converted to 4-hydroxybenzaldehyde by 4-hydroxybenzaldehyde synthase (4-HBS; EC number not identified). The 3′-hydroxylation of 4-hydroxybenzaldehyde by the C3′H will then yield 3,4-dihydroxybenzaldehyde followed by the *O*-methylation of the 3′ hydroxyl group by an *O*-methyltransferase (OMT) to produce vanillin. Nevertheless, the yield of plant-based vanillin is relatively low, and the extraction and purification present several challenges [50]. In recent decades, the increased knowledge regarding the metabolic pathway and the corresponding enzymes involved in the conversion of various natural substrates, such as ferulic acid, eugenol, isoeugenol, lignin, and glucose, to bio-vanillin has provided new opportunities for bioengineering microorganisms for vanillin biosynthesis. Several authors have reported the biotechnological production of vanillin in plants, bacteria, and yeast, as summarized in Appendix A—Table A1. However, microalgae, as potential expression systems, may offer a promising alternative to support the biotechnological production of vanillin. This review will highlight the importance of these photosynthetic hosts for vanillin production, which remain underexplored compared to non-photosynthetic microorganisms despite their potential for more sustainable and efficient production.

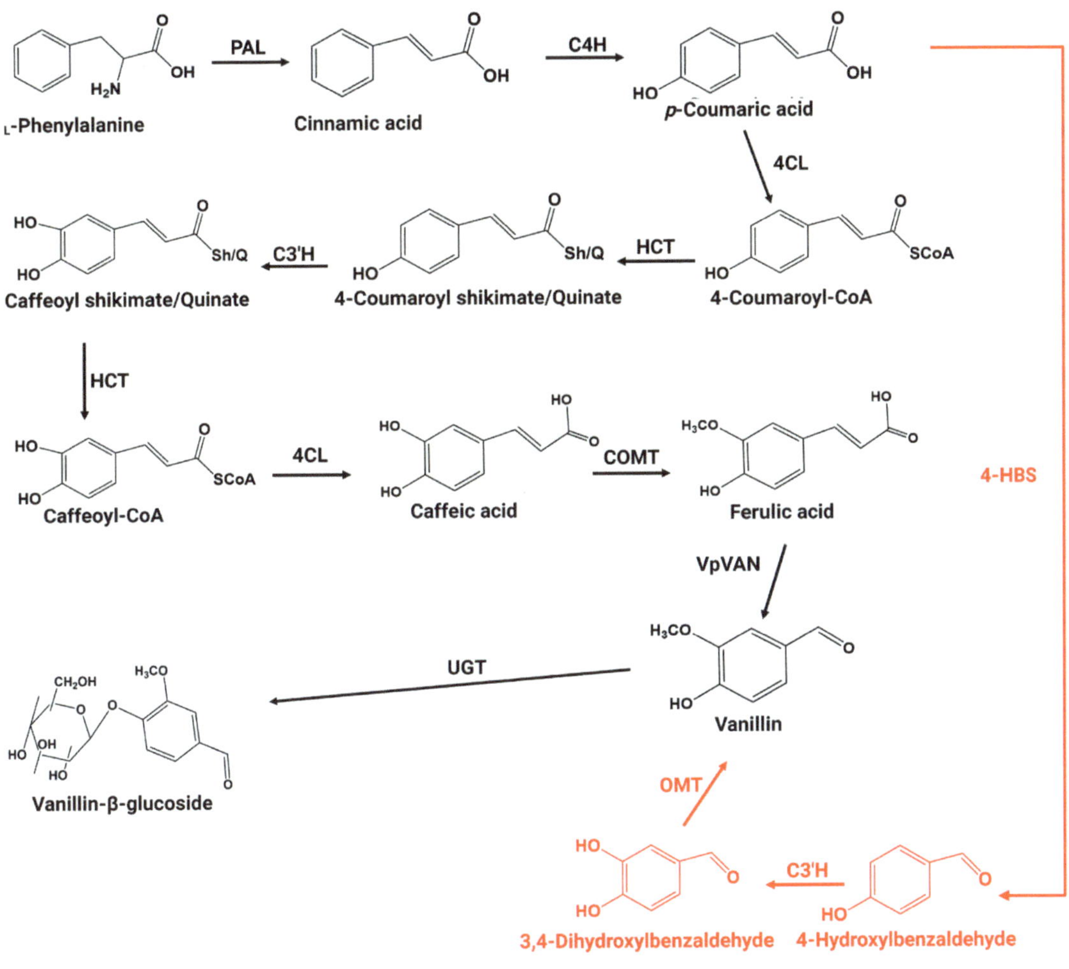

Figure 3. Vanillin and vanillin-β-D-glucoside biosynthesis pathways in *V. planifolia* pods. L-phenylalanine from the shikimate pathway will undergo a series of reactions to yield ferulic acid, which will subsequently be cleaved into vanillin by the enzyme vanillin synthetase according to Gallage, Hansen, Kannangara, Olsen, Motawia, Jørgensen, Holme, Hebelstrup, Grisoni and Møller [4]. As present in red according to Podstolski, Havkin-Frenkel, Malinowski, Blount, Kourteva and Dixon [49], 4-coumaric acid can be directly converted to 4-hydroxybenzaldehyde followed by 3′ hydroxylation and O-methylation to yield vanillin. Abbreviations: COMT: caffeate O-methyltransferase; PAL: phenylalanine ammonia lyase; OMT: O-methyltransferase; 4HBS: 4-hydroxybenzaldehyde synthase; C4H: cinnamate 4-hydroxylase; 4CL: 4-hydroxycinnamoyl-CoA ligase; HCT: hydroxycinnamoyl transferase; C3′H: coumaroyl ester 3′-hydroxylase; and VpVAN: *V. planifolia* vanillin synthase.

4. Engineering Prokaryotic and Eukaryotic Expression Systems for Sustainable Vanillin Production

Metabolic engineering aims to construct a metabolic pathway of interest in host organisms for the production of high-value-added compounds, thereby enhancing the yield and efficiency for industrial-scale production [51]. Microorganisms such as bacteria and yeast are widely used in metabolic engineering projects since they can be easily genetically modified with the availability of a wide range of molecular tools. This section summarizes the recent advancements and achievements in vanillin production using bacteria, yeast, and plant tissues. It highlights the progress in optimizing the metabolic pathways, enhancing enzyme functionality, and applying innovative genetic engineering techniques to increase the vanillin yield across these biological systems. Additionally, the section addresses the current challenges in optimizing microbial-based vanillin production, such as low yield, metabolic bottlenecks, and strain stability. Potential strategies to overcome these challenges are also explored, including advanced metabolic engineering, adaptive laboratory evolution, and process optimization to improve scalability and commercial viability.

4.1. Engineering Prokaryotic Microorganisms for Vanillin Production

4.1.1. Engineering *Escherichia coli*

E. coli has been extensively used for the heterologous production of vanillin [52]. In particular, Torre, et al. [53] immobilized the recombinant strain M109/pBB1, which was mutated at the alcohol dehydrogenase (ADH; EC 1.1.1.1) locus and expressed the genes encoding feruloyl-CoA synthetase (FCS; EC 6.2.1.34) and enoyl-CoA hydratase/aldolase (ECH; EC 4.2.1.17) from *Pseudomonas fluorescens* BF13 in a reactor. This strategy resulted in the production of 0.080 g/L of vanillin from 0.200 g/L of ferulic acid in the immobilized-cell reactor. As a result, the combination of these two genes has been widely employed in *E. coli* expression systems. This is the case for Barghini, Di Gioia, Fava and Ruzzi [12], who were able to produce 0.53 g/L of vanillin from ferulic acid by co-expressing the same two genes in *E coli* strain JM109. Further progress in vanillin production using bacteria was realized eight years later by the research of Ni, et al. [54]. In fact, for the de novo production of vanillin in *E. coli*, this research group partially mimicked the vanillin biosynthetic pathway of the plant *V. planifolia* using the *E. coli* strain genetically engineered to overproduce L-tyrosine (Figure 4B). The carbon flux from the primary metabolism of carbohydrates and simple organic compounds, such as glycerol and xyloses, was directed to vanillin via L-tyrosine. This strategy produced 97.2, 19.3, 13.3, and 24.7 mg/L of vanillin from L-tyrosine, glucose, glycerol, and xylose, respectively [54]. The conversion of protocatechuic acid to vanillic acid is a central step in the de novo biosynthesis of vanillin [55]. A central metabolism perturbation designed to increase the flux in the de novo biosynthetic pathway of vanillic acid increased the mean of the de novo vanillate yield from 132 to 205 mg/L [55]. However, protocatechuic acid, the precursor of vanillic acid, remained the dominant biosynthetic intermediate. To maximize this step of converting protocatechuic acid to vanillic acid, Kunjapur, Hyun and Prather [55] demonstrated that supplementation with methyl donor (methionine and homocysteine) improved the vanillin acid production from protocatechuic in *E. coli* strain DH10B. An artificial pathway for vanillic alcohol biosynthesis from simple carbon sources was proposed by Chen, et al. [56] using the *E. coli* BW25113 strain. This pathway required only four heterologous enzymes, namely *p*-hydroxybenzoate hydroxylase (PobA), carboxylic acid reductase (CAR), caffeate *O*-methyltransferase (COMT), and the CAR maturation factor phosphopantetheinyl transferase (Sfp) (Figure 4A), and led to 0.24 g/L of vanillyl alcohol released into the culture medium [56]. Vanillin can be converted from isoeugenol in one step via the oxidation of the side chains by the enzyme isoeugenol monooxygenase (IEM). The isoeugenol monooxygenase gene from the bacterium *Pseudomonas nitroreducens Jin1* was therefore cloned into *E. coli* [57]. This strain produced 38.34 g/L of vanillin from 49.2 g/L of isoeugenol. More recently, to optimize the yield of vanillyl alcohol production, Yang, et al. [58] mutated the shikimate dehydrogenase (*aroE*) gene in *E. coli* strain BW25113. In this study, gene *aroE* was knocked out to improve 3,4-DHBA biosynthesis of the upstream strain. Using the bioengineering *E. coli* BW25113 (YMC13/YMC14) co-culture strategy, this strain was able to produce 0.5 and 3.89 g/L of vanillic alcohol in shake-flask and fed-batch fermentations [58].

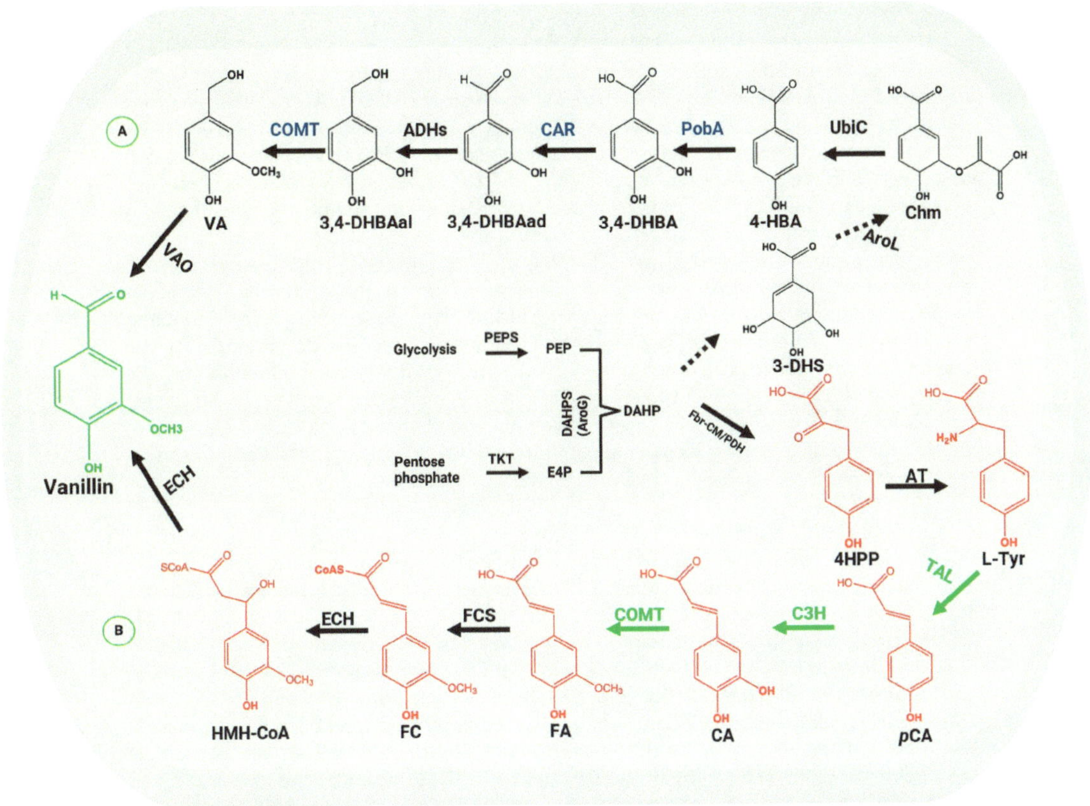

Figure 4. Engineered biosynthetic pathways for the de novo heterologous production of vanillin in *E. coli*. (**A**,**B**) represent the different metabolic pathways proposed by Chen, Shen, Wang, Wang, Zhang, Rey, Yuan and Yan [56] and Ni, Tao, Du and Xu [54]. (**A**) The arrows with black annotation in Chen's proposed pathway indicate the native pathway in *E. coli*, and the arrows blue annotation indicate the heterologous steps that have been added. (**B**) Pathway proposed by Ni, Tao, Du and Xu [54]; the green arrows contain three non-bacterial enzymes involved in the artificial pathway from tyrosine to ferulic acid, and the last black arrows of this pathway represent steps catalyzed by two bacterial enzymes involved in the conversion of ferulic acid to vanillin. Abbreviations. PEP: phosphoenolpyruvate; E4P: D-erythrose 4-phosphate; DAHP: 3-deoxy-D-arabinoheptulosonate 7-phosphate; 4-HPP: 4-hydroxyphenylpyruvate; L-Tyr: L-tyrosine; *p*CA: coumaric acid; CA: caffeic acid; FA: ferulic acid; FC: feruloyl-CoA; HMH-CoA: 4-hydroxy-3-methoxyphenyl-β-hydroxypropionyl-CoA; 3-DHS: 3-dehydroshikimic acid; Chim: chorismate; 4-HBA: 4-hydroxybenzoic acid; 3,4-DHBA: 3,4-dihydroxybenzoic acid; 3,4-DHBAad: 3,4-dihydroxybenzaldehyde; 3,4-DHBAal: 3,4-dihydroxybenzyl alcohol; VA: vanillyl alcohol; PEPS: phosphoenolpyruvate synthase; TKT: trans-ketolase; AroG: 2-dehydro-3-deoxyphosphoheptonate aldolase; AroL: shikimate kinase II; UbiC: chorismate lyase; PobA: *p*-hydroxybenzoate hydroxylase; CAR: carboxylic acid reductase; ADH: alcohol dehydrogenases; COMT: caffeate O-methyltransferase; VAO: vanillyl alcohol oxidase; AT: aminotransferase; fbr-CM/PDH: fbr-chorismate mutase/prephenate dehydrogenase; TAL: tyrosine ammonia lyase; C3H: 4-coumarate 3-hydroxylase; FCS: *trans*-feruloyl-CoA synthetase; and ECH: enoyl-CoA hydratase/aldolase.

4.1.2. Engineering *Pseudomonas* sp.

Pseudomonas sp. has also been used for the production of various plants' specialized metabolites, including vanillin. To develop an effective strain of *Pseudomonas* that is capable of producing vanillin with high yields, research from Di Gioia, Luziatelli, Negroni, Ficca, Fava and Ruzzi [29] suppressed, by directed mutagenesis, the gene (*vdh*) encoding the vanillin dehydrogenase of the *Pseudomonas fluorescens* BF13 strain. The inactivation associated with the simultaneous expression of the structural genes encoding feruloyl-CoA synthetase and enoyl-CoA hydratase/aldolase from a low-copy-number plasmid enabled this strain to produce up to 1.27 g/L of vanillin from ferulic acid [29]. Later, the non-pathogenic strain *Pseudomonas putida* KT2440 was genetically optimized to convert ferulic acid into vanillin. This genetic modification strategy aimed to enhance the efficiency of the metabolic pathway responsible for converting ferulic acid into vanillin while minimizing the formation of undesirable by-products. This was achieved by downregulating the vanillin dehydrogenase (vdh) and trans-molybdate transporter genes while enhancing the expression of feruloyl-CoA synthetase and enoyl-CoA hydratase/aldolase under the tac promoter. As a result, a yield of 1.66 g/L of vanillin was obtained, indicating the success of the genetic modifications in increasing vanillin production while minimizing by-product formation [59].

4.1.3. Engineering *Corynebacterium glutamicum*

C. glutamicum is a promising host for metabolic engineering studies, mostly for the industrial production of amino acids [60]. However, some publications have demonstrated the capacity of the genetically modified strain to metabolize vanillin. Chen, et al. [61] investigated the ability of the strain in culture to resist high concentrations of vanillin produced from lignocellulosic biomass. A transcriptome analysis revealed that, in addition to stimulating vanillin degradation enzymes, high concentrations of vanillin also caused differential regulation of antioxidant enzymes, secretory proteins, and the cell envelope [61]. These insights may be useful for the engineering of vanillin tolerance strains for suitable production of biofuels from lignocellulosic biomass. Thus, the conversion of ferulic acid to protocatechuic acid was enabled in *C. glutamicum* ATCC 21420 by cloning the vanillic acid *O*-demethylase (*vanAB*) gene of *Corynebacterium efficiens* NBRC 100395, which converts vanillic acid into protocatechuic acid [62]. This transformed strain was able to produce 1.06 g/L of protocatechuic acid from 3.10 g/L of ferulic acid. Also, an artificial biosynthetic pathway expressing the *COMT* and *CAR* genes from *Rattus norvegicus* and *Nocardia iowensis*, respectively, has recently been cloned into *C. glutamicum* [63]. The carbon flux was directed towards vanillin by the deletion of the gene *NCgl0324* encoding an aromatic aldehyde reductase. The new strain PV-IYD0324 resulting in the culture produced 0.31 g/L of vanillin from endogenous 4-hydroxylbenzoic acid.

4.1.4. Engineering *Pediococcus acidilactici*

P. acidilactici is a Gram-positive coccus used as a host for the production of aromatic compounds such as vanillin. This bacterium has several advantages in metabolic engineering, particularly growth at high temperature, pH, and osmotic pressure values [64]. The *P. acidilactici* BD16 strain transformed by heat shock to express feruloyl-CoA synthetase and the enoyl-CoA hydratase/aldolase genes from *Amycolatopsis* sp. was cultured in the presence of ferulic acid [65]. When the culture conditions were optimized, 0.47 g/L of vanillin was recovered from 0.21 g/L of ferulic acid per mg of the recombinant cell biomass. Producing vanillin from expensive substrates such as ferulic acid has lower advantages over simple and available substrates such as glucose and industrial waste. Chakraborty, et al. [66] were able to recover 4.01 g/L of vanillin by culturing the recombinant *P. acidilactici* BD16 strain in the presence of rice bran.

In summary, prokaryotic microorganisms have been widely used for vanillin production by metabolic engineering. *E. coli* seems to be the best platform so far because few studies have mentioned the presence of vanillin degradation pathways in *E. coli*. Also, up to 38.3 g/L of vanillin could be produced in the bacteria in a single step using isoeugenol monooxygenase from *Pseudomonas nitroreducens Jin1*, as shown in Appendix A—Table A1 [57].

However, the expression of vanillin biosynthesis enzymes from plants in prokaryotic systems may, in some cases, result in poor enzyme stability or activity, requiring further optimization to improve the protein folding and expression levels. This issue can potentially be addressed by using eukaryotic expression systems, which typically possess more advanced protein-folding mechanisms and post-translational modifications (PTMs), thereby enhancing enzyme functionality and stability. Post-translational modifications such as glycosylation are essential for the optimal activity of some eukaryotic enzymes, affecting their folding, stability, and ability to interact with other molecules in the biosynthetic pathway [67,68]. When these modifications are absent or insufficient, it can lead to premature degradation of the enzymes, thereby reducing their activity and, consequently, the amount of vanillin produced.

4.2. Engineering Eukaryotic Organisms for Vanillin Production

Eukaryotic organisms, used as expression systems for vanillin production, are mainly yeasts and plant tissues.

4.2.1. Engineering *Saccharomyces cerevisiae* and *Schizosaccharomyces pombe*

S. cerevisiae is a yeast that is widely used in industry for the production of high-value-added compounds due to its alcohol tolerance, well-elucidated genome, and fermentation performance [69]. According to the Evolva project, Hansen, et al. [70] studied the ability of the yeast strains *S. cerevisiae* and *S. pombe* to produce vanillin from glucose. In this study, the vanillin biosynthetic pathway was constructed using the 3-dehydroshikimate dehydratase genes from the ascomycete *Podospora pauciseta*, the carboxylic acid reductase from the genus *Nocardia*, *O*-methyltransferase from *Homo sapiens*, and UDP-glycosyltransferase from *Arabidopsis thaliana* (red arrows in Figure 5A). The efficiency of this pathway was achieved by removing the alcohol dehydrogenase and the vanillin-β-D-glucosidase genes involved in vanillin and vanillin-β-glucoside degradation [71]. Therefore, 0.065 and 0.045g/L of vanillin were produced in *S. cerevisiae* and *S. pombe*, respectively, after 48 h of culture. Later, the glycosylated vanillin produced was optimized by Brochado, et al. [72] based on an in silico metabolic engineering algorithm. Indeed, this approach enables the identification of target reactions in a metabolic network whose inactivation would lead to an improvement in vanillin production. Bioinformatics analysis has thus identified pyruvate decarboxylase and glutamate dehydrogenase genes in *S. cerevisiae* whose suppression is thought to be associated with increased glycosyltransferase activity. The experimental analysis of four strains with two single gene deletions in these genes in the same study showed that three of the strains produced 1.5 times more glycosylated vanillin than the unmutated strains [72]. Plant tissues are mainly composed of lignin, which is a raw material residue obtained from various industrial bioprocesses. Due to its aliphatic and aromatic structures, it is a widely used substrate in the flavor industry [73]. Similarly, Shen, et al. [74] tested the ability of the EMV-8 strain of *S. cerevisiae* mutated with ethyl methane sulfonate to tolerate high concentrations of vanillin during the lignin biotransformation processes. The use of ethyl methane sulfonate is an engineering strategy that increases the oxidoreductase activity and antioxidant capacity of a microorganism. Such a mutation strategy enabled the EMV-8 strain to maintain a specific growth rate in the presence of 2 g/L of vanillin, in contrast to the reference strain NAN-27, which could not grow in the same conditions because of vanillin cytotoxicity [74]. To explain the redox mechanisms that allow the EMV-8 strain to tolerate high concentrations of vanillin, Liang, et al. [75] identified the different genes expressed and their implications in the mechanisms of vanillin tolerance in *S. cerevisiae*. Thus, the metabolism of the strain under investigation was characterized in a culture by the overexpression of the mRNA-binding aldo-keto reductase (Gcy1), its Ypr1 paralog, the peroxisome membrane receptor (Pex5), and the multiprotein complex involved in specific growth (Mbf1). However, only the Gcy1 and Ypr1 proteins showed vanillin reductase activities that are dependent on NADPH and NADH [75]. To reproduce the de novo biosynthetic pathway of vanillin from tyrosine in *S. cerevisiae*, Qiu, et al. [76] developed a recombinant yeast strain by cloning five heterologous enzymes, namely tyrosine ammonia lyase (TAL) from the actinomycete *Saccharothrix espanaensis*, the combination of three hydroxylase enzymes [(4-coumarate 3-hydroxylase (C3H) from *Saccharothrix espanaensis*, 4-hydroxyphenylacetate 3-hydroxylase (4HPA3H) from *Pseudomonas aeruginosa*, and cytochrome P450 monooxygenase (CYP199A2) from *Rhodopseudomonas palustris*)], caffeate *O*-methyltransferase (COMT) from *Arabidopsis thaliana*, *trans*-feruloyl-CoA synthetase (FCS), and enoyl-CoA hydratase/aldolase (ECH) from *Streptomyces* sp. (dark blue arrows in Figure 5C). By mutating the *ARO4* and *ARO7* genes, which encode 3-deoxy-D-arabino heptulosonate-7-phosphate synthase (DAHPS) and chorismate mutase (CM), respectively, into others to limit the negative feedback of tyrosine and increase its flux to vanillin, then by substituting the rate-limiting enzyme tyrosine ammonia lyase (*sam8* gene) with 4-coumarate 3-hydroxylase (*sam5* gene) to increase the caffeic acid pool, these researchers were able to produce 0.008g/L of vanillin from 0.4 g/L of ferulic acid [76]. Most microorganisms lack the ability to accumulate aldehydes due to the presence of endogenous dehydrogenases and reductases, which facilitate their rapid metabolism and prevent their accumulation in the cells. To overcome this challenge in the case of vanillin production in yeast, Mo and Yuan [20] Mo and Yuan [20] recently developed a minimal aromatic aldehyde reduction (MARE) yeast platform for engineering vanillin production from glucose. They first used a combinatorial gene deletion approach, targeting the *adh6*, *adh7*, *gre3*, *gcy1*, *ydl124w*, *ypr1*, *ari1*, *ydr541c*, and *aad3* genes, which are critical for vanillin reduction and oxidation. Subsequently, they substituted the *hfd1* gene with *ubiC* to prevent vanillin oxidation and provide the precursor for ubiquinone synthesis, thus enhancing vanillin accumulation while limiting vanillyl alcohol production. Building upon this platform, they further optimized the co-factor supply (NADPH/SAM) and implemented dual precursor synthesis pathways along with metabolic reconfiguration using a phosphoketolase pathway (green arrows in Figure 5C). Through systematic engineering, they achieved a significant milestone of producing 0.36g/L of vanillin, demonstrating the potential of the MARE yeast platform for sustainable vanillin production. Similarly, Xin, et al. [77] meticulously engineered *Saccharomyces cerevisiae* BY4742, which carried deletions in twenty-two genes associated with vanillin reduction and oxidation [78], to facilitate the conversion of lignin-derived monomers into vanillin. They first reconstructed two pathways separately: the CoA-independent pathway, expressing *VpVAN* from *V. planifolia*, and the CoA-dependent non-β-oxidation pathway, expressing 4CL from *Petroselinum crispum* and Ech from *Pseudomonas putida* KT2440. In addition, they fused hpaC and hpaB enzymes to enable the simultaneous hydroxylation of *p*-coumaric acid and *p*-hydroxybenzaldehyde into caffeic acid and protocatechualdehyde, respectively. Furthermore, they regulated S-adenosylmethionine metabolism by expressing 3′-O-methyltransferase from *Arabidopsis thaliana* (AtCOMT), *Homo sapiens* (HsCOMT), and *Nicotiana tabacum* (NtCOMT), along with methylenetetrahydrofolate reductase from *Arabidopsis thaliana*. By also controlling the branch and shikimic acid pathways, the microbial cell factory, employing the CoA-dependent non-β-oxidation pathway, successfully produced 0.29g/L of vanillin from authentic lignocellulosic biomass hydrolysates [77].

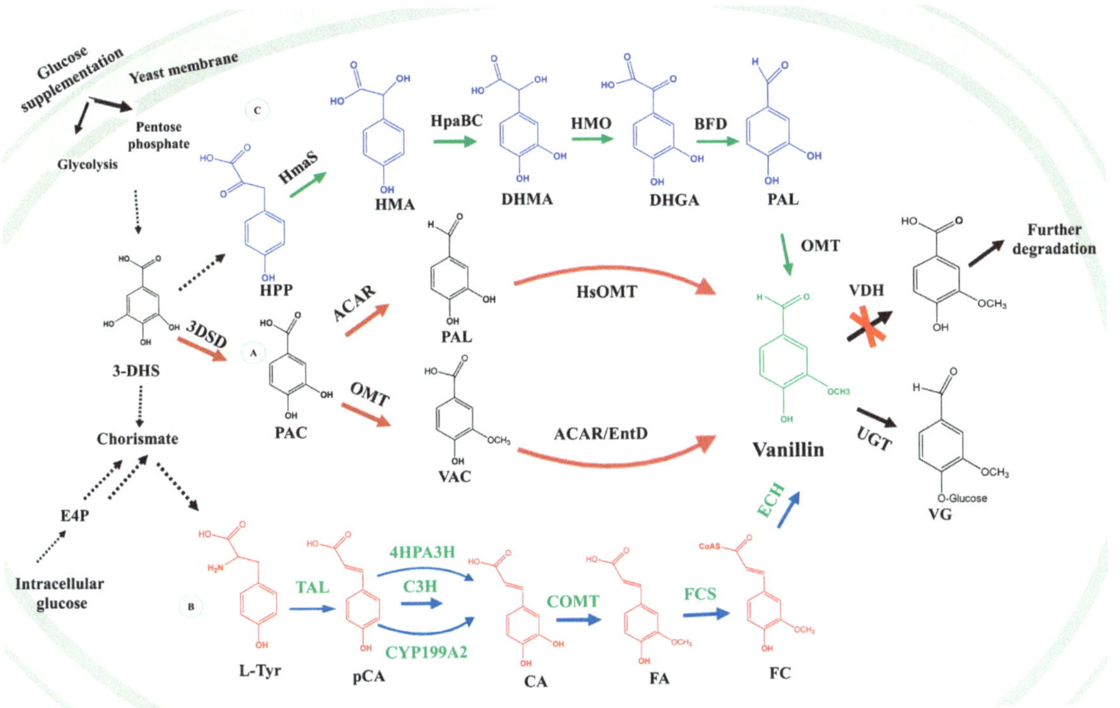

Figure 5. Biosynthesis pathways for vanillin and glycosylated vanillin production in *Saccharomyces cerevisiae* yeast [20,70,76,79]. (**A**) (Red arrows) represent the pathway proposed by Strucko, Magdenoska and Mortensen [79] and Hansen, Møller, Kock, Bünner, Kristensen, Jensen, Okkels, Olsen, Motawia and Hansen [70], while (**B**) (dark blue arrows) and (**C**) (green arrows) represent the pathways proposed by Qiu et al. (2022) and Mo and Yuan [20], respectively. Abbreviations. 3DHS: 3-dehydroshikimic acid; E4P: erythrose-4-phosphate; PCA: protocatechuic acid; L-Tyr: L-tyrosine; *p*-CA: *p*-coumaric acid; CA: caffeic acid; FA: ferulic acid; FC: feruloyl-CoA; 3,4-DHBAad: 3,4-dihydroxybenzaldehyde; PAL: protocatechuic aldehyde and VAC: vanillic acid; HPP: hydroxyphenylpyruvate; HMA: hydroxymandelate; DHMA: 3,4-dihydroxymandelate; DHGA: 3,4-dihydroxyphenylglyoxylate; 3-DSD: 3-dehydroshikimate dehydratase; ACAR: aromatic carboxylic acid reductase; EntD: phosphopantetheine transferase; HsOMT: O-methyltransferase (homo sapiens); UGT: UDP-glycosyltransferase; TAL: tyrosine ammonia-lyase; C3H: 4-coumarate 3-hydroxylase; 4HPA3H: 4-hydroxyphenylacetate-3-hydroxylase; COMT: caffeate O-methyltransferase; FCS: trans-feruloyl-CoA synthetase; ECH: enoyl-CoA hydratase/aldolase; HmaS: hydroxymandelate synthase; HpaBC: flavin-dependent monooxygenase; HMO: hydroxymandelate oxidase; BFD: benzoylformate decarboxylase; and VDH: vanillin dehydrogenase. The red cross materializes the suppression of genes involved in vanillin oxidation and reduction.

4.2.2. Engineering Plant Tissues for the Production of Vanillin

Although most of the studies are limited to the production of vanillin in microorganisms, a few have investigated plant tissues. Specifically, Chee, et al. [80] genetically transformed the pepper (*Capsicum frutescens*) cells with the vanillin synthetase gene from *Vanilla planifolia* (VpVAN) by microparticle bombardment. A high-performance liquid chromatography (HPLC) analysis of the extracts from the transformant cells in a culture led to the detection of 5.7×10^{-3} g/L of vanillin, which was sixteen times higher than the control [80]. An amount of 0.54 g of vanillin per gram of fresh callus culture was also produced from rice embryonic cells transformed by *Agrobacterium* to express the codon optimized sequence of vanillin synthetase [81]. However, it is important to note that the enzymatic activity of this enzyme is subject to several controversies, and no kinetic parameters such as catalytic activity (Kcat) and Michaelis constant (Km) have yet been reported [82,83]. The progress of plant-tissue-culture-based vanillin synthesis has not been commercially viable and has been hindered by factors such as long culture duration, low yields, and increased by-product formation [3].

5. Current Challenges and Potential Strategies for Enhancing Microbial-Based Vanillin Production Systems

Despite the notable progress in microbial-based vanillin production, several challenges remain. One of the primary hurdles involves the low conversion efficiency and yield, largely due to the cytotoxicity of vanillin and its intermediates to host cells. This toxicity often impairs cell growth and enzyme functionality, limiting production [84]. Indeed, vanillin concentrations exceeding 0.5–1 g/L have been shown to significantly inhibit microbial growth, thereby reducing the final product [70]. Additionally, by-product formation during the biosynthesis process complicates the downstream purification, reducing the overall efficiency and increasing the production costs [50]. Another challenge is the metabolic burden on microorganisms engineered for vanillin production. Complex metabolic pathways, often involving multiple genetic modifications and CRISPR-Cas9 gene knockouts, can reduce cellular fitness, leading to lower productivity [85]. This strain on cellular resources often results in slower growth rates, decreased metabolic efficiency, and suboptimal vanillin yields, further complicating the scaling of biotechnological vanillin production for industrial applications. To address these challenges, potential strategies include metabolic pathway optimization by fine-tuning the pathways through adaptive laboratory evolution or optimizing the gene expression to enhance the flux toward vanillin biosynthesis, balancing precursor availability, co-factor regeneration, and improving enzyme activity [20,77,84,86]. Another approach is tolerance engineering, where microbial strains are engineered to increase their tolerance to vanillin and its intermediates, achieved through the overexpression of efflux pumps or modification of cellular stress response mechanisms to mitigate the cytotoxic effects of vanillin on microbial cells. Efflux pumps actively transport vanillin and its toxic intermediates out of the cell, reducing the intracellular concentrations and minimizing cellular damage [3]. Additionally, enhancing the stress response pathways can bolster the cell's ability to tolerate and recover from vanillin-induced stress, leading to improved growth and productivity [75]. These strategies are crucial for overcoming cytotoxicity, which is one of the major challenges in microbial-based vanillin production [3,72,87]. Enzyme engineering, using rational or directed evolution-based methods, can improve catalytic efficiency and reduce the formation of unwanted by-products, thereby enhancing yield and purity [86,88]. Diversifying host systems by exploring non-traditional microbial hosts, such as photosynthetic microalgae, presents a promising alternative for vanillin production, potentially lowering costs and integrating CO_2 fixation, which contributes to a more sustainable bioproduction platform.

6. Green Biology: Harnessing Microalgae as Natural Producers of Vanillin Precursors and Promising Hosts for Biotechnological Vanillin Production

Metabolically engineered photosynthetic microorganisms, such as microalgae, can offer several advantages as potential hosts for vanillin production. First, their ability to perform photosynthesis allows them to utilize CO_2 as a carbon source, reducing the reliance on external feedstocks and contributing to carbon sequestration [89,90]. This makes the process more sustainable compared to non-photosynthetic microorganisms. Additionally, microalgae can produce high-value precursors and metabolites relevant to vanillin biosynthesis, potentially enhancing the overall yield. Moreover, it has been documented that the cost associated with biotechnology production in microalgae is much cheaper compared to animal cells and even lower than that of microbes and plants [91].

6.1. Microalgae as Natural Producers of Vanillin Precursors

Microalgae, with their remarkable metabolic versatility, have developed intricate biochemical pathways that enable them to produce a wide range of specialized metabolites, including valuable phenolic compounds [92]. The biosynthesis of phenolic compounds in microalgae is a complex and highly regulated process, involving several metabolic pathways that enable the production of bioactive molecules that are essential for the survival and adaptation of microalgae to their environment [93,94]. This justifies the fact that *Phaeodactylum tricornutum* was able to produce various vanillin precursors and metabolites, such as coumaric acid, caffeic acid, ferulic acid, protocatechuic acid, gallic acid, and vanillic acid, from in vitro cultures grown in natural seawater containing elevated levels of copper (Cu) and iron (Fe) metals [93]. The presence of these metals likely enhanced the biosynthetic pathways involved in the production of phenolic compounds, including those that serve as precursors for vanillin synthesis. This demonstrates the potential of diatoms like *P. tricornutum* as biotechnological platforms for producing valuable aromatic compounds such as vanillin. Moreover, Bhuvana, et al. [95], using an HPLC method, detected peaks corresponding to caffeic acid derivatives in methanolic extracts of *Nannochloropsis oculata* and *Chlorella vulgaris*. Further, *Haematococcus pluvialis* was fed with vanillin precursors such as ferulic acid, coniferyl aldehyde, and *p*-coumaric acid by Tripathi, et al. [96], resulting in the production of vanillin (0.010 g/L), vanillyl alcohol (0.003 g/L), vanillic acid (0.005 g/L), protocatechuic acid (0.001 g/L), and *p*-hydroxybenzoic acid (0.001 g/L) in free and immobilized cell cultures. Together, these data provide compelling evidence for the presence of the putative enzymes involved in the synthesis of vanillin, its precursors, and metabolites in microalgae, as evaluated and outlined in the following Table 1. Indeed, we identified potential homologs of vanillin biosynthetic enzymes in microalgae by comparing the sequences to those already characterized in *V. planifolia* (VpVAN), *Streptomyces* sp. V-1 (FCS and ECH), and *Pseudomonas nitroreducens* Jin1 (IEM). An immature *Vp*VAN candidate displaying the cathepsin propeptide inhibitor (I29), peptidase_C1, and granulin domains was identified in the protein databases of various microalgae (Table 1). Additionally, a putative FCS similar to long-chain fatty acyl-CoA synthetases was identified in these microalgae species. This protein contains domains such as the fatty acyl-CoA synthetase domain, a dimer interface polypeptide, an acyl-activating enzyme (AAE) consensus motif, as well as AMP and CoA binding sites. Since the CoA-dependent non-β-oxidation pathway involves both FCS and ECH, a putative ECH with a crotonase-like domain, a substrate binding site, oxyanion hole (OAH)-forming residues, and a trimer interface polypeptide binding site was also found. However, for the isoeugenol monooxygenase pathway, only three of these microalgae species (*C. reinhardtii*, *H. pluvialis*, and *C. vulgaris*) present a putative IEM, characterized by a retinal pigment epithelial membrane protein domain. These putative proteins in microalgae could play roles similar to those in the vanillin biosynthesis pathway in *V. planifolia* or in bacteria. Indeed, phenolic compounds are primarily synthesized through the shikimic acid pathway, a biochemical route present in most plants and microorganisms [97–99]. This pathway, which leads to the formation of phenylpropanoid acid precursors like L-phenylalanine, could serve as the upstream mechanism for the biosynthesis of various vanillin precursors. The key enzyme typically initiating phenolic compound biosynthesis is phenylalanine ammonia-lyase (PAL), which catalyzes the deamination of L-phenylalanine to form cinnamic acid, a crucial precursor of vanillin. However, recent studies have shown that not all microalgal taxa possess the PAL enzyme [94]. This absence suggests that microalgae may utilize alternative pathways for phenolic compound biosynthesis, potentially opening up other routes for the production of vanillin precursors via different enzymatic mechanisms. Such alternative strategies highlight the metabolic diversity within microalgae, offering distinct advantages for engineering approaches aimed at optimizing vanillin production.

Table 1. Putative vanillin biosynthesis and degradation enzymes identified in microalgae.

Homologues of Vanillin biosynthesis Enzymes	Conserved Domains Database	Function	Component	Process	Target Biochemical Mechanisms
Putative Vanillin synthetase	214853, 425470, 239068, 197621.	Enables cysteine-type endopeptidase activity	Extracellular space and lysosome	Involved in protein catabolic process	Retro-aldol-type reaction [4]
Putative Feruloyl-CoA synthetase	341284;	Enables acyl-activating enzyme activity	NI	Catalyze the ATP-dependent acylation of fatty acids in a two-step reaction	CoA-dependent non-β-oxidation reaction [100]
Putative Enoyl-CoA Aldolase/hydratase	474030 119339	Enables enoyl-CoA hydratase activity	Mitochondrion	Involved in fatty acid beta-oxidation	CoA-dependent non-β-oxidation reaction [100]
Putative Isoeugenol Monooxygenase	442887	Enables carotenoid dioxygenase activity	NI	Involved in carotene catabolic process	Oxidative cleavage [101]

A BlastP was performed in NCBI using the vanillin synthase from *Vanilla planifolia* (GenBank: AKG47593.1); feruloyl-CoA synthetase from *Streptomyces* sp. *V-1* (GenBank: AGR34008.1); enoyl-CoA aldolase/hydratase from *Streptomyces* sp. *V-1* (GenBank: AGR34009.1), and isoeugenol monooxygenase from *Pseudomonas nitroreducens* Jin1 (GenBank: ACP17973.1) as queries in the protein databases of four microalgae species (*C. reinhardtii*, *P. tricornutum*, *H. pluvialis*, and *C. vulgaris*). Homologous enzymes sharing the same active domain were found in these microalgae species. CDD: Conserved Domains Database; NI: not identified.

6.2. Engineering and Modulation of Robust Microalgal-Based Vanillin Production Systems

The engineering of robust microalgal-based systems for vanillin production represents a promising frontier in biotechnology, aimed at utilizing the metabolic flexibility of microalgae for sustainable vanillin biosynthesis. This development involves the integration of advanced genetic engineering techniques to optimize the metabolic pathways responsible for producing vanillin and its precursors. The key strategies include introducing or enhancing the pathways involved in the synthesis of phenylpropanoid acids such as ferulic acid, *p*-coumaric acid, and caffeic acid, which serve as essential precursors for vanillin production [7]. Genetic engineering techniques, including the CRISPR-Cas9 and transgenic approaches, are widely employed to modify the key enzymes and regulatory elements in the microalgal genome [102,103]. These modifications may enable more precise control of the metabolic flux toward vanillin production, minimizing the formation of unwanted by-products while maximizing efficiency.

A significant challenge in engineering microalgae for vanillin production is the efficient expression of heterologous genes within their complex cellular environment. Research efforts have been undertaken to introduce vanillin biosynthesis pathways into microalgae like *P. tricornutum*. In that case, the researchers attempted to propagate the full vanillin biosynthesis pathway from *V. planifolia* into *P. tricornutum* through an episomal system, incorporating multiple key enzymes required for vanillin synthesis [104]. While the pathway introduction was successful at the mRNA level, issues such as episomal instability and difficulties in protein expression hindered the actual vanillin production [105]. Indeed, episomal plasmids are commonly used in genetic engineering to introduce new pathways into host cells without altering the cell's native genome [106–108]. However, these plasmids can be unstable, especially under prolonged culture conditions or selective pressures, leading to the gradual loss of the plasmid from the cell population. To overcome these challenges, researchers can explore strategies to improve the expression of heterologous genes, such as optimizing the gene integration into stable genomic locations, enhancing the promoter efficiency, and fine-tuning the expression systems [109,110]. Advances in algal synthetic biology tools, including modular plasmid systems and improved transformation techniques, are enabling more precise and efficient engineering of microalgae. Microalgae like *C. reinhardtii* offer a highly versatile platform for genetic manipulation, with extensive molecular tools and mutant libraries available [111,112]. Both the chloroplast and nuclear genomes of *C. reinhardtii* can be engineered to express vanillin biosynthesis enzymes, providing flexibility in targeting specific metabolic compartments for efficient vanillin production. Therefore, this approach enables the genetic modification of the nuclear genome (with the genes encoding enzymes that require N-glycosylation for enhanced activity) or the chloroplast genome (with the genes encoding enzymes that do not require N-glycosylation). This facilitates the reconstruction of partial vanillin biosynthesis pathways from *V. planifolia* or known bacterial pathways, ultimately leading to the development of industrial strains of microalgae with high vanillin yields.

Furthermore, the integration of efficient downstream processing techniques is critical to developing robust microalgal vanillin production systems. The extraction and purification of vanillin from microalgal biomass must be cost-effective and efficient to maximize the yield. Innovative approaches, such as the use of biocompatible solvents, microwave-assisted extraction, pulsed electric fields, supercritical fluids, crystallization, and membrane-based separation technologies, are currently being explored to streamline the recovery of vanillin from engineered microbial and plant-based vanillin production systems [50]. These techniques can also be adapted to microalgae, further enhancing their potential as a sustainable and scalable source of vanillin. By optimizing these extraction methods, the industrial viability of microalgal-based vanillin production could be significantly improved, ensuring high-purity vanillin at reduced processing costs and energy consumption. Ultimately, a vanillin-tolerant microalgae could be optimized for resistance to redox products, enabling the accumulation of vanillin without undergoing reduction or oxidation processes. Then, the optimized designed expression system carrying the vanillin biosynthesis pathway described in *V. planifolia* or the bacterial pathway can be used to engineer the nuclear and/or chloroplast genome of the tolerant microalgae. Finally, the optimized transformants are then cultivated on a large scale to produce vanillin for the cosmetics, food, and pharmaceutical industries.

7. Conclusions and Perspectives

This review has explored the potential of leveraging biotechnological advances to engineer microbial, plant, and algal systems for sustainable vanillin production. As the consumer demand for natural and eco-friendly ingredients grows, the biotechnological synthesis of vanillin offers a promising alternative to the traditional methods reliant on vanilla beans or petrochemical routes. We discussed and highlighted the production and industrial application of vanillin by deciphering the current knowledge regarding the key enzymes involved in vanillin biosynthesis. The insights gained from microbial and plant systems have already established a foundation for understanding vanillin biosynthesis. Advances in metabolic engineering have enabled the development of genetically modified microbial strains capable of efficiently producing vanillin and its precursors, although challenges such as yield optimization and pathway stability remain.

Microalgae have emerged as a promising platform for high-value molecule production owing to their metabolic diversity and capacity for large-scale cultivation. This unique biological system not only presents opportunities for vanillin biosynthesis but also offers added environmental benefits, such as carbon dioxide capture. While substantial progress has been achieved in the metabolic engineering of microalgae for the production of high-value molecules, the full potential of microalgae for vanillin production is yet to be realized. The complexity of heterologous gene expression in these systems remains a critical challenge. Researchers must continue to refine the gene integration strategies, enhance the promoter efficiency, and fine-tune the expression systems to ensure stable and high-yield production. The scaling-up of vanillin production also poses challenges. Although laboratory-level successes have demonstrated the feasibility of non-photosynthetic microorganism systems for vanillin production, scaling these systems for commercial applications will require further optimization of the growth conditions, bioreactor designs, and downstream processing methods.

Looking ahead, several key areas will drive future research into biotechnological vanillin production using microalgae. First, a deeper understanding of microalgal metabolism, particularly in the context of vanillin biosynthesis, will enable more targeted and efficient metabolic engineering. Exploring the evolutionary diversity of microalgal species could uncover novel biosynthetic pathways and enzymes that further improve the efficiency and yield of vanillin production. Second, the integration of omics technologies, including genomics, transcriptomics, proteomics, and metabolomics, will provide valuable insights into the regulation of the vanillin biosynthesis pathways and the identification of bottlenecks that limit production. Collaboration between academia and industry will be crucial to accelerating the development of scalable, cost-effective production platforms. The growing interest in bio-based vanillin from the food, cosmetics, pharmaceutical, and agrochemical industries will drive investment in research and development. By addressing the challenges of gene expression, pathway optimization, and process scalability, it is likely that microbial and algal systems will become viable alternatives for sustainable vanillin production in the near future.

Author Contributions: Conceptualization, A.W.T. and I.D.-P.; validation, A.W.T., F.A., F.M.-M. and I.D.-P.; writing—original draft preparation, A.W.T.; writing—review and editing, A.W.T., F.A., F.M.-M. and I.D.-P.; visualization, A.W.T. and F.A.; supervision, I.D.-P.; project administration, F.M.-M. and I.D-P.; funding acquisition, I.D.-P. All authors have read and agreed to the published version of the manuscript.

Funding: This research was funded by the Université du Québec à Trois-Rivières—Annexe C program to I.D.-P.

Institutional Review Board Statement: Not applicable.

Informed Consent Statement: Not applicable.

Data Availability Statement: The original contributions presented in the study are included in the review and appendix; further inquiries can be directed to the corresponding author.

Conflicts of Interest: The authors declare no conflicts of interest.

Appendix A

Table A1. Metabolic engineering of prokaryotic and eukaryotic organisms for the production of vanillin.

Engineering Platform	Genes (Origin)	Bioengineering Strategies Transformation Method	Insertion Type	Mutations	Substrates (g/L)	Complexity	Product	Volume of Culture (mL)	Fermentation Time (h)	Yield (g/L)	Ref.
Escherichia coli BL21(DE3)	*IEM* (*Pseudomonas nitroreducens Jin1*)	Heat shock	Episomal	None	49.2 Isoeugenol	2	Vanillin	NI	16	38.3	[57]
Escherichia coli BL21(DE3)	*IEM* (*Pseudomonas putida IE27*)	NI	Genome integration	None	37.7 Isoeugenol	2	Vanillin	10	48	28.3	[113]
Escherichia coli BL21 (DE3)	*IEM* (*microbial metagenome*)	NI	NI	None	16.4 Isoeugenol	2	Vanillin	15	48	2.2	[114]
Escherichia coli JM109	*fcs* and *ech* (*Pseudomonas fluorescens BF13*)	NI	NI	None	0.64 Ferulic acid	4	Vanillin	NI	6	0.53	[12]
Escherichia coli FR13	*fcs* and *ech* (*Pseudomonas fluorescens BF13*)	NI	Genome integration	None	4.5 Ferulic acid	4	Vanillin	15	24	4.3	[115]
Escherichia coli BW25113	*PobA* (*Pseudomonas aeruginosa*), *CAR* (*Mycobacterium marinum*), *sfp* (*Bacillus subtilis*), and *comt* (*Arabidopsis thaliana*)	NI	NI	None	1 3,4-dihydroxy benzoyl alcohol.	4	Vanillyl alcohol	20	48	0.24	[56]
Escherichia coli K-12 MG1655	*CAR* (*Nocardia iowensis*), *sfp* (*Bacillus subtilis*), *comt* (*Homo sapiens*), *DSD* (*Bacillus thuringiensis*)	P1 bacteriophage transduction	Genome integration	Deletion of *yqhC-dkgA* genes	Protocatechuic acid	4	Vanillate	50	72	0.004	[55]
Pseudomonas fluorescens BF13	*fcs* and *ech* (*Pseudomonas fluorescens BF13*)	Heat shock	NI	Deletion of *vdh* gene	0.48 Ferulic acid	4	Vanillin	2000	24	1.27	[29]
Escherichia coli	*tal* (*Saccharothrix espanaensis*), *sam5* (*actinomycete*), *comt* (*Arabidopsis thaliana*), *sam8* (*Saccharothrix espanaensis*), *fcs*, and *ech* (*Streptomyces* sp.)	Co-transformation	Genome integration	Deletion of *TyrR* gene	L-Tyrosine	6	Vanillin	50	12	0.013–0.097	[54]

Table A1. *Cont.*

Engineering Platform	Genes (Origin)	Bioengineering Strategies Transformation Method	Insertion Type	Mutations	Substrates (g/L)	Complexity	Product	Volume of Culture (mL)	Fermentation Time (h)	Yield (g/L)	Ref.
Escherichia coli MG1655 RARE	*LCC (unknown prokaryote), TPADO (Comamonas* sp.*), DCDDH (Comamonas* sp.*), O-MT (Rattus norvegicus), and CAR (Nocardia iowensis).*	Heat shock	NI	None	Monomer terephthalic acid	6	Vanillin	10	24	0.11	[116]
Escherichia coli BW25113	*PobA (Pseudomonas aeruginosa), CAR (Mycobacterium marinum), sfp (Bacillus subtilis), and COMT (Arabidopsis thaliana)*	Heat shock	NI	Deletion of *aroE* gene	ND	6	Vanillyl alcohol	50 and 3000	2 and 24	3.89 and 0.55	[58]
Escherichia coli	*fcs and ech (Amycolatopsis* sp.*)*	NI	NI	None	Glucose	6	Vanillin	NI	48	0.50	[117]
Pediococcus acidilactici BD16	*fcs and ech (Amycolatopsis* sp.*)*	Heat shock	Episomal	None	0.21 Ferulic Acid per mg of recombinant cell biomass	4	Vanillin	100	0.33	0.47	[55]
Pediococcus acidilactici BD16	*fcs and ech (Pediococcus acidilactici BD16)*	Heat shock	Episomal	None	Rice bran	4	Vanillin	50	24	4.01	[56]
Pseudomonas putida KT2440	*Pp-SpuC and ATA (Chromobacterium), adh (Bacillus)*	Electroporation	Genome integration	None	Lignin-derived substrates	4	Vanillin and vanilly-lamine	50 and 100	24	0.10 and 0.13	[118]
Pseudomonas putida KT2440	*fcs and ech (Pseudomonas putida)*	Electroporation	Genome integration	Deletion of *vdh* and *modABC* genes	1.94 Ferulic acid	4	Vanillin	5	3	1.66	[59]
Corynebacterium glutamicum ATCC 21420	*vanAB (Corynebacterium efficiens NBRC 100395)*	Electroporation	Genome integration	None	3.1 Ferulic Acid	4	Protocatechuic acid	400	24	1.06	[62]
Corynebacterium glutamicum PV-IYΔ0324	*comt (Rattus norvegicus) and CAR (Nocardia iowensis)*	NI	Genome integration	Deletion of *NCgl0324* gene	Endogenous 4-Hydroxylbenzoic acid	6	Vanillin	25	48	0.31	[63]

Table A1. *Cont.*

Engineering Platform	Genes (Origin)	Bioengineering Strategies Transformation Method	Insertion Type	Mutations	Substrates (g/L)	Complexity	Product	Volume of Culture (mL)	Fermentation Time (h)	Yield (g/L)	Ref.
Amycolatopsis sp. *ATCC 39116*	*fcs and ech* (*Amycolatopsis* sp.)	Direct mycelium transformation	Genome integration	Deletion of *vdh* gene	5 Glucose	6	Vanillin	2000	20	22.3	[119]
Saccharomyces cerevisiae and Schizosaccharomyces pombe	*3DSD (Podospora pauciseta), ACAR (Nocardia* sp.*), O-MT (Homo sapiens), and UGT (Arabidopsis thaliana)*	Lithium acetate/polyethylene glycol/single carrier DNA	Genome integration	Deletion of *adh6* gene	Glucose	6	Vanillin	5	48	0.065 and 0.045	[70]
Saccharomyces cerevisiae VAN286	*UGT (Arabidopsis thaliana)*	NI	NI	Deletion of *pdc1* and *gdh1* genes	Glucose	6	Vanillin-β-D-glucoside	NI	90	0.50	[72]
Saccharomyces cerevisiae	*Sam8 (Actinomycete), sam5/ hpaB/ CYP199A2(Saccharothrix espanaensis/Pseudomonas aeruginosa/Rhodopseudomonas palustris), comt (Arabidopsis thaliana), fcs, and ech (Streptomyces* sp.*).*	Lithium acetate/polyethylene glycol/single carrier DNA	Genome integration	Deletion of *Aro4* and *Aro7* genes	0.4 ferulic acid	6	Vanillin	50	96	0.008	[76]
Saccharomyces cerevisiae S288c and CEN.PK	*ACAR (Neurospora* sp.*), EntD (Escherichia coli), UGT (Arabidopsis thaliana), OMT (Homo sapiens), and 3DSD (Podospora pauciseta)*	Lithium acetate/polyethylene glycol/single carrier DNA	Genome integration	Deletion of *Bgl1* and *adh6* genes	20 Glucose	6	Vanillin-β-D-glucoside	50	45	2	[79]

Table A1. *Cont.*

Engineering Platform	Genes (Origin)	Bioengineering Strategies Transformation Method	Insertion Type	Mutations	Substrates (g/L)	Complexity	Product	Volume of Culture (mL)	Fermentation Time (h)	Yield (g/L)	Ref.
Saccharomyces cerevisiae BY4741	*AroZ (Podospora anserina), OMT (Homo sapiens), CAR (Segniliparus Rotundus), GapC (Clostridium acetobutylicum), Pos5c (S. cerevisiae), Sfp (Bacillus subtilis), EntD (E. coli), MetK (E. coli), hpaB (Pseudomonas aeruginosa), hpaC (Salmonella enterica), Xfpk (Bifidobacterium breve), CkPta (Clostridium kluyveri), HmaS (A. orientalis), HMO (S. coelicolor A3), and BFD (P. putida KT2440)*	Electroporation	Combinaison of both episomal expression and genome integration	Deletion of adh6, adh7, gre3, gcy1, ydl124w, ypr1, ari1, ydr541c, aad3, *hfd1, AKR, and AL DR family genes*	Glucose	4	Vanillin	10	120	0.36	[20]
Saccharomyces cerevisiae BY4742	*4CL (Petroselinum crispum), Ech (Pseudomonas putida KT2440), VpVAN (V. planifolia), hpaB(Pseudomonas aeruginosa), hpaC (Salmonella enterica), PobA (Pseudomonas putida KT2440), COMT (Arabidopsis thaliana), METHER (Arabidopsis thaliana), and PsVAO (Penicillium simplicissimum)*	NI	Genome integration	Deletion of *adh6, adh7, bdh2, bdh1, hfd1, pdc5, aro10, pha2, pgm1, trp2, pad1, ahfd1, AKR, and ALDR family* genes	lignocellulosic biomass	6	Vanillin	50	24	0.29	[77]

Table A1. *Cont.*

Engineering Platform	Genes (Origin)	Bioengineering Strategies Transformation Method	Insertion Type	Mutations	Substrates (g/L)	Complexity	Product	Volume of Culture (mL)	Fermentation Time (h)	Yield (g/L)	Ref.
Oryza sativa	*VpVAN* (*V. planifolia*)	Agrobacterium-mediated transformation	Genome integration	None	Endogenous ferulic Acid	4	Vanillin	NI	NI	0.00054	[81]
Capsicum frutescens	*VpVAN* (*V. planifolia*)	Microparticle bombardment	Genome integration	None	Endogenous ferulic Acid	4	Vanillin	NI	NI	0.00057	[80]

Abbreviations: TAL (*sam 8*): tyrosine ammonia lyase; C3H (sam5): 4-coumarate 3-hydroxylase; COMT: caffeate O-methyltransferase; fcs: feruloyl-CoA synthetase; ech: enoyl-CoA hydratase/aldolase; vdh: vanillin dehydrogenase; IEM: isoeugenol monooxygenase; PobA: 4-hydroxybenzoate hydroxylase; CAR: carboxyl acid reductase; Pp-SpuC: putrescine transaminase; ATA: amine transaminase; adh: alcohol dehydrogenase; sfp: phosphopantetheinyl transferase; COMT: catechol O-methyltransferase; 3DSD: 3-dehydroshikimate dehydratase; LCC: leaf–branch compost cutinase; TPADO: terephthalate 1,2-dioxygenase; DCDDH: 1,4-dicarboxylic acid dehydrogenase; OMT: O-methyltransferase; aroE: shikimate dehydrogenase; vanAB: vanillate O-demethylase; ACAR: aryl carboxylic acid reductase; UGT: UDP-glycosyltransferase; ADH6: alcohol dehydrogenase; Pdc1: pyruvate decarboxylase; gdh1: glutamate dehydrogenase; 4HPA3H (*hpaB* and C): 4- hydroxyphenylacetate 3-hydroxylase;CYP199A2: cytochrome P450 monooxygenase; DAHP: 3-deoxy-D-arabino heptulosonate-7-phosphate synthase; CHA: chorismate mutase; EntD: phosphopantetheine transferase; BGL1: beta-glucosidase; VpVAN: vanillin synthetase; NCgl0324: aromatic aldehyde reductase; 4CL: hydroxycinnamic acid CoA ligase; METHER: methylenetetrahydrofolate reductase; PsVAO: vanillyl alcohol oxidase; bdh1 and 2: 3-hydroxybutyrate dehydrogenase1 and 2; hfd1: fatty aldehyde deshydrogenase; pha2: prephenate dehydratase 2; pgm1: phosphoglucomutase 1; trp2: DOPAchrome tautomerase; pad1: phenylacrylic acid decarboxylase; AKR: aldo-keto reductases; ALDR: adrenoleukodystrophy-related; gcy1: glycerol 2-dehydrogenase; ari1: carbonyl reductase; ydr541c: aldehyde reductase; aad3: aryl-alcohol dehydrogenase; HpaBC: flavin-dependent monooxygenase; HMO: hydroxymandelate oxidase; BFD: benzoylformate decarboxylase. NI: not identified. The numbers 2, 4, and 6 represent the level of complexity of the heterologous pathway; 2 is less complex; 4 is moderately complex; and 6 is very complex. COMT in red is to differentiate the catechol O-methyltransferase from the caffeate O-methyltransferase.

References

1. Spence, C. Odour hedonics and the ubiquitous appeal of vanilla. *Nat. Food* **2022**, *3*, 837–846. [CrossRef] [PubMed]
2. Furuya, T.; Miura, M.; Kuroiwa, M.; Kino, K. High-yield production of vanillin from ferulic acid by a coenzyme-independent decarboxylase/oxygenase two-stage process. *New Biotechnol.* **2015**, *32*, 335–339. [CrossRef] [PubMed]
3. Xu, L.; Liaqat, F.; Sun, J.; Khazi, M.I.; Xie, R.; Zhu, D. Advances in the vanillin synthesis and biotransformation: A review. *Renew. Sustain. Energy Rev.* **2024**, *189*, 113905. [CrossRef]
4. Gallage, N.J.; Hansen, E.H.; Kannangara, R.; Olsen, C.E.; Motawia, M.S.; Jørgensen, K.; Holme, I.; Hebelstrup, K.; Grisoni, M.; Møller, B.L. Vanillin formation from ferulic acid in Vanilla planifolia is catalysed by a single enzyme. *Nat. Commun.* **2014**, *5*, 4037. [CrossRef] [PubMed]
5. Taira, J.; Toyoshima, R.; Ameku, N.; Iguchi, A.; Tamaki, Y. Vanillin production by biotransformation of phenolic compounds in fungus, Aspergillus luchuensis. *AMB Express* **2018**, *8*, 40. [CrossRef]
6. Krings, U.; Berger, R.G. Biotechnological production of flavours and fragrances. *Appl. Microbiol. Biotechnol.* **1998**, *49*, 1–8. [CrossRef]
7. Gallage, N.J.; Moller, B.L. Vanillin-bioconversion and bioengineering of the most popular plant flavor and its de novo biosynthesis in the vanilla orchid. *Mol. Plant* **2015**, *8*, 40–57. [CrossRef]
8. Omokhua-Uyi, A.G.; Van Staden, J. Natural product remedies for COVID-19: A focus on safety. *S. Afr. J. Bot.* **2021**, *139*, 386–398. [CrossRef]
9. SNS, Vanillin Market Size, Share and Segmentation by Product Type (Natural, Synthetic Vanillin), by End-Use (Food & Beverage, Fragrance, Pharmaceutical), by Regions and Global Market Forecast 2023–2030. *Microorganisms*. 2022. Available online: https://www.snsinsider.com/reports/vanillin-market-2244 (accessed on 25 November 2024).
10. Hocking, M.B. Vanillin: Synthetic Flavoring from Spent Sulfite Liquor. *J. Chem. Educ.* **1997**, *74*, 1055. [CrossRef]
11. Zhou, N.; Thilakarathna, W.P.D.W.; He, Q.S.; Rupasinghe, H.P.V. A Review: Depolymerization of Lignin to Generate High-Value Bio-Products: Opportunities, Challenges, and Prospects. *Front. Energy Res.* **2022**, *9*, 758744. [CrossRef]
12. Barghini, P.; Di Gioia, D.; Fava, F.; Ruzzi, M. Vanillin production using metabolically engineered Escherichia coli under non-growing conditions. *Microb. Cell Fact.* **2007**, *6*, 13. [CrossRef] [PubMed]
13. Braga, A.; Guerreiro, C.E.; Belo, I. Generation of Flavors and Fragrances Through Biotransformation and De Novo Synthesis. *Food Bioprocess. Technol.* **2018**, *11*, 2217–2228. [CrossRef]
14. Muheim, A.; Lerch, K. Towards a high-yield bioconversion of ferulic acid to vanillin. *Appl. Microbiol. Biotechnol.* **1999**, *51*, 456–461. [CrossRef]
15. Chou, T.H.; Ding, H.Y.; Hung, W.J.; Liang, C.H. Antioxidative characteristics and inhibition of alpha-melanocyte-stimulating hormone-stimulated melanogenesis of vanillin and vanillic acid from Origanum vulgare. *Exp. Dermatol.* **2010**, *19*, 742–750. [CrossRef] [PubMed]
16. Kundu, A.; Mitra, A. Flavoring extracts of Hemidesmus indicus roots and Vanilla planifolia pods exhibit in vitro acetylcholinesterase inhibitory activities. *Plant Foods Hum. Nutr.* **2013**, *68*, 247–253. [CrossRef]
17. Ma, Q.; Liu, L.; Zhao, S.; Huang, Z.; Li, C.; Jiang, S.; Li, Q.; Gu, P. Biosynthesis of vanillin by different microorganisms: A review. *World J. Microbiol. Biotechnol.* **2022**, *38*, 40. [CrossRef]
18. Huang, Y.; Yang, Y.; Xue, J.; Liao, Y.; Fu, X.; Zhu, C.; Li, J.; Zeng, L.; Yang, Z. Biosynthetic Pathway and Bioactivity of Vanillin, a Highly Abundant Metabolite Distributed in the Root Cortex of Tea Plants (Camellia sinensis). *J. Agric. Food Chem.* **2024**, *72*, 1660–1673. [CrossRef]
19. Xu, L.; Liaqat, F.; Khazi, M.I.; Sun, J.; Zhu, D. Natural deep eutectic solvents-based green extraction of vanillin: Optimization, purification, and bioactivity assessment. *Front. Nutr.* **2024**, *10*, 1279552. [CrossRef]
20. Mo, Q.; Yuan, J. Minimal aromatic aldehyde reduction (MARE) yeast platform for engineering vanillin production. *Biotechnol. Biofuels Bioprod.* **2024**, *17*, 4. [CrossRef]
21. Jassey, V.E.J.; Hamard, S.; Lepère, C.; Céréghino, R.; Corbara, B.; Küttim, M.; Leflaive, J.; Leroy, C.; Carrias, J.-F. Photosynthetic microorganisms effectively contribute to bryophyte CO2 fixation in boreal and tropical regions. *ISME Commun.* **2022**, *2*, 64. [CrossRef]
22. Cordero, B.F.; Couso, I.; León, R.; Rodríguez, H.; Vargas, M.A. Enhancement of carotenoids biosynthesis in Chlamydomonas reinhardtii by nuclear transformation using a phytoene synthase gene isolated from Chlorella zofingiensis. *Appl. Microbiol. Biotechnol.* **2011**, *91*, 341–351. [CrossRef] [PubMed]
23. Shamriz, S.; Ofoghi, H. Expression of Recombinant PfCelTOS Antigen in the Chloroplast of Chlamydomonas reinhardtii and its Potential Use in Detection of Malaria. *Mol. Biotechnol.* **2019**, *61*, 102–110. [CrossRef] [PubMed]
24. Wang, Q.; Zhuang, J.; Ni, S.; Luo, H.; Zheng, K.; Li, X.; Lan, C.; Zhao, D.; Bai, Y.; Jia, B.; et al. Overexpressing CrePAPS Polyadenylate Activity Enhances Protein Translation and Accumulation in Chlamydomonas reinhardtii. *Mar. Drugs* **2022**, *20*, 276. [CrossRef]
25. Passariello, C.L.; Marchionni, S.; Carcuro, M.; Casali, G.; Della Pasqua, A.; Hrelia, S.; Malaguti, M.; Lorenzini, A. The Mediterranean Athlete's Nutrition: Are Protein Supplements Necessary? *Nutrients* **2020**, *12*, 3681. [CrossRef] [PubMed]
26. Mohammadi Gheisar, M.; Hosseindoust, A.; Kim, I.H. Evaluating the effect of microencapsulated blends of organic acids and essential oils in broiler chickens diet. *J. Appl. Poult. Res.* **2015**, *24*, 511–519. [CrossRef]

27. Pereira, A.M.; de Lurdes Nunes Enes Dapkevicius, M.; Borba, A.E.S. Alternative pathways for hydrogen sink originated from the ruminal fermentation of carbohydrates: Which microorganisms are involved in lowering methane emission? *Anim. Microbiome* **2022**, *4*, 5. [CrossRef]
28. Malone, S.L.; Oh, Y.; Arndt, K.A.; Burba, G.; Commane, R.; Contosta, A.R.; Goodrich, J.P.; Loescher, H.W.; Starr, G.; Varner, R.K. Gaps in network infrastructure limit our understanding of biogenic methane emissions for the United States. *Biogeosciences* **2022**, *19*, 2507–2522. [CrossRef]
29. Di Gioia, D.; Luziatelli, F.; Negroni, A.; Ficca, A.G.; Fava, F.; Ruzzi, M. Metabolic engineering of Pseudomonas fluorescens for the production of vanillin from ferulic acid. *J. Biotechnol.* **2011**, *156*, 309–316. [CrossRef]
30. Lee, J.; Cho, J.Y.; Lee, S.Y.; Lee, K.W.; Lee, J.; Song, J.Y. Vanillin protects human keratinocyte stem cells against ultraviolet B irradiation. *Food Chem. Toxicol.* **2014**, *63*, 30–37. [CrossRef]
31. Loh, Y.H.; Wu, Q.; Chew, J.L.; Vega, V.B.; Zhang, W.; Chen, X.; Bourque, G.; George, J.; Leong, B.; Liu, J.; et al. The Oct4 and Nanog transcription network regulates pluripotency in mouse embryonic stem cells. *Nat. Genet.* **2006**, *38*, 431–440. [CrossRef]
32. Paul, V.; Rai, D.C.; Ramyaa Lakshmi, T.S.; Srivastava, S.K.; Tripathi, A.D. A comprehensive review on vanillin: Its microbial synthesis, isolation and recovery. *Food Biotechnol.* **2021**, *35*, 22–49. [CrossRef]
33. Arya, S.S.; Rookes, J.E.; Cahill, D.M.; Lenka, S.K. Vanillin: A review on the therapeutic prospects of a popular flavouring molecule. *Adv. Tradit. Med.* **2021**, *21*, 1–17. [CrossRef]
34. Mani, A.; Ahamed, A.; Ali, D.; Alarifi, S.; Akbar, I. Dopamine-Mediated Vanillin Multicomponent Derivative Synthesis via Grindstone Method: Application of Antioxidant, Anti-Tyrosinase, and Cytotoxic Activities. *Drug Des. Devel. Ther.* **2021**, *15*, 787–802. [CrossRef] [PubMed]
35. Olatunde, A.; Mohammed, A.; Ibrahim, M.A.; Tajuddeen, N.; Shuaibu, M.N. Vanillin: A food additive with multiple biological activities. *Eur. J. Med. Chem. Rep.* **2022**, *5*, 100055. [CrossRef]
36. Bezerra, D.P.; Soares, A.K.; de Sousa, D.P. Overview of the Role of Vanillin on Redox Status and Cancer Development. *Oxid. Med. Cell Longev.* **2016**, *2016*, 9734816. [CrossRef]
37. Kim, H.J.; Hwang, I.K.; Won, M.H. Vanillin, 4-hydroxybenzyl aldehyde and 4-hydroxybenzyl alcohol prevent hippocampal CA1 cell death following global ischemia. *Brain Res.* **2007**, *1181*, 130–141. [CrossRef]
38. Yan, X.; Liu, D.F.; Zhang, X.Y.; Liu, D.; Xu, S.Y.; Chen, G.X.; Huang, B.X.; Ren, W.Z.; Wang, W.; Fu, S.P.; et al. Vanillin Protects Dopaminergic Neurons against Inflammation-Mediated Cell Death by Inhibiting ERK1/2, P38 and the NF-κB Signaling Pathway. *Int. J. Mol. Sci.* **2017**, *18*, 389. [CrossRef]
39. Kim, J.H.; Lee, H.-O.; Cho, Y.-J.; Kim, J.; Chun, J.; Choi, J.; Lee, Y.; Jung, W.H. A vanillin derivative causes mitochondrial dysfunction and triggers oxidative stress in Cryptococcus neoformans. *PLoS ONE* **2014**, *9*, e89122. [CrossRef]
40. Hariono, M.; Abdullah, N.; Damodaran, K.; Kamarulzaman, E.E.; Mohamed, N.; Hassan, S.S.; Shamsuddin, S.; Wahab, H.A. Potential new H1N1 neuraminidase inhibitors from ferulic acid and vanillin: Molecular modelling, synthesis and in vitro assay. *Sci. Rep.* **2016**, *6*, 38692. [CrossRef]
41. Zhao, D.; Jiang, Y.; Sun, J.; Li, H.; Huang, M.; Sun, X.; Zhao, M. Elucidation of The Anti-Inflammatory Effect of Vanillin In Lps-Activated THP-1 Cells. *J. Food Sci.* **2019**, *84*, 1920–1928. [CrossRef]
42. Yun, H.M.; Kim, E.; Kwon, Y.J.; Park, K.R. Vanillin Promotes Osteoblast Differentiation, Mineral Apposition, and Antioxidant Effects in Pre-Osteoblasts. *Pharmaceutics* **2024**, *16*, 485. [CrossRef] [PubMed]
43. Gendron, D. Vanillin: A Promising Biosourced Building Block for the Preparation of Various Heterocycles. *Front. Chem.* **2022**, *10*, 949355. [CrossRef] [PubMed]
44. Lou, M.; Li, S.; Jin, F.; Yang, T.; Song, R.; Song, B. Pesticide Engineering from Natural Vanillin: Recent Advances and a Perspective. *Engineering* **2024**. [CrossRef]
45. Jiang, W.; Chen, X.; Feng, Y.; Sun, J.; Jiang, Y.; Zhang, W.; Xin, F.; Jiang, M. Current Status, Challenges, and Prospects for the Biological Production of Vanillin. *Fermentation* **2023**, *9*, 389. [CrossRef]
46. Funk, C.; Brodelius, P.E. Phenylpropanoid Metabolism in Suspension Cultures of Vanilla planifolia Andr. 1: IV. Induction of Vanillic Acid Formation. *Plant Physiol.* **1992**, *99*, 256–262. [CrossRef]
47. Schoch, G.; Goepfert, S.; Morant, M.; Hehn, A.; Meyer, D.; Ullmann, P.; Werck-Reichhart, D. CYP98A3 from Arabidopsis thaliana is a 3′-hydroxylase of phenolic esters, a missing link in the phenylpropanoid pathway. *J. Biol. Chem.* **2001**, *276*, 36566–36574. [CrossRef]
48. Kundu, A. Vanillin biosynthetic pathways in plants. *Planta* **2017**, *245*, 1069–1078. [CrossRef]
49. Podstolski, A.; Havkin-Frenkel, D.; Malinowski, J.; Blount, J.W.; Kourteva, G.; Dixon, R.A. Unusual 4-hydroxybenzaldehyde synthase activity from tissue cultures of the vanilla orchid Vanilla planifolia. *Phytochemistry* **2002**, *61*, 611–620. [CrossRef]
50. Liaqat, F.; Xu, L.; Khazi, M.I.; Ali, S.; Rahman, M.U.; Zhu, D. Extraction, purification, and applications of vanillin: A review of recent advances and challenges. *Ind. Crops Prod.* **2023**, *204*, 117372. [CrossRef]
51. Zhao, M.L.; Cai, W.S.; Zheng, S.Q.; Zhao, J.L.; Zhang, J.L.; Huang, Y.; Hu, Z.L.; Jia, B. Metabolic Engineering of the Isopentenol Utilization Pathway Enhanced the Production of Terpenoids in Chlamydomonas reinhardtii. *Mar. Drugs* **2022**, *20*, 577. [CrossRef]
52. Converti, A.; Aliakbarian, B.; Domínguez, J.M.; Bustos Vázquez, G.; Perego, P. Microbial production of biovanillin. *Braz. J. Microbiol.* **2010**, *41*, 519–530. [CrossRef] [PubMed]
53. Torre, P.; De Faveri, D.; Perego, P.; Ruzzi, M.; Barghini, P.; Gandolfi, R.; Converti, A. Bioconversion of ferulate into vanillin by Escherichia coli strain JM109/pBB1 in an immobilized-cell reactor. *Ann. Microbiol.* **2004**, *54*, 517–527.

54. Ni, J.; Tao, F.; Du, H.; Xu, P. Mimicking a natural pathway for de novo biosynthesis: Natural vanillin production from accessible carbon sources. *Sci. Rep.* **2015**, *5*, 13670. [CrossRef] [PubMed]
55. Kunjapur, A.M.; Hyun, J.C.; Prather, K.L. Deregulation of S-adenosylmethionine biosynthesis and regeneration improves methylation in the E. coli de novo vanillin biosynthesis pathway. *Microb. Cell Fact.* **2016**, *15*, 61. [CrossRef] [PubMed]
56. Chen, Z.; Shen, X.; Wang, J.; Wang, J.; Zhang, R.; Rey, J.F.; Yuan, Q.; Yan, Y. Establishing an Artificial Pathway for De Novo Biosynthesis of Vanillyl Alcohol in Escherichia coli. *ACS Synth. Biol.* **2017**, *6*, 1784–1792. [CrossRef]
57. Wang, Q.; Wu, X.; Lu, X.; He, Y.; Ma, B.; Xu, Y. Efficient Biosynthesis of Vanillin from Isoeugenol by Recombinant Isoeugenol Monooxygenase from Pseudomonas nitroreducens Jin1. *Appl. Biochem. Biotechnol.* **2021**, *193*, 1116–1128. [CrossRef]
58. Yang, M.; Meng, H.; Li, X.; Wang, J.; Shen, X.; Sun, X.; Yuan, Q. Coculture engineering for efficient production of vanillyl alcohol in Escherichia coli. *aBIOTECH* **2022**, *3*, 292–300. [CrossRef]
59. Graf, N.; Altenbuchner, J. Genetic engineering of Pseudomonas putida KT2440 for rapid and high-yield production of vanillin from ferulic acid. *Appl. Microbiol. Biotechnol.* **2014**, *98*, 137–149. [CrossRef]
60. Park, S.H.; Kim, H.U.; Kim, T.Y.; Park, J.S.; Kim, S.-S.; Lee, S.Y. Metabolic engineering of Corynebacterium glutamicum for L-arginine production. *Nat. Commun.* **2014**, *5*, 4618. [CrossRef]
61. Chen, C.; Pan, J.; Yang, X.; Guo, C.; Ding, W.; Si, M.; Zhang, Y.; Shen, X.; Wang, Y. Global Transcriptomic Analysis of the Response of Corynebacterium glutamicum to Vanillin. *PLoS ONE* **2016**, *11*, e0164955. [CrossRef]
62. Okai, N.; Masuda, T.; Takeshima, Y.; Tanaka, K.; Yoshida, K.I.; Miyamoto, M.; Ogino, C.; Kondo, A. Biotransformation of ferulic acid to protocatechuic acid by Corynebacterium glutamicum ATCC 21420 engineered to express vanillate O-demethylase. *AMB Express* **2017**, *7*, 130. [CrossRef]
63. Kim, H.S.; Choi, J.A.; Kim, B.Y.; Ferrer, L.; Choi, J.M.; Wendisch, V.F.; Lee, J.H. Engineered Corynebacterium glutamicum as the Platform for the Production of Aromatic Aldehydes. *Front. Bioeng. Biotechnol.* **2022**, *10*, 880277. [CrossRef] [PubMed]
64. Mora, D.; Fortina, M.G.; Parini, C.; Manachini, P.L. Identification of Pediococcus acidilactici and Pediococcus pentosaceus based on 16S rRNA and ldhD gene-targeted multiplex PCR analysis. *FEMS Microbiol. Lett.* **1997**, *151*, 231–236. [CrossRef] [PubMed]
65. Kaur, B.; Chakraborty, D.; Kumar, B. Metabolic engineering of Pediococcus acidilactici BD16 for production of vanillin through ferulic acid catabolic pathway and process optimization using response surface methodology. *Appl. Microbiol. Biotechnol.* **2014**, *98*, 8539–8551. [CrossRef] [PubMed]
66. Chakraborty, D.; Selvam, A.; Kaur, B.; Wong, J.W.C.; Karthikeyan, O.P. Application of recombinant Pediococcus acidilactici BD16 (fcs+/ech+) for bioconversion of agrowaste to vanillin. *Appl. Microbiol. Biotechnol.* **2017**, *101*, 5615–5626. [CrossRef]
67. Scheper, A.F.; Schofield, J.; Bohara, R.; Ritter, T.; Pandit, A. Understanding glycosylation: Regulation through the metabolic flux of precursor pathways. *Biotechnol. Adv.* **2023**, *67*, 108184. [CrossRef]
68. Bryan, L.; Clynes, M.; Meleady, P. The emerging role of cellular post-translational modifications in modulating growth and productivity of recombinant Chinese hamster ovary cells. *Biotechnol. Adv.* **2021**, *49*, 107757. [CrossRef]
69. Liu, Z.L.; Ma, M.; Song, M. Evolutionarily engineered ethanologenic yeast detoxifies lignocellulosic biomass conversion inhibitors by reprogrammed pathways. *Mol. Genet. Genom.* **2009**, *282*, 233–244.
70. Hansen, E.H.; Møller, B.L.; Kock, G.R.; Bünner, C.M.; Kristensen, C.; Jensen, O.R.; Okkels, F.T.; Olsen, C.E.; Motawia, M.S.; Hansen, J. De novo biosynthesis of vanillin in fission yeast (Schizosaccharomyces pombe) and baker's yeast (Saccharomyces cerevisiae). *Appl. Env. Microbiol.* **2009**, *75*, 2765–2774. [CrossRef]
71. Moehs, C.P.; Allen, P.V.; Friedman, M.; Belknap, W.R. Cloning and expression of solanidine UDP-glucose glucosyltransferase from potato. *Plant J.* **1997**, *11*, 227–236. [CrossRef]
72. Brochado, A.R.; Matos, C.; Møller, B.L.; Hansen, J.; Mortensen, U.H.; Patil, K.R. Improved vanillin production in baker's yeast through in silico design. *Microb. Cell Factories* **2010**, *9*, 84. [CrossRef] [PubMed]
73. Zamzuri, N.A.A.; Abd-Aziz, S. Biovanillin from agro wastes as an alternative food flavour. *J. Sci. Food Agric.* **2013**, *93*, 429–438. [CrossRef] [PubMed]
74. Shen, Y.; Li, H.; Wang, X.; Zhang, X.; Hou, J.; Wang, L.; Gao, N.; Bao, X. High vanillin tolerance of an evolved Saccharomyces cerevisiae strain owing to its enhanced vanillin reduction and antioxidative capacity. *J. Ind. Microbiol. Biotechnol.* **2014**, *41*, 1637–1645. [CrossRef] [PubMed]
75. Liang, Z.; Wang, X.; Bao, X.; Wei, T.; Hou, J.; Liu, W.; Shen, Y. Newly identified genes contribute to vanillin tolerance in Saccharomyces cerevisiae. *Microb. Biotechnol.* **2021**, *14*, 503–516. [CrossRef] [PubMed]
76. Qiu, D.; Wang, M.; Zhou, C.; Zhao, J.; Zhang, G. De novo biosynthesis of vanillin in engineered Saccharomyces cerevisiae. *Chem. Eng. Sci.* **2022**, *263*, 118049. [CrossRef]
77. Xin, X.; Zhang, R.-K.; Liu, S.-C.; He, Z.-J.; Liu, R.-Y.; Lan, H.-N.; Liu, Z.-H.; Li, B.-Z.; Yuan, Y.-J. Engineering yeast to convert lignocellulose into vanillin. *Chem. Eng. J.* **2024**, *485*, 149815. [CrossRef]
78. Chen, R.; Gao, J.; Yu, W.; Chen, X.; Zhai, X.; Chen, Y.; Zhang, L.; Zhou, Y.J. Engineering cofactor supply and recycling to drive phenolic acid biosynthesis in yeast. *Nat. Chem. Biol.* **2022**, *18*, 520–529. [CrossRef]
79. Strucko, T.; Magdenoska, O.; Mortensen, U.H. Benchmarking two commonly used Saccharomyces cerevisiae strains for heterologous vanillin-β-glucoside production. *Metab. Eng. Commun.* **2015**, *2*, 99–108. [CrossRef]
80. Chee, M.J.; Lycett, G.W.; Khoo, T.J.; Chin, C.F. Bioengineering of the Plant Culture of Capsicum frutescens with Vanillin Synthase Gene for the Production of Vanillin. *Mol. Biotechnol.* **2017**, *59*, 1–8. [CrossRef]

81. Arya, S.S.; Mahto, B.K.; Sengar, M.S.; Rookes, J.E.; Cahill, D.M.; Lenka, S.K. Metabolic Engineering of Rice Cells with Vanillin Synthase Gene (VpVAN) to Produce Vanillin. *Mol. Biotechnol.* **2022**, *64*, 861–872. [CrossRef]
82. Diamond, A.; Barnabé, S.; Desgagné-Penix, I. Is a spice missing from the recipe? The intra-cellular localization of vanillin biosynthesis needs further investigations. *Plant Biol.* **2022**, *25*, 3–7. [CrossRef] [PubMed]
83. Yang, H.; Barros-Rios, J.; Kourteva, G.; Rao, X.; Chen, F.; Shen, H.; Liu, C.; Podstolski, A.; Belanger, F.; Havkin-Frenkel, D.; et al. A re-evaluation of the final step of vanillin biosynthesis in the orchid Vanilla planifolia. *Phytochemistry* **2017**, *139*, 33–46. [CrossRef] [PubMed]
84. Liu, Y.; Sun, L.; Huo, Y.-X.; Guo, S. Strategies for improving the production of bio-based vanillin. *Microb. Cell Factories* **2023**, *22*, 147. [CrossRef] [PubMed]
85. Mao, J.; Zhang, H.; Chen, Y.; Wei, L.; Liu, J.; Nielsen, J.; Chen, Y.; Xu, N. Relieving metabolic burden to improve robustness and bioproduction by industrial microorganisms. *Biotechnol. Adv.* **2024**, *74*, 108401. [CrossRef]
86. Chen, Q.; Jiang, Y.; Kang, Z.; Cheng, J.; Xiong, X.; Hu, C.Y.; Meng, Y. Engineering a Feruloyl–Coenzyme A Synthase for Bioconversion of Phenylpropanoid Acids into High-Value Aromatic Aldehydes. *J. Agric. Food Chem.* **2022**, *70*, 9948–9960. [CrossRef]
87. Pattrick, C.A.; Webb, J.P.; Green, J.; Chaudhuri, R.R.; Collins, M.O.; Kelly, D.J. Proteomic Profiling, Transcription Factor Modeling, and Genomics of Evolved Tolerant Strains Elucidate Mechanisms of Vanillin Toxicity in Escherichia coli. *mSystems* **2019**, *4*. [CrossRef]
88. Fujimaki, S.; Sakamoto, S.; Shimada, S.; Kino, K.; Furuya, T. Engineering a coenzyme-independent dioxygenase for one-step production of vanillin from ferulic acid. *Appl. Environ. Microbiol.* **2024**, *90*, e0023324. [CrossRef]
89. Zeng, X.; Danquah, M.K.; Chen, X.D.; Lu, Y. Microalgae bioengineering: From CO2 fixation to biofuel production. *Renew. Sustain. Energy Rev.* **2011**, *15*, 3252–3260. [CrossRef]
90. Kselíková, V.; Singh, A.; Bialevich, V.; Čížková, M.; Bišová, K. Improving microalgae for biotechnology—From genetics to synthetic biology—Moving forward but not there yet. *Biotechnol. Adv.* **2022**, *58*, 107885. [CrossRef]
91. Hammerschmidt, N.; Tscheliessnig, A.; Sommer, R.; Helk, B.; Jungbauer, A. Economics of recombinant antibody production processes at various scales: Industry-standard compared to continuous precipitation. *Biotechnol. J.* **2014**, *9*, 766–775. [CrossRef]
92. Cichonski, J.; Chrzanowski, G. Microalgae as a Source of Valuable Phenolic Compounds and Carotenoids. *Molecules* **2022**, *27*, 8852. [CrossRef] [PubMed]
93. Rico, M.; López, A.; Santana-Casiano, J.M.; Gonzàlez, A.G.; Gonzàlez-Dàvila, M. Variability of the phenolic profile in the diatom Phaeodactylum tricornutum growing under copper and iron stress. *Limnol. Oceanogr.* **2013**, *58*, 144–152. [CrossRef]
94. Del Mondo, A.; Sansone, C.; Brunet, C. Insights into the biosynthesis pathway of phenolic compounds in microalgae. *Comput. Struct. Biotechnol. J.* **2022**, *20*, 1901–1913. [CrossRef] [PubMed]
95. Bhuvana, P.; Sangeetha, P.; Anuradha, V.; Ali, M.S. Spectral characterization of bioactive compounds from microalgae: N. oculata and C. vulgaris. *Biocatal. Agric. Biotechnol.* **2019**, *19*, 101094. [CrossRef]
96. Tripathi, U.; Ramachandra Rao, S.; Ravishankar, G.A. Biotransformation of phenylpropanoid compounds to vanilla flavor metabolites in cultures of Haematococcus pluvialis. *Process Biochem.* **2002**, *38*, 419–426. [CrossRef]
97. Liyanage, N.S.; Awwad, F.; Gonçalves dos Santos, K.C.; Jayawardena, T.U.; Mérindol, N.; Desgagné-Penix, I. Navigating Amaryllidaceae alkaloids: Bridging gaps and charting biosynthetic territories. *J. Exp. Bot.* **2024**, erae187. [CrossRef]
98. Desgagné-Penix, I. Biosynthesis of alkaloids in Amaryllidaceae plants: A review. *Phytochem. Rev.* **2021**, *20*, 409–431. [CrossRef]
99. Tzin, V.; Galili, G. New Insights into the Shikimate and Aromatic Amino Acids Biosynthesis Pathways in Plants. *Mol. Plant* **2010**, *3*, 956–972. [CrossRef]
100. Yang, W.; Tang, H.; Ni, J.; Wu, Q.; Hua, D.; Tao, F.; Xu, P. Characterization of two Streptomyces enzymes that convert ferulic acid to vanillin. *PLoS ONE* **2013**, *8*, e67339. [CrossRef]
101. Ryu, J.-Y.; Seo, J.; Park, S.; Ahn, J.-H.; Chong, Y.; Sadowsky, M.J.; Hur, H.-G. Characterization of an Isoeugenol Monooxygenase (Iem) from *Pseudomonas nitroreducens* Jin1 That Transforms Isoeugenol to Vanillin. *Biosci. Biotechnol. Biochem.* **2013**, *77*, 289–294. [CrossRef]
102. Jeong, B.-r.; Jang, J.; Jin, E. Genome engineering via gene editing technologies in microalgae. *Bioresour. Technol.* **2023**, *373*, 128701. [CrossRef] [PubMed]
103. Patel, V.K.; Das, A.; Kumari, R.; Kajla, S. Recent progress and challenges in CRISPR-Cas9 engineered algae and cyanobacteria. *Algal Res.* **2023**, *71*, 103068. [CrossRef]
104. Slattery, S.S.; Diamond, A.; Wang, H.; Therrien, J.A.; Lant, J.T.; Jazey, T.; Lee, K.; Klassen, Z.; Desgagné-Penix, I.; Karas, B.J.; et al. An Expanded Plasmid-Based Genetic Toolbox Enables Cas9 Genome Editing and Stable Maintenance of Synthetic Pathways in Phaeodactylum tricornutum. *ACS Synth. Biol.* **2018**, *7*, 328–338. [CrossRef]
105. Diamond, A.; Diaz-Garza, A.M.; Li, J.; Slattery, S.S.; Merindol, N.; Fantino, E.; Meddeb-Mouelhi, F.; Karas, B.J.; Barnabé, S.; Desgagné-Penix, I. Instability of extrachromosomal DNA transformed into the diatom Phaeodactylum tricornutum. *Algal Res.* **2023**, *70*, 102998. [CrossRef]
106. Roig-Merino, A.; Urban, M.; Bozza, M.; Peterson, J.D.; Bullen, L.; Büchler-Schäff, M.; Stäble, S.; van der Hoeven, F.; Müller-Decker, K.; McKay, T.R.; et al. An episomal DNA vector platform for the persistent genetic modification of pluripotent stem cells and their differentiated progeny. *Stem Cell Rep.* **2022**, *17*, 143–158. [CrossRef]

107. Wade-Martins, R.; James, M.R.; Frampton, J. Long-term stability of large insert genomic DNA episomal shuttle vectors in human cells. *Nucleic Acids Res.* **1999**, *27*, 1674–1682. [CrossRef]
108. Chen, Y.; Partow, S.; Scalcinati, G.; Siewers, V.; Nielsen, J. Enhancing the copy number of episomal plasmids in Saccharomyces cerevisiae for improved protein production. *FEMS Yeast Res.* **2012**, *12*, 598–607. [CrossRef]
109. Suttangkakul, A.; Sirikhachornkit, A.; Juntawong, P.; Puangtame, W.; Chomtong, T.; Srifa, S.; Sathitnaitham, S.; Dumrongthawatchai, W.; Jariyachawalid, K.; Vuttipongchaikij, S. Evaluation of strategies for improving the transgene expression in an oleaginous microalga Scenedesmus acutus. *BMC Biotechnol.* **2019**, *19*, 4. [CrossRef]
110. Cagney, M.H.; O'Neill, E.C. Strategies for producing high value small molecules in microalgae. *Plant Physiol. Biochem.* **2024**, *214*, 108942. [CrossRef]
111. Einhaus, A.; Baier, T.; Kruse, O. Molecular design of microalgae as sustainable cell factories. *Trends Biotechnol.* **2024**, *42*, 728–738. [CrossRef]
112. Naduthodi, M.I.S.; Claassens, N.J.; D'Adamo, S.; van der Oost, J.; Barbosa, M.J. Synthetic Biology Approaches To Enhance Microalgal Productivity. *Trends Biotechnol.* **2021**, *39*, 1019–1036. [CrossRef] [PubMed]
113. Yamada, M.; Okada, Y.; Yoshida, T.; Nagasawa, T. Vanillin production using Escherichia coli cells over-expressing isoeugenol monooxygenase of Pseudomonas putida. *Biotechnol. Lett.* **2008**, *30*, 665–670. [CrossRef]
114. Zhao, L.; Xie, Y.; Chen, L.; Xu, X.; Zhao, C.X.; Cheng, F. Efficient biotransformation of isoeugenol to vanillin in recombinant strains of Escherichia coli by using engineered isoeugenol monooxygenase and sol-gel chitosan membrane. *Process Biochem.* **2018**, *71*, 76–81. [CrossRef]
115. Luziatelli, F.; Brunetti, L.; Ficca, A.G.; Ruzzi, M. Maximizing the Efficiency of Vanillin Production by Biocatalyst Enhancement and Process Optimization. *Front. Bioeng. Biotechnol.* **2019**, *7*, 279. [CrossRef] [PubMed]
116. Sadler, J.C.; Wallace, S. Microbial synthesis of vanillin from waste poly(ethylene terephthalate). *Green Chem.* **2021**, *23*, 4665–4672. [CrossRef] [PubMed]
117. Yoon, S.-H.; Li, C.; Kim, J.-E.; Lee, S.-H.; Yoon, J.-Y.; Choi, M.-S.; Seo, W.-T.; Yang, J.-K.; Kim, J.-Y.; Kim, S.-W. Production of Vanillin by Metabolically Engineered Escherichia coli. *Biotechnol. Lett.* **2005**, *27*, 1829–1832. [CrossRef]
118. Manfrão-Netto, J.H.C.; Lund, F.; Muratovska, N.; Larsson, E.M.; Parachin, N.S.; Carlquist, M. Metabolic engineering of Pseudomonas putida for production of vanillylamine from lignin-derived substrates. *Microb. Biotechnol.* **2021**, *14*, 2448–2462. [CrossRef]
119. Fleige, C.; Meyer, F.; Steinbüchel, A. Metabolic Engineering of the Actinomycete Amycolatopsis sp. Strain ATCC 39116 towards Enhanced Production of Natural Vanillin. *Appl. Environ. Microbiol.* **2016**, *82*, 3410–3419. [CrossRef]

BioChem

MDPI

Systematic Review

Appropriate Prescription of Non-Steroidal Anti-Inflammatory Drugs in Geriatric Patients—A Systematic Review

Carolina Costa [1,2], Diana Soares [1], Ana Borges [3], Ana Gonçalves [4], José Paulo Andrade [2,5] and Hugo Ribeiro [2,6,7,8,*]

[1] Health Family Unit Barão do Corvo, Local Health Unit Gaia and Espinho, 4434-502 Vila Nova de Gaia, Portugal; ana.gomes.costa@ulsge.min-saude.pt (C.C.); diana.arnelas.soares@ulsge.min-saude.pt (D.S.)
[2] Faculty of Medicine, University of Porto, 4200-319 Porto, Portugal; jandrade@med.up.pt
[3] Health Family Unit Covelo, Local Health Unit S. João, 4200-319 Porto, Portugal; raquelborges1608@gmail.com
[4] Health Family Unit Santo André de Canidelo, Local Health Unit Gaia and Espinho, 4434-502 Vila Nova de Gaia, Portugal; anaduartemg@gmail.com
[5] RISE@Health, 4200-319 Porto, Portugal
[6] Community Support Team in Palliative Care Gaia, Local Health Unit Gaia and Espinho, 4434-502 Vila Nova de Gaia, Portugal
[7] Faculty of Medicine, University of Coimbra, 3000-548 Coimbra, Portugal
[8] Center for Innovative Biomedicine and Biotechnology—CIMAGO, 3000-548 Coimbra, Portugal
* Correspondence: hribeiroff@gmail.com

Abstract: The elderly population is growing worldwide. Non-steroidal anti-inflammatory drugs (NSAIDs) are commonly prescribed, but their adverse events can pose significant risks. Different NSAID molecules can exhibit varying risk profiles. This study aims to evaluate the cardiovascular, gastrointestinal, and renal safety profiles of ibuprofen, naproxen, acemetacin, diclofenac, celecoxib, and etoricoxib in elderly patients. A comprehensive literature search was conducted in PubMed and Cochrane Library. For the selection of articles, we used Medical Subject Headings (MeSH) terms "aged" sequentially and together with "ibuprofen", "diclofenac", "naproxen", "acemetacin", "celecoxib", and "etoricoxib". To assess the quality and interest of the articles, four independent reviewers screened titles and abstracts to identify potentially eligible studies. Strength of Recommendation Taxonomy (SORT) was used to rate the quality of individual studies and to establish recommendation strengths (RS). From 2086 articles identified, 39 studies met the inclusion criteria. Twenty studies analyzed cardiovascular safety, fourteen gastrointestinal safety, and four renal safety. When CV risk is the main concern celecoxib or naproxen are a good first choice (RS B). In high GI risk addition of PPI to naproxen or celecoxib use should be recommended (RS A). When renal function is on focus, celecoxib remains as first line of therapy (RS A). Diclofenac in the geriatric population should be avoided (RS B). Celecoxib is a good choice for elderly patients for whom it is difficult to direct pain treatment based on a single known risk factor (RS B).

Keywords: elderly; anti-inflammatory agents; non-steroidal; pharmacology; clinical; physiological effects of drugs

1. Introduction

Over the last decades, population aging has become evident, with a continuous increase in the elderly population worldwide [1]. Europeans are living longer than ever before. EU-27's median age is projected to increase by 4.5 years over the next three decades, reaching 48.2 years by 2050 [2]. Portugal shows the same tendency, with the population aged 65 years or older representing 24% of the national population in 2023 [3].

Aging is a normal process, during which humans experience a decline in organic, functional, and social status. This leads to enhanced vulnerability and increased risk of various aging-related diseases that become more disabling, chronic, degenerative, and are frequently associated with pain [4,5]. Geriatric patients present characteristics make them twice more likely to have adverse drug events [6]. This should be taken into account when prescribing different drugs [7].

Chronic pain is one of the most common conditions encountered in clinical practice, particularly in elderly patients, and is associated with substantial impairment, depression, anxiety, and isolation [8]. As stated in the consensus-based definition by the International Association for Hospice & Palliative Care, palliative care can be defined as "active holistic care of individuals with serious health-related suffering due to severe illness, with aims to improve the quality of life of patients, their families and their caregivers". Many age-related diseases lack curative treatments. Effective pain management in the elderly is critical to improving these patients' quality of life, functional capacity, and performance status [9]. As such, pain treatment is a palliative measure.

Non-steroidal anti-inflammatory drugs (NSAIDs) are widely prescribed for their analgesic, anti-inflammatory, and antipyretic properties. NSAIDs exert their effects primarily through inhibiting COX-1 and COX-2 enzymes, which play a crucial role in synthesizing prostaglandins and thromboxane. With pain and inflammation being common in older adults, these drugs are frequently used and difficult to manage, attending to the various morbidities that are common in this population [10]. In fact, NSAIDs are one of the most commonly responsible medicines to cause adverse drug events, accounting for approximately 11% of the preventable drug-related hospital admissions [6].

Side effects of NSAIDs are known and described in the literature, namely significant risks concerning cardiovascular (CV), gastrointestinal (GI), and renal impairment [11]. Nevertheless, different types of molecules are available, with pharmacokinetic characteristics that can make them more or less suitable for each situation [5].

Naproxen is a reversible inhibitor of COX isozymes and has a higher potency for COX-1 than COX-2. Ibuprofen inhibits COX-1 and COX-2 with comparable potency while diclofenac and celecoxib have intermediate COX-2 selectivity. Etoricoxib is considered a highly selective COX-2 inhibitor due to its greater potential for sparing COX-1 activity [12]. These affinity differences impact their potential for adverse effects: COX-1 is constitutively present in the gastroduodenal mucosa and mediates its cytoprotection. Therefore, COX-1 inhibition entails greater risks of adverse GI effects. COX-2 plays an important role in the regulation of renal function, is inducible, and is primarily involved in the inflammatory response, being that its inhibition is associated with CV toxicity [13].

Naproxen, ibuprofen and diclofenac are three of the most used NSAIDs worldwide [14]. According to INFARMED—National Authority for Medicines and Health Products—celecoxib and etoricoxib are the selective cyclooxygenase-2 (COX-2) inhibitors available in oral formulation in Portugal. Acemetacin showed promising results in terms of safety profile due to its phase 2 hepatic biotransformation and less renal elimination [5].

Portuguese national clinical guidelines regarding the use of NSAIDs and COX-2 inhibitors identify naproxen as the NSAID of choice for patients with a high CV risk or patients on secondary prevention. Selective COX-2 inhibitors are described as those with the most negligible GI impact but greater CV risk. However, this document was last updated in 2013 [15]. Both last recommendations from the International Association on the Study of Pain (IASP) and the United Kingdom National Health System (NHS) suggest starting with naproxen or ibuprofen plus proton pump inhibitor (PPI) in patients with CV risk [16,17]. Nevertheless, a recent study suggests that the risk of GI, CV, and renal adverse drug reactions is significantly reduced with the use of selective COX-2 inhibitors compared to non-selective NSAIDs [18].

Understanding the safety profiles of these NSAIDs is crucial for guiding clinical decision-making in the elderly, where balancing therapeutic benefits with potential risks is paramount.

Bearing this in mind, the main purpose of this work is to perceive the CV, GI, and renal safety profiles of ibuprofen, naproxen, acemetacin, diclofenac, celecoxib, and etoricoxib in elderly patients. Therefore, we performed a systematic review aimed at systematizing the available data about the safety profiles of these drugs to provide clinicians with evidence-based information regarding the preferred NSAID options for this population. The authors intend to contribute to a more informed, patient-adapted choice, while minimizing potential adverse events and increasing the efficiency of care.

2. Methods

A comprehensive literature search was conducted in PubMed and the Cochrane Library. For the selection of articles, we used Medical Subject Headings (MeSH) terms "aged" (A person 65 years of age or older) sequentially and together with "ibuprofen", "diclofenac", "naproxen", "acemetacin", "celecoxib", and "etoricoxib", published from January 2003 to January 2023.

Studies were included if they met the following criteria: participants were individuals aged 65 years or older, or the study included a subgroup analysis focusing on the elderly. Compared with non-use or placebo, the interventions involved using ibuprofen, naproxen, acemetacin, diclofenac, celecoxib, or etoricoxib. The outcomes assessed, either as primary or secondary, were renal, GI, and CV adverse events. We included meta-analyses, systematic reviews, and original research articles reporting observational studies or randomized controlled trials (RCTs).

Exclusion criteria were applied to studies in languages other than Portuguese or English, studies focusing on pain in a surgical context, and studies using the selected NSAIDs administered by routes other than oral.

To assess the quality and interest of the articles, four independent reviewers screened titles and abstracts to identify potentially eligible studies. Full-text articles were retrieved and assessed for eligibility based on the inclusion criteria and contributions with relevant information to the aim of this review.

Strength of Recommendation Taxonomy (SORT) was used to rate the quality of individual studies and establish recommendation strengths [19]. In line with the defined inclusion criteria on study design, only studies classified as level one and two were selected. Level of evidence one means the study is a meta-analysis or a systematic review with consistent findings or a high-quality randomized control trial. Level of evidence two means the study is a meta-analysis or a systematic review of lower quality or clinical trials with inconsistent findings, cohort study, or case-control study. Regarding the recommendation strength, C is assigned to expert opinions, bench research, a consensus guideline, clinical experience, usual practice, or a case series study; A is assigned to Cochrane reviews with a clear recommendation, a clinical evidence rating of beneficial, consistent findings from at least two good-quality randomized control trials or two good-quality diagnostic cohort studies, or a systematic review/meta-analysis of same [19].

This review follows the Preferred Reporting Items for Systematic Reviews and Meta-Analyses (PRISMA) guidelines for transparent reporting of systematic reviews [20].

3. Results

Overall, 3423 abstracts were obtained, 2308 from PubMed and 1115 from Cochrane Library. After removing duplicates, a total of 2086 studies were screened. Of those, 1513 studies were excluded based on titles not meeting inclusion criteria. Of 469 abstracts, 233 were selected for full-length reading based on the potentially relevant results listed in the abstract. One was not retrieved due to incapacity of access.

As shown in Figure 1, after full reading, a final group of 39 studies was reached and those were included in this systematic review. Selection was based on a clear focus on our study population, a comparison between drugs of interest or placebo, and with main outcomes being CV, GI or renal safety.

Twenty studies analyzed CV safety (Table 1). One study considered CV and GI safety simultaneously and fourteen considered GI safety only (Table 2). The remaining four studies investigated the renal safety of the NSAIDs of interest (Table 3). There were no studies about acemetacin safety that fulfilled the inclusion criteria.

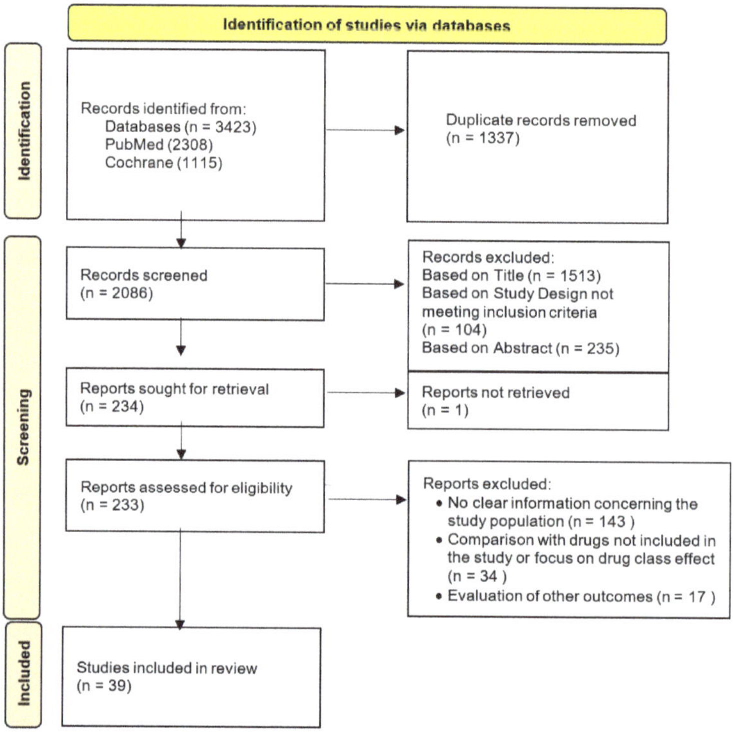

Figure 1. Preferred Reporting Items for Systematic Review and Meta-Analysis (PRISMA) protocols used in this study.

Some of the final selected articles provide data supporting a higher CV and GI risk in the elderly population. In the MEDAL program study, patients were randomized to etoricoxib 60 or 90 mg or diclofenac 150 mg once daily [21]. In two different analyses of the MEDAL program, Krum et al. aimed to assess the impact of these drugs on the risk of congestive heart failure (CHF) and their hypertensive effects. Despite the used drugs, an age of at least 65 was strongly associated with CHF hospitalization and increased blood pressure [22,23].

In a cohort of patients taking either celecoxib or naproxen, the strongest predictor of thrombotic cardiac events was an age of 65 years or older [24]. The same age group in an RCT of patients taking celecoxib 200–400 mg bid showed higher absolute risk for CV events [25].

Laine L. et al. identified age $\geq$ 65 years as a significant predictor of discontinuation of diclofenac or etoricoxib due to GI side effects [26]. Chan FK. et al. and Lai KC. et al. found age as an independent risk factor for ulcer recurrence [27,28]. On the other hand, Kyerematеng K. et al. and Hsiang KW. et al. did not find any difference in the overall rate or type of adverse events between elderly and younger participants, did not find age > 70 years to be a risk factor for symptomatic ulcers in patients taking the studied NSAIDs [29,30].

In terms of CV safety, in the Solomon DH et al. study, naproxen showed a protective effect against CV events, when compared to diclofenac and ibuprofen [31]. Schneeweiss S. et al.'s results suggest an equivalent risk for acute myocardial infarction (AMI) between ibuprofen and celecoxib, but a more favorable CV safety profile of naproxen compared to celecoxib [32]. On the other hand, Lévesque LE et al. found a neutral effect of naproxen on CV risk, same as celecoxib [33]. Likewise, Hudson M. et al. found no statistically significant differences between celecoxib, naproxen, and ibuprofen in the risk of CHF [34], and Mamdani M. et al. found no statistically significant differences between celecoxib and naproxen for AMI [35].

In turn, diclofenac demonstrated an increased risk of major adverse CV events when compared to ibuprofen, naproxen, and non-use [36]. It also showed a higher risk for thromboembolic events, namely stroke and venous thromboembolism [37,38].

CV risks of etoricoxib are comparable to those of diclofenac [21].

Evaluating GI safety, according to Schneeweiss S. et al. celecoxib produces a significant reduction in GI complications compared with naproxen, diclofenac, and ibuprofen [32]. In spite of that, celecoxib, as all NSAIDs, does not appear to be free of risks, showing a modest but still statistically significant risk of upper GI events as compared with non-use as stated in Chang CH et al.'s work [39]. In RCTs, diclofenac showed a worst GI safety profile when compared to celecoxib, even when added to PPI [40]. Etoricoxib may offer a GI safety advantage over diclofenac [21].

Concerning renal safety, a single 50 mg daily dose of diclofenac demonstrates a significant reduction in glomerular filtration rate, urine flow, and excretion rates of sodium and potassium. Pre-treatment with diuretics and angiotensin-converting enzyme inhibitors (drugs frequently used by elderly patients) exacerbated the renal impairment caused by diclofenac [41]. Higher doses of NSAIDs were associated with a higher risk of acute kidney failure (AKF). Both celecoxib and naproxen show a dose-dependent increase in the risk of AKF. However, the risk increase is more pronounced with higher doses of naproxen [42].

Table 1. Main characteristics of the systemically reviewed studies on CV safety of NSAIDs on a geriatric population.

References	Year	Study Design	Population n (%)	Studied Drug of Interest (Dosage)	Main Findings in Population of Interest	SORT Evidence Level
Cannon CP et al. [21]	2006	RCT	Overall = 34,701 ≥65 y = 14,397 (41.5%)	Diclofenac (150 mg/day) Etoricoxib (60 mg or 90 mg/day)	For elderly patients, the study suggests that the CV risk of etoricoxib is comparable to diclofenac.	1
Bertagnolli MM et al. [25]	2009	RCT	Overall = 2035 ≥65 y = ND	Celecoxib (200 mg 2id or 400 mg bid)	Patients aged 65 or older showed higher absolute risk for CV events.	1
Krum H, et al. [22]	2009	RCT	Overall = 34,695 ≥65 y = 14,386 (41.5%)	Diclofenac (150 mg id) Etoricoxib (60 or 90 mg id)	Older age increased the risk of CHF hospitalization.	1
Krum H, et al. [23]	2009	RCT	Overall = 23,498 ≥65 y = 9998 (42.5%)	Diclofenac (150 mg id) Etoricoxib (60 or 90 mg id)	Age at least 65 years was strongly associated with increased BP.	1
Mamdani M et al. [35]	2003	Retrospective Cohort Study	Overall = 593,808 ≥65 y = 593,808 (100%)	Celecoxib (ND) Naproxen (ND)	No significant differences in AMI rates were observed between celecoxib and naproxen. The findings do not support a short-term reduced risk of AMI with naproxen.	2
Lévesque LE et al. [33]	2005	Retrospective Cohort Study	Overall = 113,927 ≥65 y = 113,927 (100%)	Celecoxib (≤200 mg or >200 mg/day) Naproxen (ND)	The study found no increased risk among users of celecoxib, regardless of the dose prescribed. Naproxen was not associated with an increased CV risk or benefit.	2
Hudson M et al. [43]	2005	Retrospective Cohort Study	Overall = 18,503 ≥65 y = 18,503 (100%)	Ibuprofen (ND) Naproxen (ND)	Trend toward an increase in the rate of recurrent AMI with longer duration of ibuprofen exposure. Trend toward a lower rate of recurrent AMI in patients taking naproxen and AAS compared to AAS alone.	2
Solomon DH, et al. [31]	2006	Longitudinal CohortStudy	Overall = 74,838 ≥65 y = 74,838 (100%)	Celecoxib low ≤ 200 mg, high > 200 mg Ibuprofen, Diclofenac and Naproxen (low ≤ 75% of the maximum anti-inflammatory dosage, high > 75%)	Naproxen showed a protective effect against CV events, particularly compared to diclofenac and ibuprofen.Diclofenac was associated with a higher risk of AMI, though not significantly different for overall CV events.	2

Table 1. *Cont.*

References	Year	Study Design	Population n (%)	Studied Drug of Interest (Dosage)	Main Findings in Population of Interest	SORT Evidence Level
Lévesque LE et al. [44]	2006	Population-Based Cohort Study	Overall = 113,927 ≥65 y = 113,927 (100%)	Celecoxib (ND)	The risk increase for first-time use of celecoxib was not statistically significant.	2
Rahme E et al. [45]	2007	Retrospective Cohort Study	Overall = 283,799 ≥65 y = 283,799 (100%)	Celecoxib (≤200 mg/day vs. >200 mg/day OR 400 mg/day vs. >400 mg/day) Diclofenac (≤150 mg/day vs. >150 mg/day) and Ibuprofen (≤1600 mg/day vs. >1600 mg/day)	Study suggests that celecoxib does not confer a higher CV risk compared to ibuprofen and diclofenac.	2
Hudson M et al. [34]	2007	Case-Control Study	Overall = 42,560 ≥65 y = 42,560 (100%)	Celecoxib (≤200 mg id and >200 mg/day) Diclofenac (≤100 mg id and >100 mg/day) Ibuprofen (≤1200 mg id and >1200 mg/day) Naproxen (≤500 mg id and >500 mg/day)	Risks of recurrent CHF in patients exposed to naproxen, diclofenac, and ibuprofen were not significantly different compared with those of patients exposed to celecoxib at target doses.	2
Abraham NS et al. [46]	2007	Retrospective Cohort Study	Overall = 384,322 ≥65 y = 384,322 (100%)	Celecoxib (200 mg median daily dose) Ibuprofen (1800 mg median daily dose) Naproxen (1000 mg median daily dose)	Celecoxib was associated with a non-significant increase in the risk of AMI or cerebrovascular accident compared to naproxen. Ibuprofen showed a slightly higher risk of AMI compared to naproxen but had a similar risk for cerebrovascular accident compared to naproxen.	2
Solomon DH, et al. [47]	2008	Retrospective Cohort Study	Overall = 76,082 ≥65 y = 76,082 (100%)	Celecoxib (ND) Diclofenac (ND) Ibuprofen (ND) Naproxen (ND)	Ibuprofen appeared to confer an increased CVR in patients aged ≥80 years.	2
Cunnington M et al. [24]	2008	Retrospective Cohort Study	Overall = 80,826 ≥65 y = 39,048 (48.3%)	Celecoxib (ND) Naproxen (ND)	One of the strongest predictors of AMI/ ischemic stroke risk was age >65 years.	2
Mangoni AA, et al. [48]	2010	Retrospective Case-Control Study	Overall = 138,774 ≥65 y = 138,774 (100%)	Celecoxib (ND) Diclofenac (ND) Ibuprofen (ND) Naproxen (ND)	Neither naproxen or celecoxib showed a significant association with the risk of ischemic or hemorrhagic stroke. Low to moderate use of ibuprofen and diclofenac may be associated with a reduced risk of ischemic stroke.	2
Mangoni AA, et al. [49]	2010	Retrospective Case-Control Study	Overall = 138,774 ≥65 y = 138,774 (100%)	Celecoxib (ND) Diclofenac (ND) Ibuprofen (ND) Naproxen (ND)	There was an increase in the risk of AMI with increasing supplies of the individual NSAIDs naproxen and ibuprofen, but not for diclofenac.Analysis of the individual NSAIDs showed a reduced risk of cardiac arrest associated with low exposure to diclofenac and an increased risk of arrhythmias associated with moderate-high exposure to naproxen.	2
Schjerning et al. [50]	2011	Retrospective Cohort Study	Overall = 83,677 ≥65 y = ND	Celecoxib (ND) Diclofenac (ND) Ibuprofen (ND) Naproxen (ND)	Age-stratified analysis found that patients ≥80 years of age had a higher risk of death during the first week of treatment when taking diclofenac compared to younger patients.	2
Caughey GE, et al. [38]	2011	Retrospective Cohort Study	Overall = 162,065 ≥65 y = 162,065 (100%)	Diclofenac (ND) Ibuprofen (ND)	An increased risk of stroke was observed with diclofenac. Ibuprofen was not associated with ischemic or hemorrhagic stroke.	2
Lee T, et al. [37]	2016	Case-control Study	Overall = 24,079 ≥70 y = 15,435 (64.1%)	Celecoxib (ND) Diclofenac (ND) Ibuprofen (ND) Naproxen (ND)	Among current users, risk of venous thromboembolism was increased in diclofenac and ibuprofen. However, such association was not observed among current naproxen users.	2

Table 1. *Cont.*

References	Year	Study Design	Population n (%)	Studied Drug of Interest (Dosage)	Main Findings in Population of Interest	SORT Evidence Level
Schmidt M, et al. [36]	2018	Emulated Trial	Overall = 7,608,766 ≥70 y = 779,250 (10.2%)	Diclofenac (ND) Ibuprofen (ND) Naproxen (ND)	The increased risk of major adverse CV events with diclofenac initiation was consistent in the elderly subgroups compared to ibuprofen, naproxen, and non-use.	2

AAS: Acetylsalicylic Acid; AMI: Acute myocardial infarction; BP: Blood Pressure; CHF: Cardiac Heart Failure; CI: Cardiovascular; OA: Osteoarthritis; ND: No Data; NSAID: Non-steroidal Anti-Inflammatory Drugs; RCT: Randomized Controlled Trial.

Table 2. Main characteristics of the systemically reviewed studies on GI safety of NSAIDs on a geriatric population.

References	Year	Study Design	Population n (%)	Studied Drugs of Interest (Dosage)	Main Findings in Population of Interest	SORT Evidence Level
Goldstein JL et al. [51]	2003	RCT	Overall = 186 ≥65 y = 186 (100%)	Naproxen (500 mg bid)	Naproxen use was associated with a significantly higher incidence of gastroduodenal ulcers and erosions compared to placebo.	1
Laine L et al. [52]	2004	RCT	Overall = 1519 ≥65 y = 467 (30.7%)	Ibuprofen (800 mg bid)	No significant treatment-by-subgroup interaction was seen for ulcer risk factors, including age ≥ 65 years. Rates of discontinuation and gastroduodenal ulcers and erosions were significantly higher in the ibuprofen group compared with placebo group.	1
Chan FK et al. [27]	2004	RCT	Overall = 287 ≥65 y = ND	Celecoxib (200 mg bid) Diclofenac (75 mg bid)	Age ≥ 75 years was an independent riskfactor predicting ulcer recurrence.	1
Lai KC et al. [28]	2005	RCT	Overall = 242 ≥65 y = 81 (33.5%)	Celecoxib (200 mg/day) Naproxen (750 mg/day)	Celecoxib was not inferior to lansoprazole plus naproxen in preventing GI ulcer complications, including those aged 65 or older. Age 65 years or more was an independent risk factor for ulcer recurrence.	1
Laine L et al. [53]	2008	Prospective Cohort Trial	Overall = 34,701 ≥65 y = 14,396 (41.5%)	Diclofenac (150 mg/day) Etoricoxib (60 or 90 m/day)	Older age was a significant predictor for lower GI clinical events, with a 2-fold increase in risk.	1
Laine L e at. [26]	2010	RCT	Overall = 34,701 ≥65 y = 14,227 (41.0%)	Diclofenac (150 mg/day) Etoricoxib (60 or 90 mg/day)	Age ≥ 65 years was a significant predictor of discontinuation due to dyspepsia, clinical events, and complicated events.	1
Arber N et al. [54]	2012	RCT	Overall = 3588 ≥65 y = 29.7% (APC Study); 37.9% (PreSAP study)	Celecoxib (200 mg/400 mg bid or 400 mg id)	Age ≥65 years was associated with an increased risk of GI events in both celecoxib and placebo groups. The noninferiority of celecoxib to placebo was not established.	1
Kellner HL et al. [40]	2012	RCT	Overall = 2446 ≥65 y = 2446 (100%)	Celecoxib (200 mg bid) Diclofenac (75 mg bid)	Celecoxib was associated with significantly fewer clinically significant upper and lower GI events compared with diclofenac plus omeprazole in patients aged ≥65 years.	1
Couto A et al. [55]	2018	RCT	Overall = 818 ≥65 y = 358 (43.7%)	Naproxen (220 mg bid)	The rate of reported GI AE was comparable in the naproxen and placebo groups for all age groups.	1
Kyeremateng K et al. [29]	2019	Retrospective Pooled Analysis of RCTs	Overall = 1494 ≥65 y = 544 (36.4%)	Naproxen (440 mg/day) Ibuprofen (600–1200 mg/day)	In ≥65 years there was no difference in the overall rate or type of AE compared with younger participants. There were no significant differences in overall adverse events between naproxen, placebo or ibuprofen.	1
Biskupiak JE et al. [56]	2006	Retrospective Cohort Study	Overall = 60,212 ≥65 y = ND	Ibuprofen (ND) Naproxen (ND)	Elderly patients had a higher incidence of serious GI toxicities compared to younger adults both with naproxen and ibuprofen.	2

Table 2. *Cont.*

References	Year	Study Design	Population n (%)	Studied Drugs of Interest (Dosage)	Main Findings in Population of Interest	SORT Evidence Level
Schneeweiss S et al. [32]	2006	Population-Based Cohort Study	Overall = 49,711 ≥65 y = 49,711 (100%)	Celecoxib (ND) Diclofenac (ND) Naproxen (ND) Ibuprofen (ND)	Celecoxib produces a significant short-term reduction in GI complications compared with all nonselective NSAIDs combined. Diclofenac appears to have the least favorable safety profile among individual NSAIDs. The risk of AMI was significantly higher for diclofenac compared to celecoxib, while ibuprofen did not show a significant difference in AMI risk when compared to celecoxib. Naproxen may have a more favorable CV safety profile compared to celecoxib, indicated by the lowest risk difference for MI among the NSAIDs studied, although this difference was not statistically significant.	2
Hsiang KW et al. [30]	2010	Prospective Observational Cohort Study	Overall = 933 ≥70 y = 490 (49.3%)	Celecoxib (100–400 mg/day) Etoricoxib (60 mg/day)	Age >60 or >70 years was not a risk factor for symptomatic ulcers and ulcer complications in COX-2 inhibitor users. Use of COX-2 inhibitors was well tolerated in regard to the upper GI tract in patients aged >70 years.	2
Sostek MB et al. [57]	2011	Open-label, Multicenter, Phase III Study	Overall = 239 ≥65 y = 78 (32.6%)	Naproxen (500 mg bid)	Twice-daily dosing with naproxen/esomeprazole was not associated with any new or unexpected safety issues. Subgroup analysis suggest that the safety profile of naproxen/esomeprazole is similar in <65 vs. ≥65 years old patients.	2
Chang CH et al. [39]	2011	Case-Crossover Study	Overall = 40,635 ≥65 y = ND	Celecoxib (ND)	Celecoxib was associated with higher risk of upper GI events vs. non-use. Elderly were at higher risk for upper GI adverse events.	2

AE: Adverse Events; CI: Cardiovascular; GI: Gastrointestinal; PPI: Proton Pump Inhibitors; OA: Osteoarthritis; ND: No Data; NSAID: Non-steroidal Anti-Inflammatory Drugs; RCT: Randomized Controlled Trial; AMI: Acute myocardial infarction.

Table 3. Main characteristics of the systemically reviewed studies on renal safety of NSAIDs on a geriatric population.

References	Year	Study Design	Population n (%)	Studied Drug (Dosage)	Main Findings in Population of Interest	SORT Evidence Level
Juhlin T et al. [41]	2004	RCT	Overall = 14 ≥65 y = 14 (100%)	Diclofenac (50 mg id)	Diclofenac caused significant reductions in GFR, urine flow, and excretion rates of sodium and potassium. Pre-treatment with diuretics and ACE-inhibitors exacerbated this impact.	2
Schneider V et al. [42]	2006	Case-Control Study	Overall = 121,722 ≥65 y = 121,722 (100%)	Celecoxib (low dose 200 mg id // high dose >200 mg id) Naproxen (low dose 750 mg id // high dose >750 mg id)	Both celecoxib and naproxen show a dose-dependent increase in the risk of AKI. However, the risk increase is more pronounced with higher doses of naproxen. The study suggests that celecoxib is associated with a lower risk of AKI compared to naproxen, especially at lower doses.	2
Winkelmayer WC et al. [58]	2008	Retrospective Cohort Study	Overall = 183,446 ≥65 y = 183,446 (100%)	Celecoxib (ND) Diclofenac (ND) Naproxen (ND) Ibuprofen (ND)	Ibuprofen seemed to have a greater rate of AKI compared with celecoxib.	2
Benson P et al. [59]	2012	Prospective observational study	Overall = 44 ≥65 y = 44 (100%)	Celecoxib (400 mg bid)	High-dose celecoxib was relatively well-tolerated by elderly patients, with stable renal function and minor electrolyte alterations.	2

ACE: Angiotensin-converting enzyme; AKI: Acute kidney injury; GFR: Glomerular filtration rate; ND: No Data; NSAID: Non-steroidal anti-inflammatory drugs; RCT: Randomized controlled trial.

4. Discussion

Recommendations about NSAID prescriptions are frequently based on COX selectivity, but there are marked differences in each NSAID's molecular and chemical properties, which confer different pharmacological results [5].

Age is widely described in the literature as an important CV risk factor, with a high prevalence of hypertension, ischemic heart disease, atrial fibrillation, and heart failure (level of evidence 1). Reduced gastric and intestinal motility, decreased splanchnic blood flow, impaired mucosal protective mechanisms and enzyme secretion, hypochlorhydria, and mucosal atrophy are some of the age-related GI changes [60,61].

Our results show that older age acts as an independent CV risk factor and that the elderly have a higher baseline risk for GI complications (level of evidence 2). Both can be exacerbated by NSAID use. Even though two studies do not find age as a risk factor for symptomatic peptic ulcer disease (PUD) [29,30], a manifestation of PUD in the elderly is often atypical [60]. Also, these studies, like many others in this matter, did not account for *H. pylori* infection, one of the most frequent causes of PUD [60], which might be an important confounder.

Most existing orientations are targeted at the adult population and usually divide recommendations into CV risk vs. GI risk to direct the choice of NSAID [15–17]. Nevertheless, the elderly frequently fulfill both risks plus the expected decline in kidney function, making drug selection particularly challenging [5].

Both the United Kingdom NHS and IASP recommend starting with naproxen or ibuprofen plus PPI [16,17]. IASP proposes low doses of celecoxib as an alternative option in high CV risk patients [16].

In the results of this systematic review, two studies support the potential superiority of naproxen in terms of CV risk [31,32]. When analyzing these publications, no statistical significance is reached on one of them [32], and in the other, the doses of the drugs used in the analysis are not clearly presented [31]. Other studies suggest a neutral impact of naproxen in thromboembolic events [37,48]. Once again, these studies also do not clearly present the doses used for each drug, which can be an important bias. Consequently, in light of the results gathered in this systematic review, opting for first-line naproxen in the elderly high CV risk population is not supported by scientific evidence (level of evidence 1). When comparing ibuprofen with naproxen, our results suggest a modest inferiority of ibuprofen to naproxen based on two studies that document a higher risk for AMI and venous thromboembolic events [37,46]. Therefore, ibuprofen should be considered as a second-line therapy compared to naproxen in elderly patients with high CV risk (level of evidence 1).

When celecoxib is only compared with naproxen, it shows no inferiority regarding AMI, CHF, and cerebrovascular events risk [34,35,46,48]. Taking this into consideration, in high CV risk elderly patients, in studies that compare one to another, celecoxib shows a similar safety profile to naproxen (level of evidence 2).

It is important to acknowledge that all the mentioned studies are based on prescription records, and lack direct measures of adherence, dosage, duration of treatment or nonprescription NSAID use.

As previously mentioned, COX-2 inhibitors are designed to prevent GI issues, and therefore, celecoxib is expected, right from the start, to show a better safety profile when it comes to GI events. Despite this theoretical advantage, data from our review continue to find that the use of celecoxib is not without harm, showing a modest risk of superior GI effects when compared with placebo (level of evidence 2) [39].

However, when compared to diclofenac, naproxen, and ibuprofen, celecoxib showed a better GI safety profile (level of evidence 2), including greater safety compared to diclofenac in combination with PPIs [32,40]. When compared only with naproxen plus lansoprazole, celecoxib alone was not inferior in preventing GI ulcer complications [28]. Regarding the use of naproxen alone, it was associated with a significantly higher incidence of gastroduodenal ulcers and erosions compared to placebo (level of evidence 1). Consequently, in light of the results gathered in this systematic review, we consider that celecoxib has a better GI safety profile (level of evidence 2), followed by naproxen plus PPI (level of evidence 2).

Celecoxib demonstrates a better safety profile when compared with naproxen and ibuprofen (level of evidence 2).

A study comparing naproxen and celecoxib concluded that both show a dose-dependent increase in the risk of AKF. Still, the risk increase was more pronounced with higher doses of naproxen [42]. Celecoxib also had a lower rate of AKF compared to ibuprofen [58].

When celecoxib was compared with placebo, it showed a stable renal function and minor electrolyte alterations [59].

All these studies advocate for a better safety profile for celecoxib (level of evidence 1).

Acemetacin was one of the drugs this systematic review intended to study. A promising safety profile was expected due to its specific pharmacokinetic characteristics. Being a prodrug and bearing a weak reduction of prostaglandin synthesis might reduce the damage of the GI mucosa [62] (level of evidence 3). Of all the drugs included in our study, acemetacin has the lowest renal excretion, which may be better suited for the elderly with compromised renal function [5] (level of evidence 3). Despite its theoretical potential, no papers were identified that evaluated its use in the elderly.

Diclofenac presents a lower renal excretion compared with most NSAIDs and a phase 2 liver biotransformation, making it theoretically a good choice for elderly patients [5]. Nevertheless, IASP and NHS advise against the use of diclofenac in high CV risk patients [16,17]. In our findings, diclofenac showed an overall poorer CV safety profile compared to naproxen and ibuprofen (level of evidence 2). Diclofenac also showed a worse GI safety profile when compared to celecoxib, even when added to PPI (level of evidence 2) [32,40]. Additionally, a single low dose of 50 mg daily had a significant impact on renal function (level of evidence 2) [41].

Findings on etoricoxib were also very scarce. Of the studies retrieved in this systematic review, the MEDAL trial was the only one that allowed comparison between this drug and the other NSAIDs studied [21]. It suggests that the cardiovascular risks of etoricoxib are comparable to those of diclofenac (level of evidence 2), which, as discussed above, showed a poorer CV safety profile.

A prior pharmacovigilance and literature review study [18] concluded that the risk of GI, CV, and renal adverse drug reactions was significantly lower with the use of selective COX-2 inhibitors compared to non-selective NSAIDs. Although this is one of the populations with higher needs and prescriptions for NSAIDs, few studies focus on geriatric patients.

Our study has some limitations. The search was limited to certain databases, which might have excluded relevant studies that were not indexed. A global assessment of the results obtained leads us to reflect on their heterogeneity. The multiplicity of study types, drug comparisons, doses, and treatment periods limit the achievement of solid recommendations. Also, the determination of the strength of recommendations is affected by inherent subjectivity.

5. Conclusions

When CV risk is the main concern, according to our findings, celecoxib or naproxen are a suitable choice (recommendation strength B). Acemetacin can be considered (recommendation strength C). Etoricoxib and diclofenac should be avoided in these patients (recommendation strength B).

Regarding the GI safety profile, celecoxib remains the first-line choice, as does naproxen plus PPI (recommendation strength B).

When renal function is in focus, celecoxib remains the first-line therapy (recommendation strength A). Naproxen can be considered a second-line therapy in these patients (recommendation strength B). Acemetacin can also be considered (recommendation strength C). Ibuprofen and diclofenac should be used with caution in these patients (recommendation strength B).

These results will assist in making tailored treatment decisions for patients, reducing the risk of adverse events, and enhancing care efficiency.

Author Contributions: Conceptualization, C.C. and H.R.; methodology, C.C., J.P.A. and H.R.; validation, A.B., A.G., J.P.A. and H.R.; formal analysis, C.C., D.S., A.B. and A.G.; investigation, C.C., D.S., A.G. and A.B.; resources, H.R.; data curation, C.C. and D.S.; writing—original draft preparation, C.C.; writing—review and editing, J.P.A. and H.R.; visualization, D.S., A.B. and A.G.; supervision, H.R.; project administration, J.P.A.; funding acquisition, H.R. All authors have read and agreed to the published version of the manuscript.

Funding: This research received no specific grant from any funding agency in the public, commercial, or not-for-profit sectors.

Data Availability Statement: Dataset available on request from the authors.

Conflicts of Interest: The authors declare no conflict of interest.

References

1. Jain, N.; Kourampi, I.; Umar, T.P.; Almansoor, Z.R.; Anand, A.; Ur Rehman, M.E.; Jain, S.; Reinis, A. Global population surpasses eight billion: Are we ready for the next billion? *AIMS Public Health* **2023**, *10*, 849–866. [CrossRef] [PubMed]
2. European Commission. Eurostat Statistics Explained, July 2020, Ageing Europe—Statistics on Population Developments. 2020. Available online: https://ec.europa.eu/eurostat/statistics-explained/index.php?title=Ageing_Europe_-_introduction (accessed on 1 May 2024).
3. PORDATA. População Residente: Total e por Grandes Grupos Etários. 2024. Available online: https://www.pordata.pt/pt/estatisticas/populacao/populacao-residente/populacao-residente-por-sexo-e-grupo-etario (accessed on 1 May 2024).
4. Tenchov, R.; Sasso, J.M.; Wang, X.; Zhou, Q.A. Aging Hallmarks and Progression and Age-Related Diseases: A Landscape View of Research Advancement. *ACS Chem. Neurosci.* **2024**, *15*, 1–30. [CrossRef] [PubMed]
5. Ribeiro, H.; Rodrigues, I.; Napoleão, L.; Lira, L.; Marques, D.; Veríssimo, M.; Andrade, J.P.; Dourado, M. Non-steroidal anti-inflammatory drugs (NSAIDs), pain and aging: Adjusting prescription to patient features. *Biomed. Pharmacother.* **2022**, *150*, 112958. [CrossRef] [PubMed]
6. Zazzara, M.B.; Palmer, K.; Vetrano, D.L.; Carfì, A.; Onder, G. Adverse drug reactions in older adults: A narrative review of the literature. *Eur. Geriatr. Med.* **2021**, *12*, 463–473. [CrossRef]
7. Fialová, D.; Laffon, B.; Marinković, V.; Tasić, L.; Doro, P.; Sóos, G.; Mota, J.; Dogan, S.; Brkic, J.; Teixeira, J.P.; et al. Medication use in older patients and age-blind approach: Narrative literature review (insufficient evidence on the efficacy and safety of drugs in older age, frequent use of PIMs and polypharmacy, and underuse of highly beneficial nonpharmacological strategies). *Eur. J. Clin. Pharmacol.* **2019**, *75*, 451–466.
8. Azevedo, L.F.; Costa-Pereira, A.; Mendonça, L.; Dias, C.C.; Castro-Lopes, J.M. Epidemiology of chronic pain: A population-based nationwide study on its prevalence, characteristics and associated disability in Portugal. *J. Pain* **2012**, *13*, 773–783. [CrossRef]
9. Reid, M.C.; Eccleston, C.; Pillemer, K. Management of chronic pain in older adults. *BMJ* **2015**, *350*, h532. [CrossRef]
10. Sohail, R.; Mathew, M.; Patel, K.K.; Reddy, S.A.; Haider, Z.; Naria, M.; Habib, A.; Abdin, Z.U.; Chaudhry, W.R.; Akbar, A. Effects of Non-steroidal Anti-inflammatory Drugs (NSAIDs) and Gastroprotective NSAIDs on the Gastrointestinal Tract: A Narrative Review. *Cureus* **2023**, *15*, e37080. [CrossRef] [PubMed]
11. Szeto, C.C.; Sugano, K.; Wang, J.G.; Fujimoto, K.; Whittle, S.; Modi, G.K.; Chen, C.-H.; Park, J.-B.; Tam, L.-S.; Vareesangthip, K.; et al. Non-steroidal anti-inflammatory drug (NSAID) therapy in patients with hypertension, cardiovascular, renal or gastrointestinal comorbidities: Joint APAGE/APLAR/APSDE/APSH/APSN/PoA recommendations. *Gut* **2020**, *69*, 617–629. [CrossRef]
12. Brune, K.; Patrignani, P. New insights into the use of currently available non-steroidal anti-inflammatory drugs. *J. Pain Res.* **2015**, *8*, 105–118. [CrossRef]
13. Zarghi, A.; Arfaei, S. Selective COX-2 Inhibitors: A Review of Their Structure-Activity Relationships. *Iran. J. Pharm. Res.* **2011**, *10*, 655–683. [PubMed]
14. Conaghan, P.G. A turbulent decade for NSAIDs: Update on current concepts of classification, epidemiology, comparative efficacy, and toxicity. *Rheumatol. Int.* **2012**, *32*, 1491–1502. [CrossRef] [PubMed]
15. Saúde, D.G.d. Anti-Inflamatórios não Esteroides Sistémicos em Adultos: Orientações para a Utilização de Inibidores da COX-2. 2013. Available online: https://normas.dgs.min-saude.pt/wp-content/uploads/2019/09/anti-inflamatorios-nao-esteroides-sistemicos-em-adultos-orientacoes-para-a-utilizacao-de-inibidores-da-cox.pdf (accessed on 1 May 2024).
16. Minhas, D.; Nidhaan, A.; Husni, M.E. Recommendations for the Use of Nonsteroidal Anti-inflammatory Drugs and Cardiovascular Disease Risk: Decades Later, Any New Lessons Learned? *Rheum. Dis. Clin. N. Am.* **2023**, *49*, 179–191. [CrossRef] [PubMed]
17. Service N-NH. Oral Non-Steroidal Antiinflammatory (NSAID) Guidelines. 2020. Available online: https://www.nhsggc.org.uk/media/254733/oral-non-steroidal-anti-inflammatory-nsaid-guideline.pdf (accessed on 1 May 2024).

18. Monteiro, C.; Silvestre, S.; Duarte, A.P.; Alves, G. Safety of Non-Steroidal Anti-Inflammatory Drugs in the Elderly: An Analysis of Published Literature and Reports Sent to the Portuguese Pharmacovigilance System. *Int. J. Environ. Res. Public Health* **2022**, *19*, 3541. [CrossRef] [PubMed]
19. Ebell, M.H.; Siwek, J.; Weiss, B.D.; Woolf, S.H.; Susman, J.; Ewigman, B.; Bowman, M. Strength of recommendation taxonomy (SORT): A patient-centered approach to grading evidence in the medical literature. *J. Am. Board Fam. Pract.* **2004**, *17*, 59–67. [CrossRef] [PubMed]
20. Moher, D.; Liberati, A.; Tetzlaff, J.; Altman, D.G. Preferred reporting items for systematic reviews and meta-analyses: The PRISMA statement. *PLoS Med.* **2009**, *6*, e1000097. [CrossRef] [PubMed]
21. Cannon, C.P.; Curtis, S.P.; Bolognese, J.A.; Laine, L. Clinical trial design and patient demographics of the Multinational Etoricoxib and Diclofenac Arthritis Long-term (MEDAL) study program: Cardiovascular outcomes with etoricoxib versus diclofenac in patients with osteoarthritis and rheumatoid arthritis. *Am. Heart J.* **2006**, *152*, 237–245. [CrossRef] [PubMed]
22. Krum, H.; Curtis, S.P.; Kaur, A.; Wang, H.; Smugar, S.S.; Weir, M.R.; Laine, L.; Brater, D.C.; Cannon, C. Baseline factors associated with congestive heart failure in patients receiving etoricoxib or diclofenac: Multivariate analysis of the MEDAL program. *Eur. J. Heart Fail.* **2009**, *11*, 542–550. [CrossRef] [PubMed]
23. Krum, H.; Swergold, G.; Curtis, S.P.; Kaur, A.; Wang, H.; Smugar, S.S.; Weir, M.; Laine, L.; Brater, D.C.; Cannon, C. Factors associated with blood pressure changes in patients receiving diclofenac or etoricoxib: Results from the MEDAL study. *J. Hypertens.* **2009**, *27*, 886–893. [CrossRef]
24. Cunnington, M.; Webb, D.; Qizilbash, N.; Blum, D.; Mander, A.; Funk, M.J.; Weil, J. Risk of ischaemic cardiovascular events from selective cyclooxygenase-2 inhibitors in osteoarthritis. *Pharmacoepidemiol. Drug Saf.* **2008**, *17*, 601–608. [CrossRef]
25. Bertagnolli, M.M.; Eagle, C.J.; Zauber, A.G.; Redston, M.; Breazna, A.; Kim, K.; Tang, J.; Rosenstein, R.B.; Umar, A.; Bagheri, D.; et al. Five-year efficacy and safety analysis of the Adenoma Prevention with Celecoxib Trial. *Cancer Prev. Res.* **2009**, *2*, 310–321. [CrossRef] [PubMed]
26. Laine, L.; Curtis, S.P.; Cryer, B.; Kaur, A.; Cannon, C.P. Risk factors for NSAID-associated upper GI clinical events in a long-term prospective study of 34 701 arthritis patients. *Aliment. Pharmacol. Ther.* **2010**, *32*, 1240–1248. [CrossRef] [PubMed]
27. Chan, F.K.; Hung, L.C.; Suen, B.Y.; Wong, V.W.; Hui, A.J.; Wu, J.C.; Leung, W.K.; Lee, Y.T.; To, K.F.; Chung, S.C.S.; et al. Celecoxib versus diclofenac plus omeprazole in high-risk arthritis patients: Results of a randomized double-blind trial. *Gastroenterology* **2004**, *127*, 1038–1043. [CrossRef] [PubMed]
28. Lai, K.C.; Chu, K.M.; Hui, W.M.; Wong, B.C.; Hu, W.H.; Wong, W.M.; Lam, S.-K. Celecoxib compared with lansoprazole and naproxen to prevent gastrointestinal ulcer complications. *Am. J. Med.* **2005**, *118*, 1271–1278. [CrossRef]
29. Kyeremateng, K.; Troullos, E.; Paredes-Diaz, A. Safety of naproxen compared with placebo, ibuprofen and acetaminophen: A pooled analysis of eight multiple-dose, short-term, randomized controlled studies. *Curr. Med. Res. Opin.* **2019**, *35*, 1671–1676. [CrossRef]
30. Hsiang, K.W.; Chen, T.S.; Lin, H.Y.; Luo, J.C.; Lu, C.L.; Lin, H.C.; Lee, K.C.; Chang, F.Y.; Lee, S.D. Incidence and possible risk factors for clinical upper gastrointestinal events in patients taking selective cyclooxygenase-2 inhibitors: A prospective, observational, cohort study in Taiwan. *Clin. Ther.* **2010**, *32*, 1294–1303. [CrossRef] [PubMed]
31. Solomon, D.H.; Avorn, J.; Stürmer, T.; Glynn, R.J.; Mogun, H.; Schneeweiss, S. Cardiovascular outcomes in new users of coxibs and nonsteroidal antiinflammatory drugs: High-risk subgroups and time course of risk. *Arthritis Rheum.* **2006**, *54*, 1378–1389. [CrossRef] [PubMed]
32. Schneeweiss, S.; Solomon, D.H.; Wang, P.S.; Rassen, J.; Brookhart, M.A. Simultaneous assessment of short-term gastrointestinal benefits and cardiovascular risks of selective cyclooxygenase 2 inhibitors and nonselective nonsteroidal antiinflammatory drugs: An instrumental variable analysis. *Arthritis Rheum.* **2006**, *54*, 3390–3398. [CrossRef]
33. Lévesque, L.E.; Brophy, J.M.; Zhang, B. The risk for myocardial infarction with cyclooxygenase-2 inhibitors: A population study of elderly adults. *Ann. Intern. Med.* **2005**, *142*, 481–489. [CrossRef]
34. Hudson, M.; Rahme, E.; Richard, H.; Pilote, L. Risk of congestive heart failure with nonsteroidal antiinflammatory drugs and selective Cyclooxygenase 2 inhibitors: A class effect? *Arthritis Rheum.* **2007**, *57*, 516–523. [CrossRef]
35. Mamdani, M.; Rochon, P.; Juurlink, D.N.; Anderson, G.M.; Kopp, A.; Naglie, G.; Austin, P.C.; Laupacis, A. Effect of selective cyclooxygenase 2 inhibitors and naproxen on short-term risk of acute myocardial infarction in the elderly. *Arch. Intern. Med.* **2003**, *163*, 481–486. [CrossRef] [PubMed]
36. Schmidt, M.; Sørensen, H.T.; Pedersen, L. Diclofenac use and cardiovascular risks: Series of nationwide cohort studies. *BMJ* **2018**, *362*, k3426. [CrossRef] [PubMed]
37. Lee, T.; Lu, N.; Felson, D.T.; Choi, H.K.; Dalal, D.S.; Zhang, Y.; Dubreuil, M. Use of non-steroidal anti-inflammatory drugs correlates with the risk of venous thromboembolism in knee osteoarthritis patients: A UK population-based case-control study. *Rheumatology* **2016**, *55*, 1099–1105. [CrossRef]
38. Caughey, G.E.; Roughead, E.E.; Pratt, N.; Killer, G.; Gilbert, A.L. Stroke risk and NSAIDs: An Australian population-based study. *Med. J. Aust.* **2011**, *195*, 525–529. [CrossRef]
39. Chang, C.H.; Chen, H.C.; Lin, J.W.; Kuo, C.W.; Shau, W.Y.; Lai, M.S. Risk of hospitalization for upper gastrointestinal adverse events associated with nonsteroidal anti-inflammatory drugs: A nationwide case-crossover study in Taiwan. *Pharmacoepidemiol. Drug Saf.* **2011**, *20*, 763–771. [CrossRef]

40. Kellner, H.L.; Li, C.; Essex, M.N. Efficacy and safety of celecoxib versus diclofenac and omeprazole in elderly arthritis patients: A subgroup analysis of the CONDOR trial. *Curr. Med. Res. Opin.* **2012**, *28*, 1537–1545. [CrossRef]
41. Juhlin, T.; Björkman, S.; Höglund, P. Cyclooxygenase inhibition causes marked impairment of renal function in elderly subjects treated with diuretics and ACE-inhibitors. *Eur. J. Heart Fail.* **2005**, *7*, 1049–1056. [CrossRef] [PubMed]
42. Schneider, V.; Lévesque, L.E.; Zhang, B.; Hutchinson, T.; Brophy, J.M. Association of selective and conventional nonsteroidal antiinflammatory drugs with acute renal failure: A population-based, nested case-control analysis. *Am. J. Epidemiol.* **2006**, *164*, 881–889. [CrossRef]
43. Hudson, M.; Baron, M.; Rahme, E.; Pilote, L. Ibuprofen may abrogate the benefits of aspirin when used for secondary prevention of myocardial infarction. *J. Rheumatol.* **2005**, *32*, 1589–1593.
44. Lévesque, L.E.; Brophy, J.M.; Zhang, B. Time variations in the risk of myocardial infarction among elderly users of COX-2 inhibitors. *CMAJ* **2006**, *174*, 1563–1569. [CrossRef]
45. Rahme, E.; Watson, D.J.; Kong, S.X.; Toubouti, Y.; LeLorier, J. Association between nonnaproxen NSAIDs, COX-2 inhibitors and hospitalization for acute myocardial infarction among the elderly: A retrospective cohort study. *Pharmacoepidemiol. Drug Saf.* **2007**, *16*, 493–503. [CrossRef] [PubMed]
46. Abraham, N.S.; El-Serag, H.B.; Hartman, C.; Richardson, P.; Deswal, A. Cyclooxygenase-2 selectivity of non-steroidal anti-inflammatory drugs and the risk of myocardial infarction and cerebrovascular accident. *Aliment. Pharmacol. Ther.* **2007**, *25*, 913–924. [CrossRef] [PubMed]
47. Solomon, D.H.; Glynn, R.J.; Rothman, K.J.; Schneeweiss, S.; Setoguchi, S.; Mogun, H.; Avorn, J.; Sturmer, T. Subgroup analyses to determine cardiovascular risk associated with nonsteroidal antiinflammatory drugs and coxibs in specific patient groups. *Arthritis Rheum.* **2008**, *59*, 1097–1104. [CrossRef] [PubMed]
48. Mangoni, A.A.; Woodman, R.J.; Gilbert, A.L.; Knights, K.M. Use of non-steroidal anti-inflammatory drugs and risk of ischemic and hemorrhagic stroke in the Australian veteran community. *Pharmacoepidemiol. Drug Saf.* **2010**, *19*, 490–498. [CrossRef]
49. Mangoni, A.A.; Woodman, R.J.; Gaganis, P.; Gilbert, A.L.; Knights, K.M. Use of non-steroidal anti-inflammatory drugs and risk of incident myocardial infarction and heart failure, and all-cause mortality in the Australian veteran community. *Br. J. Clin. Pharmacol.* **2010**, *69*, 689–700. [CrossRef]
50. Schjerning Olsen, A.M.; Fosbøl, E.L.; Lindhardsen, J.; Folke, F.; Charlot, M.; Selmer, C.; Laberts, M.; Olesen, J.B.; Køber, L.; Hansen, P.R.; et al. Duration of treatment with nonsteroidal anti-inflammatory drugs and impact on risk of death and recurrent myocardial infarction in patients with prior myocardial infarction: A nationwide cohort study. *Circulation* **2011**, *123*, 2226–2235. [CrossRef]
51. Goldstein, J.L.; Kivitz, A.J.; Verburg, K.M.; Recker, D.P.; Palmer, R.C.; Kent, J.D. A comparison of the upper gastrointestinal mucosal effects of valdecoxib, naproxen and placebo in healthy elderly subjects. *Aliment. Pharmacol. Ther.* **2003**, *18*, 125–132. [CrossRef]
52. Laine, L.; Maller, E.S.; Yu, C.; Quan, H.; Simon, T. Ulcer formation with low-dose enteric-coated aspirin and the effect of COX-2 selective inhibition: A double-blind trial. *Gastroenterology* **2004**, *127*, 395–402. [CrossRef]
53. Laine, L.; Curtis, S.P.; Langman, M.; Jensen, D.M.; Cryer, B.; Kaur, A.; Cannon, C.P. Lower gastrointestinal events in a double-blind trial of the cyclo-oxygenase-2 selective inhibitor etoricoxib and the traditional nonsteroidal anti-inflammatory drug diclofenac. *Gastroenterology* **2008**, *135*, 1517–1525. [CrossRef]
54. Arber, N.; Lieberman, D.; Wang, T.C.; Zhang, R.; Sands, G.H.; Bertagnolli, M.M.; Hawk, E.T.; Eagle, C.; Coindreau, J.; Zauber, A.; et al. The APC and PreSAP trials: A post hoc noninferiority analysis using a comprehensive new measure for gastrointestinal tract injury in 2 randomized, double-blind studies comparing celecoxib and placebo. *Clin. Ther.* **2012**, *34*, 569–579. [CrossRef]
55. Couto, A.; Troullos, E.; Moon, J.; Paredes-Diaz, A.; An, R. Analgesic efficacy and safety of non-prescription doses of naproxen sodium in the management of moderate osteoarthritis of the knee or hip. *Curr. Med. Res. Opin.* **2018**, *34*, 1747–1753. [CrossRef] [PubMed]
56. Biskupiak, J.E.; Brixner, D.I.; Howard, K.; Oderda, G.M. Gastrointestinal complications of over-the-counter nonsteroidal antiinflammatory drugs. *J. Pain Palliat. Care Pharmacother.* **2006**, *20*, 7–14. [CrossRef] [PubMed]
57. Sostek, M.B.; Fort, J.G.; Estborn, L.; Vikman, K. Long-term safety of naproxen and esomeprazole magnesium fixed-dose combination: Phase III study in patients at risk for NSAID-associated gastric ulcers. *Curr. Med. Res. Opin.* **2011**, *27*, 847–854. [CrossRef]
58. Winkelmayer, W.C.; Waikar, S.S.; Mogun, H.; Solomon, D.H. Nonselective and cyclooxygenase-2-selective NSAIDs and acute kidney injury. *Am. J. Med.* **2008**, *121*, 1092–1098. [CrossRef]
59. Benson, P.; Yudd, M.; Sims, D.; Chang, V.; Srinivas, S.; Kasimis, B. Renal effects of high-dose celecoxib in elderly men with stage D2 prostate carcinoma. *Clin. Nephrol.* **2012**, *78*, 376–381. [CrossRef] [PubMed]
60. Dumic, I.; Nordin, T.; Jecmenica, M.; Stojkovic Lalosevic, M.; Milosavljevic, T.; Milovanovic, T. Gastrointestinal Tract Disorders in Older Age. *Can. J. Gastroenterol. Hepatol.* **2019**, *2019*, 6757524. [CrossRef]
61. Newton, J.L. Effect of age-related changes in gastric physiology on tolerability of medications for older people. *Drugs Aging* **2005**, *22*, 655–661. [CrossRef]
62. Chávez-Piña, A.E.; McKnight, W.; Dicay, M.; Castañeda-Hernández, G.; Wallace, J.L. Mechanisms underlying the anti-inflammatory activity and gastric safety of acemetacin. *Br. J. Pharmacol.* **2007**, *152*, 930–938. [CrossRef]

BioChem

Review

Protein Ligases: Nature's Gift for Protein/Peptide Synthesis

Yvonne Ritsema [1,2], Huapeng Li [1,2] and Qingfei Zheng [1,2,*]

[1] Department of Medicinal Chemistry and Molecular Pharmacology, College of Pharmacy, Purdue University, West Lafayette, IN 47907, USA
[2] Institute for Cancer Research, Purdue University, West Lafayette, IN 47907, USA
* Correspondence: zhengqf@purdue.edu; Tel.: +1-765-494-1149

Abstract: Proteins are structurally and functionally diverse biomacromolecules that serve a variety of essential activities to ensure complex biological homeostasis. The desire to elucidate and enhance these biological functions has been at the forefront of research for many decades. However, generating active proteins via recombinant expression or through chemical total synthesis each has limitations in terms of yield and functionality. Nature has provided a solution to this problem through evolving protein ligases that catalyze the formation of amide bonds between peptides/proteins, which can be exploited by protein engineers to develop robust functional proteins. Here, we summarize the biochemical mechanisms and applications of multiple cysteine-based protein ligases, especially focusing on how they have been utilized for protein therapeutics and engineering, as well as how they inspired chemists to develop efficient methodologies for protein synthesis (e.g., native chemical ligation).

Keywords: protein ligase; peptide/protein synthesis; protein engineering; protein therapy; synthetic biology

1. Introduction

Proteins are responsible for a wide range of biological functions in cells, acting as enzymes, transporters, signaling molecules, etc. There is a structure–function relationship that exists within proteins, meaning specific amino acids or post-translational modifications (PTMs) can alter the behavior of proteins and exert various biological impacts [1]. Therefore, decades of research have gone into elucidating their structures and functions and utilizing this information to develop more robust biomolecules and introduce them as therapeutic options. Protein/peptide-based therapeutics is currently a multi-billion-dollar industry that has produced life-saving treatments for diverse diseases like diabetes and cancer; however, the production of these proteins often requires synthetic techniques that have inherent limitations [2,3].

The introduction of recombinant protein expression in *E. coli*, starting in 1982, has been revolutionary in the production of proteins with desirable qualities, which soon became the major method for producing recombinant proteins in large quantities [3,4]. The alteration of amino acid sequences to modify protein structures has led to the synthesis of recombinant proteins that have increased pharmacokinetic properties, stability, activity, etc. [5]. However, a major limitation of bacterial expression is the unavailability of PTMs found in more complex eukaryotic systems which are vital for fully functional proteins. Many proteins require PTMs for proper folding, so, oftentimes, recombinant bacterial expression leads to low yields or dysfunctional eukaryotic proteins. The development of protein expression approaches using insect or mammalian cells can overcome this drawback and successfully introduce diverse PTMs onto the proteins of interest. Even though the PTMs are available, it is still very difficult to have precise control over site-specific modifications, which can be critical to elucidate specific PTMs on biological activity [6–8]. Therefore, the addition of site-specific PTMs can require the use of chemical protein/peptide synthesis.

Chemical synthesis via solid-phase peptide synthesis (SPPS) has allowed for the generation of peptides carrying site-specific PTMs, which represents an advantage over recombinant expression. Although SPPS has been optimized for faster production and simpler purification of peptides, the generation of large peptides/proteins presents a major challenge for this technique. Every amino acid coupling causes the overall yield to take a hit, so SPPS is currently limited to linear peptides containing ~40–50 amino acids [8,9]. To combat the size limitation of SPPS and PTM control issues found in recombinant expression, researchers have turned to the use of ligation reactions. Both chemical and enzymatic, these reactions allow for the fusing of peptides/proteins to incorporate exogenous functional groups or improve their structure–function relationship, without causing any side effects (e.g., denaturing the protein structures). Moreover, these ligation approaches enable the preparation of proteins/peptides that contain non-canonical structures (e.g., D- and β-amino acids) and unnatural post-translational modifications.

In this review, we will discuss nature's gift to the field of protein synthesis and engineering through the availability of protein ligases that catalyze these ligation reactions both in vitro and in vivo. Although isolated from different hosts and require different recognition sequences, the protein ligases outlined here all proceed via a catalytic cysteine residue. The resulting thioester generated between the enzyme and peptide/protein substrate can be intercepted by an incoming primary amine of another peptide/protein donor to complete the ligation reaction. These cysteine-based protein ligases have been extensively optimized and proven to be powerful tools to produce therapeutic options and facilitate the incorporation of unnatural amino acids into proteins. In a similar catalytic manner, nature has also provided the cysteine-based mechanism of intein splicing, which potentially serves as the inspiration for chemical ligation methods, thus demonstrating the breadth of opportunities enzymatic protein ligation has given to the research world.

2. Sortase A

2.1. Mechanism and Engineering

Sortase A (SrtA) is a transpeptidase enzyme that is originally found in the cell wall of *Staphylococcus aureus* and naturally functions to anchor proteins to the cell wall [10–13]. This enzyme was discovered in 1999 by the Schneewind lab, which allowed for extensive study into its mechanism of action as a protein ligase [14,15]. Sortase A recognizes the sequence LPXTG and its catalytic cysteine residue acts a nucleophile to attack the carbonyl group of the amide bond between threonine (T) and glycine (G) [10,11]. This forms a thioacyl intermediate between sortase A and threonine and releases a C-terminal glycine. The enzyme then catalyzes the ligation between a primary amine donor on an N-terminal glycine to the LPXT-SrtA complex. Due to its nucleophilicity, the amine group attacks the carbonyl group of the thioacyl intermediate to release sortase A and ligates the two peptides through the formation of a new amide bond (Figure 1A). Seeing as though the primary amine donor must be on a glycine residue which is identical to the product released in the first step of catalysis, this ligation reaction is reversible. In addition, wild-type sortase A is Ca^{2+}-dependent and has low catalytic efficiency, which requires a high enzyme to substrate concentration and thus limits its applicability for enzymatic protein ligation.

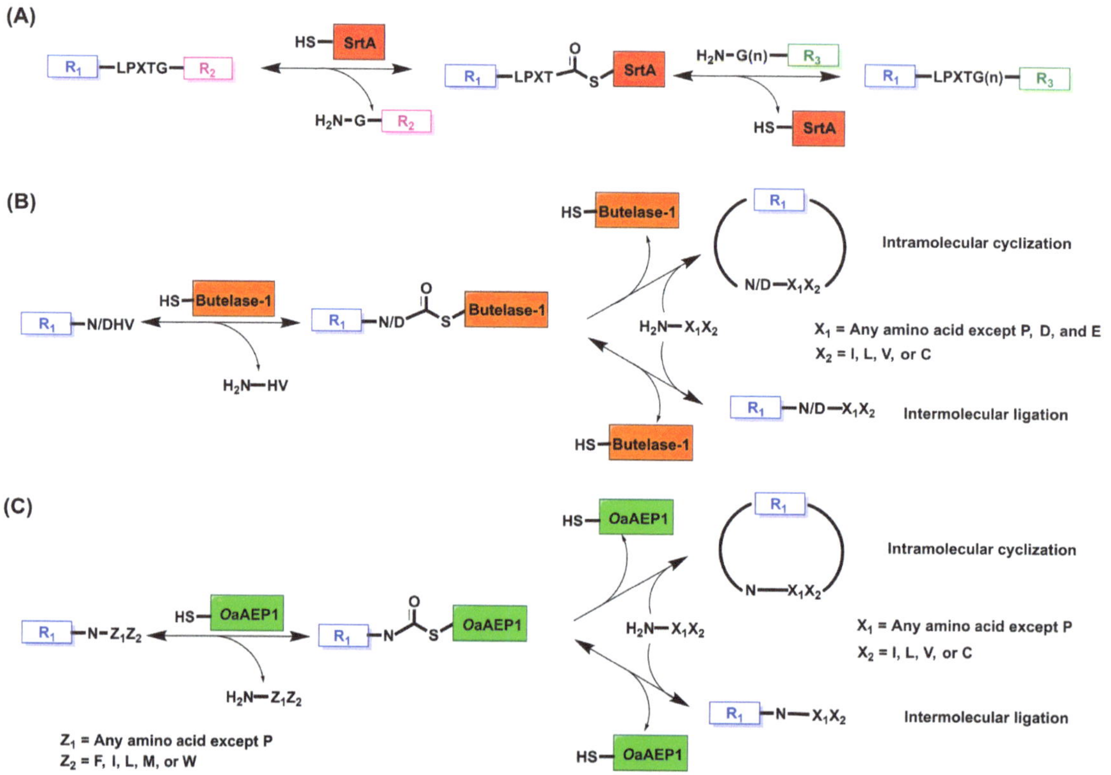

Figure 1. Enzymatic peptide/protein ligation to produce native peptide/protein backbones. (**A**) Sortase A recognizes an LPXTG sequence and attacks the carbonyl group between T and G via its catalytic cysteine. This produces a thioester intermediate that can be intercepted by a primary amine on G to undergo an aminolysis reaction. R_1 and R_2 are peptide or protein sequences. (**B**) Butelase-1 recognizes an N/DHV sequence and attacks via its catalytic cysteine to produce a thioester with N/D. Butelase-1 can catalyze either a reversible intermolecular ligation or irreversible intramolecular cyclization with a primary amine. R_1 is a peptide or protein sequence. X_1 can be any amino acid except P, D, and E. X_2 can be amino acids I, L, V, or C. (**C**) OaAEP1 recognizes an NZ_1Z_2 sequence, where Z_1 is any amino acid except for P and Z_2 can be amino acids F, I, L, M, or W. OaAEP1 can catalyze either intermolecular ligation or intramolecular cyclization with a primary amine via its catalytic cysteine residue. X_1 can be any amino acid except P. X_2 can be amino acids I, L, V, or C.

Since its initial application in enzymatic protein ligation, sortase A has been engineered to enhance its efficiency. Sortase A isolated from *S. aureus* has several setbacks, including Ca^{2+} dependency and low catalytic efficiency, which required optimization for ideal use [16,17]. To improve its catalytic efficiency, the 94R/D160N/D165A/K190E/K196T pentamutant of native sortase A, also known as eSrt, was developed by Chen et al. in 2015 [16,18]. Compared to the wild-type SrtA, eSrt has a 120-fold increased catalytic efficiency for LPETG coupling with a polyglycine peptide donor [18]. The catalytic efficiency of eSrt then allows for the usage of this enzyme with limited amounts of glycine substrate, which was a previous limitation for the wild-type enzyme [18]. Further improvements have been made to eSrt to eliminate the Ca^{2+} dependency that remained with this mutant. Hirakawa et al. identified two mutations, E105K and E108A, in the Ca^{2+} binding site that allowed sortase to have enzymatic function in the absence of Ca^{2+} [19]. However, these double mutations alone were not sufficient enough to promote ligation as they decreased the enzyme's catalytic turnover rate [19]. This led to the successful development of the P94R/E105K/E108A/D160N/D165A/K190E/K196T heptamutant, which exhibits increased catalytic efficiency and Ca^{2+} independence compared to the wild type [19,20].

Sortase A has a fairly tight specificity as it requires the recognition sequence of LPXTG; therefore, studies have engineered the enzyme to reprogram and broaden its specificity. The F40 sortase mutant was developed via phage display and contains mutations at the T164 and V168-Q172 residues [21,22]. Several amino acids with small side chains, specifically alanine, are recognized by the F40 sortase, thus enhancing its promiscuity [21]. The sequence APXTG showed preferential coupling over LPXTG, which is significant as this sequence is endogenously found in histone H3 [21,22]. Unfortunately, the F40 sortase remains Ca^{2+}-dependent and exhibits slow kinetics, resulting in limitations to its application for enzymatic protein ligation [21–23]. Other studies have focused on reprogramming the eSrt mutant to maintain its catalytic efficiency while broadening its substrate specificity to recognize both LAXTG and LPXSG substrates [13,24].

2.2. Applications

2.2.1. Histone Modification In Vitro and In Cellulo

Sortase A has demonstrated, both in vitro and in cellulo, the capability of modifying the N-terminal tails of histone H3 [22,23]. The natural amino acid sequence of histone H3 tail includes the short sequence APATGG, which allows for the use of engineered F40 sortase [21,23]. To introduce scarless H3 tails in vitro, the F40 sortase was further mutated to cW11 sortase, which possesses the additional mutations at D160, K190, and K196 [22]. The feature that makes cW11 sortase most distinct is that its backbone is cyclized, as this significantly enhances the stability of the enzyme [22,25]. cW11 sortase is able to modify H3 in several ways, one of those being the ability to ligate N-terminal tails and install PTMs onto H3 tailless nucleosomes [22]. The expression of tailless H3 and addition of difficult PTMs, such as ubiquitination, by cW11 sortase accelerate assays examining nucleosome enzymatic and binding activities [22,26,27]. In addition to introducing full tails, cW11 sortase can also produce asymmetrically ligated H3 tails, allowing one to study the effects of site-specific PTMs in various combinations [22]. Ligation is only part of the enzymatic function of cW11 sortase, as it can also cleave H3 tails at the recognition sequence [22]. This allows for the isolation and characterization of H3 tails after treatment with various "reader" proteins and "writer" enzymes to quantify their effects on histone modifications [22]. Collectively, the applicability of cW11 sortase to modify histone H3 allows for the introduction, modification, and isolation of H3′s modified N-terminal tails.

Sortase A has also been used for the semisynthesis of histones H2B and H4 in vitro to investigate the mechanism of histone deacetylation by sirtuins and histone deacetylases (HDACs) [28,29]. The N-terminal tail on histone H2B contains the sequence HPDTG, which exhibits similarities to the recognition sequence LPXTG. To accommodate these sequence changes, Wang et al. developed W4 sortase by the site-specific mutagenesis of F40 sortase [28]. The full-length semisynthesis of H2B with site-specific N-terminal acetylation was catalyzed by W4 sortase, and these versions of histone H2B were incorporated into the recombinant nucleosome core particles [28]. Histone H4 also contains a sequence similar to LPXTG, i.e., LARRGG, so Xiao et al. hypothesized that semisynthetic H4 could be produced by sortase-mediated ligation (SML) with eSrt(2A-9) [29]. Although the substitution of alanine (A) at the P2 position was tolerated by eSrt(2A-9), the variant was incapable of tolerating arginine (R) at the P4 position, so the endogenous H4 sequence was modified to LARCGG since cysteine can be converted to an arginine analog [29]. The semisynthesis of histone H2B and H4 by sortase variants allowed both studies to probe the selectivity and kinetics of HDACs and sirtuins for specific acetylated residues [28,29].

The modification of histone H3 has been further expanded from in vitro to in cellulo through the use of SrtA for sortase-mediated metathesis (SMM) (Figure 2) [23]. Truncated histone H3 was synthetically produced with an LPATGG recognition sequence for 6M-Srt to cleave between T and G, and thus produce an α-amine glycine substrate [23]. To expand the substrate scope, a peptide also carrying the LPATGG sequence was introduced so that it could also be recognized by sortase A and provide the LPAT sequence needed for ligation [23]. This allows for the addition of a cell-penetrating peptide (CPP) for in cellulo delivery and α-amine glycine on histone H3, which is lacking the endogenous tail [23]. The use of SMM allows for the delivery of histone H3 tails with site-specific modifications such as PTMs to study their impact on epigenetics in live cells [23].

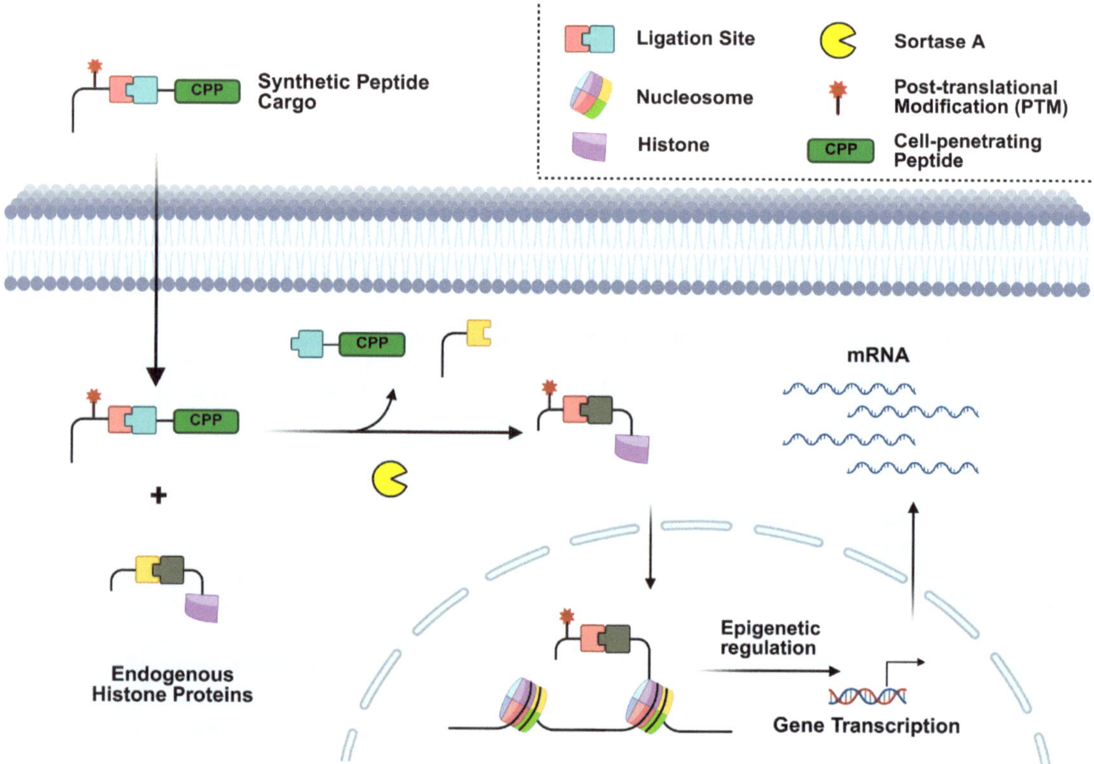

Figure 2. Sortase-mediated metathesis (SMM) of histone N-terminal tails. The histone H3 N-terminal tail has been engineered with an LPATGG recognition motif for sortase A to deliver site specific post-translational modifications (PTMs) or small molecules. Control over the delivery of these modifications allows for the study of their biological implications and therapeutic applications.

2.2.2. Protein Engineering Through "Sortagging"

Protein ligation using sortase A creates proteins with terminal modifications; however, the combined use of sortase A and click chemistry allows for N-to-N and C-to-C fused peptides/proteins [30,31]. Click chemistry is a highly chemoselective reaction between two functional groups, such as azide and alkyne, which can be utilized for bio-orthogonal transformations [32,33]. Ploegh et al. in 2011 demonstrated the combination of SrtA-mediated ligation with click chemistry, coined as "sortagging", using ubiquitin as the model protein [31]. Specifically, peptides with the recognition sequence LPETGG were synthesized with either an azide or cyclooctyne click handle [31]. N-terminal glycine on ubiquitin acted as the substrate for the SML reaction for each peptide; as a result, the incubation of both peptides produced a click reaction that formed a ubiquitin dimer [31]. Given that the click handles were present on the N-terminus of each peptide, the resulting ubiquitin dimer was an N-to-N fused protein [31]. This was extended to the production of C-to-C fused anti-GFP VHH antibody dimers, which were converted with a ~90% success rate and maintained their functionality [31]. This demonstrates the ability of "sortagging" to produce protein dimers that retain the identity of either the N- or C-terminus that is critical for function.

To improve the efficacy of cancer immunotherapy treatments, "sortagging" has been employed for the utilization of cytotoxic T lymphocytes (CTLs) that employ apoptosis to eliminate targeted cells [34,35]. He et al. set out to track CTL interaction with tumors from 4T1, a breast cancer cell line, by first utilizing click chemistry to attach the sortase A recognition sequence to each partner [34]. A thiol–maleimide click reaction was applied to attach sortase A to the CTL and a SPAAC reaction utilizing azide and dibenzocyclooctyne (DBCO) handles affixed an N-terminal polyglycine substrate to the tumor cell [34]. The introduction of a 5-Tamra-LPETG peptide provided the substrate for SML between sortase A on a CTL and the α-amine glycine donor from the tumor cell [34]. The resulting product of this ligation reaction is a fluorescently labeled tumor cell, thus allowing the tracking of CTL and tumor cell interactions in vivo [34]. To further validate this model, the authors were able to incubate this system with antitumor therapeutics, pexidartinib and cisplatin, to track their interaction with the tumor microenvironment (TME) [34]. This represents a more robust approach toward monitoring the TME, which will help in optimizing clinical immunotherapy treatments with CTLs.

2.2.3. Bacterial Cell Wall Engineering

Owing to the native environment of sortase A, it has been utilized in modifying cell wall proteins in *S. aureus* [36]. Various cell surface proteins in *S. aureus* express the LPETG recognition sequence for sortase A, allowing for the engineering of these proteins to contain small molecules such as fluorophores, biotin, etc. [36]. Nelson et al. were the first to demonstrate surface cell engineering in bacteria by tracking fluorescent labeling and biotinylation of cell wall extracts via epifluorescence microscopy and MALDI-TOF MS, respectively [36]. These authors were also the first to introduce "sortagging" in *S. aureus* [36]. Through the utilization of an N-terminal azide group on an LPETG peptide and fluorophore-containing cyclooctyne probe, a [3 + 2] cycloaddition linked the fluorophore to the sortase A substrate. SML between the LPETG peptide and *S. aureus* cell wall proteins produced fluorescently labeled proteins that could be analyzed using epifluorescence microscopy [36]. This opens the horizon for the highly specific introduction of small molecules and engineering of bacterial proteins through "sortagging".

Sortase A-mediated labeling in *S. aureus* has also been employed to attach photosensitizers to peptidoglycan [37]. An eosin photosensitizer was delivered on an LPETG peptide that binds to vancomycin, a glycopeptide antibiotic that binds to the D-Ala-D-Ala motif on the cell wall of *S. aureus* [37]. Sortase A then cleaves LPET with the photosensitizer to ligate it onto a peptidoglycan that can then be photoinduced to produce reactive oxygen species (ROS) and kill the bacterial cells [37]. SML then provides a potential mechanism to therapeutically target *S. aureus* as it is found in a variety of human afflictions including pneumonia and toxic shock syndrome [38,39].

3. Butelase

3.1. Mechanism and Engineering

Butelase-1 is an asparaginyl endopeptidase (AEP) isolated from the tropical plant *Clitoria ternatea* (*C. ternatea*) [40,41]. This ligase recognizes the motif Asn/Asp(Asx)-His-Val on a peptide and its catalytic cysteine residue acts as a nucleophile to produce an N-to-S acyl shift at the Asx carbonyl group [41–43], as a result forming a thioester intermediate between butelase-1 and Asx while releasing the His-Val dipeptide [42]. The thioester intermediate then undergoes an S-to-N acyl shift triggered by a nucleophilic N-terminal amine group to produce the ligated product [42]. This N-terminal amine group can either come from an incoming peptide in a reversible intermolecular reaction or N-terminus of the Asx peptide for an irreversible intramolecular reaction (Figure 1B) [42,44]. Therefore, one of the most impressive applications of butelase-1 is its ability to cyclize peptides with high efficiency as it has this native cyclase activity [40]. The first study to isolate and characterize butelase-1 demonstrated cyclization yields of over 95% for peptides with high catalytic efficiency; in addition, butelase-1 has demonstrated a 100-fold decrease in enzyme concentration needed for cyclization compared to sortase A [40–42].

Despite its successful applications as a peptide ligase and cyclase, the utilization of butelase-1 has been hindered by its difficult isolation from its native plant source and challenging recombinant expression in bacteria. Traditionally, butelase-1 has been isolated from the pods of *C. ternatea* and purified in a four-step process with a low yield of 5 mg kg^{-1} of the plant [45–47]. Examination of several parts of *C. ternatea* by Tam et al. revealed that butelase-1 extracted from shoots not only had higher ligation efficiency but was also able to be purified in a three-step process with a three-fold-greater yield than previously reported [45]. Extensive research has gone into improving the recombinant expression of butelase-1 in bacteria, mainly focusing on the promotion of disulfide bonds and improving its zymogen activation [45,48]. Advances in the recombinant expression of butelase-1 have found success through the employment of Rosetta-gami B (DE3) and SHuffle T7, two strains of *E. coli* that have mutated *trxB* and *gor* genes [46,48]. The mutation of these genes for reductase proteins may help facilitate the folding of butelase-1 as it naturally contains three disulfide bonds in its fully folded form [46,48]. AEPs found in plants are synthesized as zymogens which undergo irreversible autocleavage of their prodomains in acidic conditions to become active [49]. Several studies have worked to improve the activation of butelase-1 zymogen both in vitro and in vivo, which has led to the development of the engineered variant butelase AY [46,49,50]. Butelase AY contains mutations V237A and T238Y, which decrease the precipitation of the enzyme during its activation step compared to the WT and in turn improve its ligation efficiency and recombinant expression in *E. coli* RIL cells [50].

3.2. Applications

The Tam lab introduced the idea of unnatural amino acid incorporation using butelase-1 by demonstrating its ability to cyclize D-amino-acid-containing peptides [47,51]. As long as the Asx residue at the P1 position remained in the L form, butelase-1 was able to tolerate the majority of D-amino acids in the P1″ and P2″ positions [47]. The cyclization of bioactive peptides, such as D-θ-defensin, although distinctly slower than the L form, was successful and displayed antimicrobial activity against several bacterial strains [47]. The ability of butelase-1 to incorporate unnatural amino acids has been further explored with the engineered variant of the enzyme, butelase AY [50]. At the second position on the acceptor peptide, the unnatural amino acids methyl alanine (MetAla), 2-aminoisobutyric acid (Aib), and norvaline (NorVal) were all recognized [50]. However, NorVal had the best yield with 95% product conversion after 24 h [50]. Taken together, the utilization of butelase-1, both native and engineered, to develop peptides with unnatural amino acids opens the door for its potential use in bio-orthogonal systems and peptide therapeutics.

Expanding on the concept of butelase-1 employment for therapeutic purposes, this ligase has assisted in the synthesis of anticancer and antibacterial peptides [48,52]. Using a variant of butelase-1, engineered to improve its recombinant expression, Hu et al. were able to cyclize the peptide lunasin, which has shown anticancer activity in mouse and mammalian cells [48,53]. By redesigning the sequence of lunasin to contain the appropriate recognition amino acids for butelase-1, the peptide was cyclized with an over 90% yield [48]. Cyclized lunasin was then less susceptible to degradation by pepsin and trypsin, and better inhibited the proliferation of HepG2 cells compared to the linear peptide [48]. Butelase-1 mediated cyclization has also been employed to improve the stability of antimicrobial peptides for better target engagement [54]. Aside from cyclization, butelase-1 has been utilized as a ligase to develop peptide dendrimers, branched macromolecules, to act against antibiotic-resistant bacterial strains [52]. A thiodepsipeptide, where the thioester group was inserted between amino acids N and V, was utilized to increase the effectiveness of the ligation reaction between this peptide and a lysyl dendrimeric scaffold [52]. The synthesis of di-, tetra-, and octameric branched dendrimers was successful, and they displayed antimicrobial activities against several drug-resistant bacterial strains [52]. Butelase-1 ultimately possesses the potential to not only improve known peptide therapeutics but also allow for the development of peptide therapeutics for unmet clinical needs.

4. *Oa*AEP1

4.1. Mechanism and Engineering

Similar to butelase-1, *Oa*AEP1 is a cysteine-based AEP isolated from *Oldenlandia affinis* which recognizes an Asn residue at the ligation site and proceeds through the same mechanism for protein ligation (Figure 1C) [54–56]. However, *Oa*AEP1 has distinct differences from butelase-1 in that it has decreased catalytic efficiency but has better recombinant expression in bacteria [51,57]. When determining the crystal structure of *Oa*AEP1, Tam et al. discovered a C247A mutation that increased the activity of the enzyme by 160-fold [51,58,59]. C247 is positioned near the substrate pocket, so substitution for a smaller side chain with alanine increases the acceptability of the incoming nucleophilic amine [57]. Also, the hydrophobic nature of alanine was hypothesized to reduce the capability of water to hydrolyze the thioester intermediate before the ligation reaction [57]. Early recombinant expression of *Oa*AEP1 demonstrated a specificity for the C-terminal NGL motif and N-terminal GL nucleophile for intramolecular cyclization [55]. This poses a problem as the ligation reaction then becomes reversible as the released peptide can act as a nucleophile for a subsequent ligation reaction. The C247A mutation helps alleviate some of this reversibility problem and increases the promiscuity of *Oa*AEP1 by demonstrating the ability to accept a GV nucleophile, and it produces an NGV product that has poor ligation efficiency [59,60]. However, when using OaAEP1 variants with a tolerated substrate, it is oftentimes necessary to introduce metal cations to quench the reversible byproduct [61,62]. This *Oa*AEP1 variant is further being engineered to broaden the acceptance for various substrates and nucleophiles. One study has developed *Oa*AEP1-C247A-aa55-351, which can recognize NAL and accept RL nucleophiles with high catalytic efficiency [62]. This variant also resolves other issues such as purification, low pH requirements for activation, etc., that exist with *Oa*AEP1 C247A [62].

4.2. Applications

Due to the ability of *Oa*AEP1 to act as a cyclase, it has been sought out for its ability to perform head-to-tail cyclization. Oftentimes, this enzyme has been used for in vitro ligations; however, one study was able to recombinantly express a truncated version of *Oa*AEP1-C247A in *E. coli* to perform in vivo cyclization [63]. Specifically, *Oa*AEP1-C247A was co-expressed with its substrate murine dihydrofolate reductase (mDHFR), which contained prodomains, to reveal an N-terminal GL nucleophile and C-terminal NGL recognition sequence [63]. Cyclized mDHFR was purified and obtained in a similar yield to the linear mDHFR from *E. coli* while displaying increased thermal stability and no loss of enzymatic function [63]. Aside from producing macrocycles itself, the ligation capabilities of *Oa*AEP1-C247A have been used to facilitate intramolecular cyclization for the creation of phage-displayed peptide libraries [64]. An NGL substrate containing a chloroacetyl group on its N-terminus was introduced to *Oa*AEP1-C247A, which catalyzed its ligation with a phage displaying a GL-containing peptide [64]. This peptide also contains a cysteine which undergoes an S_N2 reaction with the chloroacetyl group for intramolecular cyclization [64]. Using the transcription factor TEAD4 as a proof of concept, the authors were able to perform several rounds of selection and enriched for several cyclic peptides, one of which had a K_D value 16-fold lower than its linearized counterpart [64]. These studies highlight the promise of *Oa*AEP1-C247A in producing and assisting in the enrichment of macrocycles for highly specific targets.

*Oa*AEP1 has also contributed to the development of noncanonical amino acid incorporation by accepting peptides containing unnatural or D-amino acids [65–67]. Lang et al. used genetic code expansion (GCE) to incorporate an azide glycylglycine moiety linked via an isopeptide bond to lysine—AzGGisoK—on an internal residue in ubiquitin, which then underwent a Staudinger reduction to become GGisoK [65]. A desthiobiotin-NGLH probe, dtb-NGLH, provided the substrate for the ligation reaction catalyzed by *Oa*AEP1-C247A to label the ubiquitin molecule with ~95% yield in a 1 h incubation period [65]. The authors further explored the capabilities of *Oa*AEP1-247A by dual-labeling ubiquitin, with its N-terminus acquiring an alkyne probe for proceeding click reactions, and catalyzing the ubiquitination and SUMOylation of POI with the GGisoK residue [65,66]. In this study, although the unnatural amino acid (UAA) was present, it was GG that served as the nucleophile. Rehm et al. expanded on the OaAEP1-catalyzed ligation with a UAA nucleophile by integrating an Ac-G[Dap]LGV peptide where the diaminopropionic acid (Dap) residue acted as the amine nucleophile for an NGL substrate [67]. High-yield product conversion was seen at a more neutral pH both with and without the Ni^{2+} quencher; furthermore, Ac-G[d-Dap]LGV and Ac-vGl[d-Dap]G also had high product yields, demonstrating the compatibility of this ligase with D-amino acids [67]. As Leu was important for ligation efficiency, the Dap-Leu motif was incorporated into a model protein, MCoTI-II, and had an 83% product conversion when ligated with an NGL-His_6-carrying eGFP protein [67]. *Oa*AEP1-catalyzed ligation of UAAs expands the capabilities of protein synthesis for site-specific side chain modifications and the incorporation of biochemical probes for techniques like bio-orthogonal reactions.

5. Transglutaminase

5.1. Mechanism and Engineering

Transglutaminase 2 (TGM2) is a member of the transglutaminase protein family, which are enzymes that facilitate transamidation reactions in the presence of Ca^{2+} [68,69]. This enzymatic reaction was first discovered in 1957 by Sarkar et al., who discovered that in the presence of Ca^{2+}, amines were being incorporated into soluble proteins found in guinea pig livers [70]. Decades later, it has been elucidated that TGM2 is the enzyme responsible for this crosslinking, which occurs between the γ-carboxamide group of glutamine and a primary amine [68,71,72]. TGM2 has a conserved catalytic cysteine, i.e., C277, which exerts a nucleophilic attack on the γ-carboxamide group of glutamine to produce a thioester intermediate and release ammonia [68,69]. The thioester intermediate is then attacked by a primary amine, such as lysine, to release the enzyme and create an isopeptide bond between glutamine and lysine (Figure 3A) [68]. TGM2 is heavily reliant on the surrounding concentration of Ca^{2+}, as mutagenesis studies have shown that a loss of any of its Ca^{2+} binding sites reduces or eliminates the transamidation reaction [69]. However, there is a microbial version of TGM isolated from *Streptomyces mobaraense* that is Ca^{2+}-independent while still retaining the ability to generate isopeptide bonds [73].

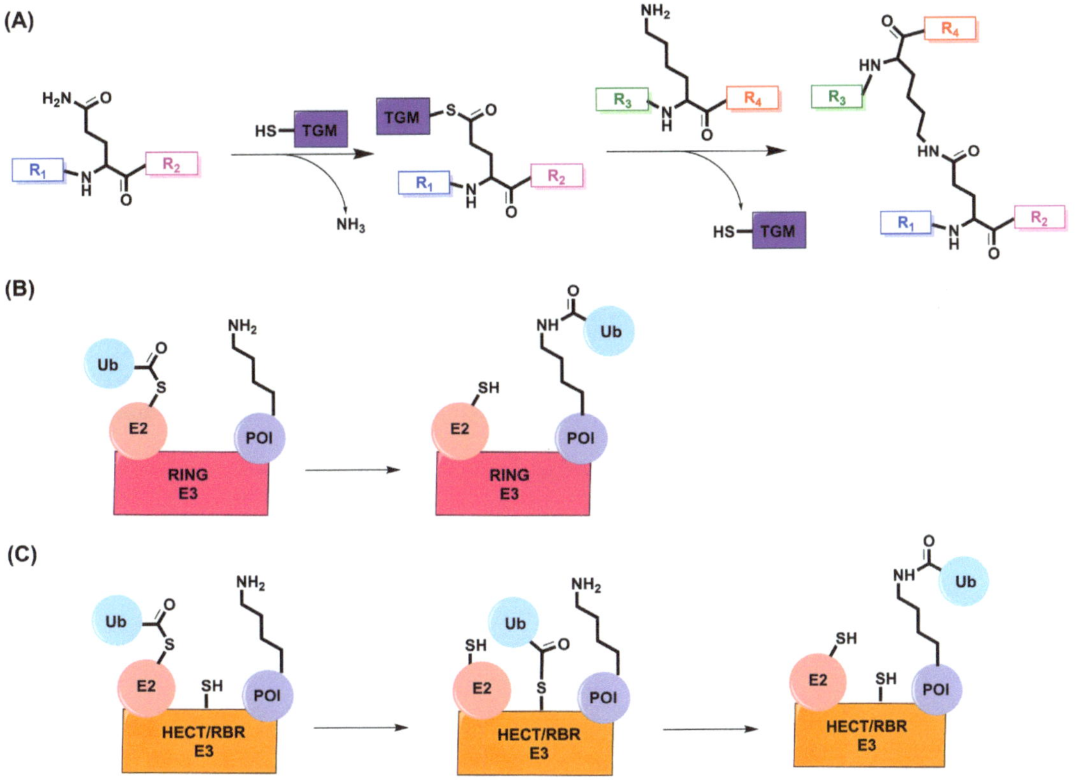

Figure 3. Enzymatic peptide ligation through isopeptide bond formation. (**A**) A catalytic cysteine on TGM attacks the γ-carboxamide group of glutamine to produce a thioester intermediate and release ammonia. The thioester intermediate undergoes an aminolysis reaction with a primary amine on lysine to release the enzyme and create an isopeptide bond. R_1–R_4 are peptide or protein sequences. (**B**) One-step transfer of ubiquitin from E2 onto a protein substrate by RING E3 ligases. The thioester intermediate between E2 and ubiquitin undergoes an aminolysis reaction with lysine on the protein substrate. (**C**) Two-step transfer of ubiquitin from E2 onto a protein substrate by HECT and RBR E3 ligases. A catalytic cysteine on HECT/RBR attacks the thioester to transfer ubiquitin onto the C-terminus of the E3 ligase. The thioester intermediate between HECT/RBR and ubiquitin undergoes an aminolysis reaction with lysine on the protein substrate. POI: protein of interest.

Microbial TGM (mTGM) is highly commercially available due to its Ca^{2+} independence and low cost but exhibits a short half-life at high temperatures, approximately 2 min at 60 °C [74]. This is a major problem considering mTGM is widely used in the food industry for production of meat, cheese, etc., and these processes require several hours at temperatures of 60 °C and above [74,75]. Many early studies looked to random mutagenesis to identify residues important to the thermostability of mTGM; however, it was the utilization of saturation mutagenesis of these hotspots in the N-terminus that produced a triple mutant with improved thermostability [76,77]. S23V-Y24N-K294L mTGM exhibited a 12-fold-higher half-life at 60 °C compared to the wildtype, which has been further improved into FRAPD-TGm1 via addition of the mutations S2P and S199A to increase enzymatic activity by approximately 115% [76–78]. The Liu lab has spent several years creating even more robust versions of this mutant (FRAPD-TGm2, FRAPD-TGm2A, and FRAPD-TGm3), with the latter two having half-lives of over 120 min above 60 °C [74,79]. FRAPD-TGm2 contains E28T-A265P-A287P mutations, granting it a two-fold-higher activity over FRAPD-TGm1 [79]. Adding mutations S116A and S179L for increased flexibility generated FRAPD-TGm2A and resulted in an additional 1.84-fold-higher activity over FRAPD-TGm2 at 79.15 U/mg [74]. To increase the surface negative charge, FRAPD-TGm3 was generated from FRAPD-TGm2 with N96E-S144E-N163D-R183E-R208E-K325E mutations and demonstrated the highest specific activity at 83.7 U/mg [79].

Aside from thermostability, mTGM has also been engineered to increase its substrate tolerance as adjacent amino acids to glutamine have been shown to impact the transamidation reaction. It has been demonstrated that the presence of arginine is not well tolerated by the wild-type mTGM, which has made this a target for increased substrate specificity [80]. One study mutated FRAPD-TGm2 and determined that the mutation Y278E shifted the substrate preference of GGGGQR from 0.05 to 0.93 and increased its activity by 1.15-fold, signifying that it is possible for transamidation to occur in the presence of arginine [81]. However, this study took place under virtual mutagenesis and molecular docking and this mutation was not stable on the physical enzyme, which limits its application and requires further optimization [81].

5.2. Applications

Although most of the attention received for TGM focuses on its inhibition for the treatment of various diseases, this enzyme has several applications as a tool for protein ligation by forming isopeptide bonds [82]. TGM has been employed in the production of site-specific antibody–drug conjugates (ADCs) utilizing various glutamine or lysine residues on the antibody [83–85]. ADCs present a promising therapeutic option for cancer by combining the specificity of antibodies for tumor cells with the cytotoxicity of small-molecule drugs; however, traditional syntheses of ADCs produce heterogenous mixtures with little control over the site of conjugation [86]. Although there are many native glutamine and lysine residues in IgG antibodies, only very specific residues are capable of undergoing a transamidation reaction with TGM, which allows for site-specific ADCs (Figure 4) [83]. For example, among the ~60 glutamine residues found in IgG1, only Q295 in the Fc domain has the ability to be modified by mTGM [85,87]. However, a nearby N-glycan at residue N297 prevents mTGM from accessing Q295, which has been the focus of several studies to remove this glycan and make Q295 available for modification [82,88]. This was first accomplished by using an amidase, PNGase, to convert the N297 into D297, which did allow for transamidation at Q295 but with alteration to the structure and function of IgG1 [82,88]. Other work has utilized an endoglycosidase, EndoS2, to trim the glycan down to its core, which then allows for mTGM to modify Q295 without compromising the IgG1 antibody [83]. An alternative approach has also been to generate mutant mTGM that has acquired specificity for Q295 in the presence of N-glycated N297 [84]. Studies have also shown that despite the inability of TGM to modify ~80 native lysine residues on IgG1, there is a C-terminal lysine, K447, that can act as an acyl acceptor [85]. Normally, this lysine is cleaved by carboxypeptidase B when expressed in HEK 293 cells; however, the addition of amino acids following K447 blocks its cleavage and allows for TGM-mediated transamidation of this residue [85,89]. Taken together, the ability to use glutamine or lysine residues on IgG antibodies for site-specific TGM-mediated conjugation signifies an important route for ADC production. Referred to as "meat glue", TGM also has a role in the food industry as it is capable of crosslinking meat pieces in combination with binding agents such as sodium caseinate and κ-carrageenan [90,91]. These crosslinks improve emulsion stability and meat gelation, which results in better meat quality and texture and reduces cooking loss [91,92]. To increase the nutritional value of meat products, TGM has been used to crosslink meat with plant-based products [93,94].

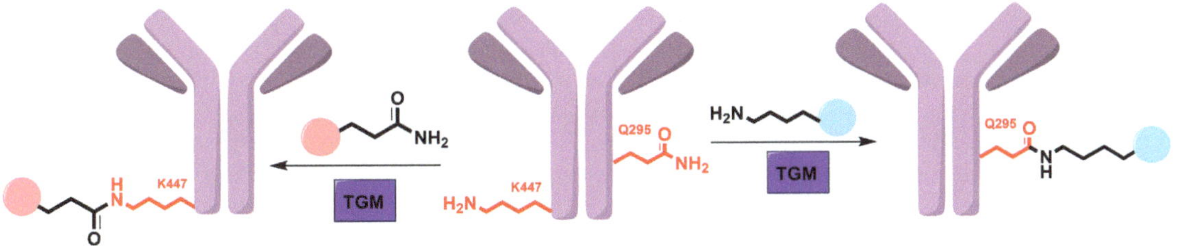

Figure 4. The production of site-specific ADCs via an isopeptide bond formation catalyzed by TGM. Despite ~60 glutamine residues existing on IgG antibodies, studies have shown that only Q295 on the Fc domain can act as an acyl donor for a ligation reaction by TGM. Similarly, K447 located on the C-terminus of the Fc domain is the only lysine residue that can act as an acyl acceptor. Drugs bearing either a primary amine or glutamine residue can then be conjugated to these residues in a site-specific manner.

6. E3 Ligase

6.1. Mechanism

In order to regulate the expression of proteins, cells have developed refined machinery through the ubiquitin proteasome pathway (UPP). Proteins marked for degradation are modified with a small molecule, ubiquitin, through an isopeptide bond between the ε-amino group on lysine and the C-terminus of ubiquitin [95,96]. Ubiquitin is passed down through E1, E2, and E3 ligases; however, it is the role of E3s to recruit E2s and facilitate the transfer of ubiquitin onto the POI [97,98]. There are over 600 E3 ligases, which can be classified into three major families: RING (really interesting new gene), HECT (homologous to E6AP C-terminus), and RBR (RING-between-RING) [99,100]. RING E3s are the largest family and can either act as monomers or in complex with several adaptor proteins for the transfer of ubiquitin [99,100]. RING E3s act in a one-step transfer as they bind both the substrate and E2–ubiquitin complex at the same time; as a result, the thioester intermediate between E2 and ubiquitin can be attacked by lysine for an aminolysis reaction (Figure 3B) [101]. HECT E3s catalyze this reaction in a two-step process by first binding to the E2–ubiquitin complex on its N-terminus, which is attacked by the C-terminal catalytic cysteine on HECT to form an E3–ubiquitin complex via a transthiolation reaction [99,101]. This complex is then positioned closely to the substrate for the aminolysis reaction to transfer ubiquitin onto the protein [99]. RBR E3s are considered a hybrid between RING and HECT seeing as though their RING1 domain binds E2, similar to RING E3s, and their RING2 domain has a catalytic cysteine for the transthiolation reaction seen in HECT E3s (Figure 3C) [101,102]. Seeing as though E3s are responsible for binding the substrate for ubiquitination, these ligases are responsible for substrate specificity through the recognition of different degrons and post-translational modifications [103].

6.2. Applications

Besides the in vitro applications of E3 for protein engineering [17], one area of drug discovery that has been rapidly developing is the production of proteolysis targeting chimeras (PROTACs) that hijack the UPP to selectively degrade proteins of interest (POIs), which is a representative application example of E3-mediated isopeptide bond formation in vivo. PROTACs consist of a warhead that will bind to a POI, an E3 ligand, and a linker connecting the two pieces together [104,105]. Several PROTACs have been developed for cancer therapeutics; however, their efficacy remains limited by aspects such as off-target degradation, large molecular weight, and varying expression of E3 ligases in different tumor cells [105–109]. To overcome these limitations, attention has been turned to the combination of click chemistry with PROTACs to deliver small pieces of the PROTAC into cells with an in situ assembly [110,111]. Wang et al. developed a click-release PROTAC (*cr*PROTAC) which used $\alpha_v\beta_3$ integrin, as it is highly expressed on cancer cells, to deliver a click handle for the bio-orthogonal reaction in tumor cells [112]. A tetrazide (Tz) group was attached to an $\alpha_v\beta_3$ integrin binding peptide, *c* (RGDyK), which, when inside the cell, reacted with a trans-cyclooctene (TCO) containing PROTAC [112]. The ARV-771 PROTAC was used to recruit the Von Hippel–Lindau (VHL) E3 ligase from the RING family to selectively degrade BRD4 in HeLa cells [112]. The authors found that not only did dosing this *cr*PROTAC induce increased activity in cancer cell lines over normal cell lines, but this PROTAC also acted in a prodrug in that BRD4 was only degraded if both click handles were present to activate the PROTAC [112]. ARV-771 has also been employed in other studies developing tissue-specific delivery and click-mediated activation of PROTACs for cancer therapeutics [113]. Yu et al. employed nanoparticles (NPs) that exposed a DBCO click handle only in the tumor microenvironment, which underwent a click reaction with an azide-modified polymeric PROTAC (POLY-PROTAC) [113]. By undergoing this click reaction in cancer-specific tissues, the ARV-771 PROTAC demonstrated 3.9-fold-higher accumulation in an MDA-MB-231 tumor xenograft model, which correlated with higher activity and decreased tumor volume [113]. These PROTACs still have limitations in that they depend on an endogenous E3 ligase, VHL, which is susceptible to resistance and has varying expression in different tissues. ClickRNA-PROTACs overcome this limitation by exhibiting spatiotemporal control of degradation in selected tumor cells. Li et al. designed an IGF2 mRNA transcript that encodes for the E3 ligase SIAH1 fused to a SNAPTag which covalently attaches to a benzylguanine (BG) group containing DBCO [114]. To complete the PROTAC, DBCO undergoes a click reaction with an azide group attached to the POI. IGF2 mRNA is specific to adrenocortical carcinoma, so this clickRNA-PROTAC can act as a targeted therapy option [114]. Collectively, these studies show the potential for targeted cancer treatments by utilizing bio-orthogonal chemistry to overcome the limitations of traditional PROTACs for therapeutic applications.

Another route to improve the tissue specificity of PROTACs is to create antibody-based PROTACs (AbTACs) where the chosen antibody will bind to tumor-specific antigens [115,116]. Once bound to the antigen, the AbTAC is internalized and releases the PROTAC, which has been demonstrated using a BRD4 PROTAC fused with an antibody specific for HER2+ cell lines [116,117]. Another feature of AbTACs that addresses the limitations of traditional PROTACs is the ability to improve the pharmacokinetic parameters of high-affinity degraders that have low bioavailability [115,118]. One of the earliest AbTAC studies came from Dragovich et al., where they synthesized a BET PROTAC, GNE-987, that had picomolar potency in acute myeloid leukemia (AML) cell lines; however, when transitioning to in vivo animal studies, this PROTAC had poor pharmacokinetic properties for oral or intravenous administration [118]. The conjugation of GNE-987 with the CLL1-targeting antibody via a cleavable disulfide linkage allowed for BET-mediated degradation by GNE-987 to shrink tumor volumes in AML mouse models, thus demonstrating how antibody conjugation can enhance PROTAC development for clinical therapeutics [115,118]. James Well's lab has also developed AbTACs to target membrane proteins and membrane-bound E3 ligases, such as RNF43, and bring them in close proximity to facilitate degradation [119,120].

7. Intein Splicing as the Inspiration for Native Chemical Ligation (NCL)

The existence of inteins was first discovered in 1990 by two independent groups, Kane et al. and Hirata et al., and for the past several decades, they have been heavily studied for their roles in protein ligation [121,122]. Inteins are believed to act as self-catalyzing enzymes which are able to undergo *cis*- or *trans*-splicing to fuse together their external protein sequences known as the N- and C-exteins [123,124]. Canonical intein splicing is a multistep process, starting with either cysteine or serine on the N-intein, which acts as a nucleophile and attacks the adjacent carbonyl group linked to the N-extein [125–128]. This produces a reversible N-to-S or N-to-O acyl shift to produce a linear intermediate. A trans(thio)esterification step, promoted by either cysteine or serine on the C-extein, links the N- and C-exteins together. To resolve the branched intermediate that is a product of this reaction, there is a rate-limiting irreversible cyclization of a conserved C-terminal asparagine. A succinimide is produced release the connected exteins from the intein. The succinimide formed on the intein undergoes a hydrolysis reaction to restore the intein. Finally, the exteins undergo an acyl shift, either S to N or O to N, to create the native peptide bond and complete the ligation reaction (Figure 5A).

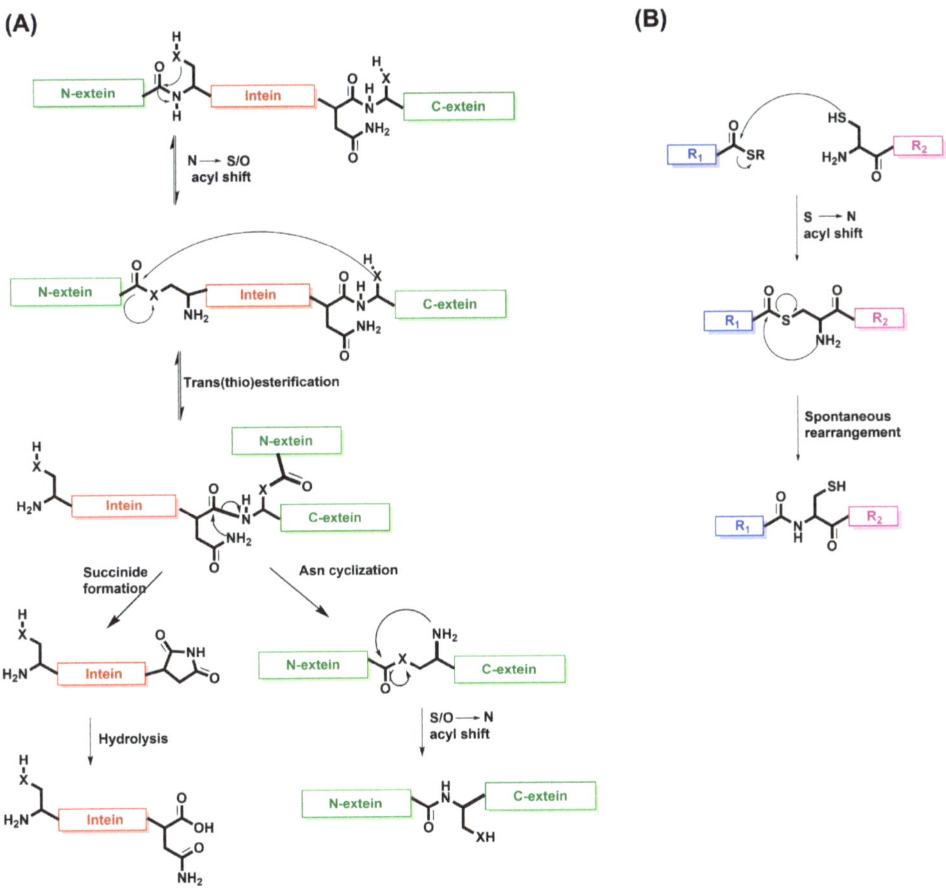

Figure 5. The mechanism of intein splicing as the inspiration behind NCL. (**A**) The multistep mechanism of class 1 canonical intein splicing. Cysteine or serine on the N-intein attacks the adjacent

carbonyl group linked to the N-extein. A reversible N-to-S or N-to-O acyl shift produces a linear intermediate that undergoes a trans(thio)esterification step promoted by either cysteine or serine on the C-extein. The irreversible cyclization of a conserved C-terminal asparagine releases intein from the fused N- and C-exteins. The resulting succinimide is hydrolyzed to restore the intein. The exteins undergo an acyl shift, either S to N or O to N, to create the native peptide bond and complete the ligation reaction. X represents either S or O. (**B**) The mechanism of NCL. A cysteine residue attacks the carbonyl group of a C-terminal thioester on a separate peptide. The thioester intermediate undergoes a spontaneous S-to-N acyl shift to form a native peptide bond and the ligated product. R_1 and R_2 are peptide or protein sequences

When comparing the mechanism of intein splicing with that of native chemical ligation (NCL), there are many similarities, which suggest that nature inspired the design of this chemical ligation. The concept of NCL to produce a peptide with a native backbone structure was first introduced in the pioneering study published by the Kent lab in 1994 [129]. In this reaction, an activated cysteine residue attacks the carbonyl group of a C-terminal thioester belonging to another peptide. This produces a thioester intermediate between the peptides that undergoes a spontaneous S-to-N acyl shift to form a native peptide bond that results in the ligated product (Figure 5B). The mechanism of NCL looks very similar to the transthioesterification and subsequent S-to-N acyl shift that occur with intein splicing. Therefore, this demonstrates how nature has gifted the world of protein engineering with a mechanism that can be implemented into chemical peptide synthesis [130].

8. Summary and Outlook

Given the vast amount of roles proteins play in biological homeostasis, decades of research have gone into understanding their functions and harnessing them to improve the research world. However, the generation of these proteins by both recombinant expression and chemical synthesis has proven to be difficult. The limitations involved in recombinant expression in bacteria often involve an absence or lack of control over PTMs, which are important for the function and folding of proteins, as a result leading to low yields or dysfunctional proteins. Chemical synthesis via SPPS retains the ability to induce site-specific modifications; however, issues with yield only allow the production of ~40–50 amino-acid-containing linear peptide sequences. Notably, nature has given us a gift through the availability of protein ligases which can catalyze the fusion of peptides/proteins synthesized either recombinantly or chemically. In this review, we highlighted protein ligases that all contain a catalytic cysteine residue to facilitate the formation of both native peptide and isopeptide bonds between peptides/proteins.

Although these protein ligases can all be isolated from natural sources, many of them have undergone extensive optimization. Studies focusing on sortase A and *Oa*AEP1 have generated several different mutants to improve their catalytic efficiencies and substrate specificities, while butelase-1 and mTGM have been engineered to improve recombinant expression and thermal stability, respectively. Engineered variants of these ligases have expanded the toolbox for many chemical biology applications, diversifying the field of proteomics. For example, butelase-1 and *Oa*AEP1 have been integral in unnatural amino acid incorporation and protein cyclization, while sortase A and E3 ligases can be combined with click chemistry to develop potent and selective cancer therapeutic options. Further expansion of this synthetic toolbox has come from the implementation of NCL for the chemical ligation of proteins, which has a natural counterpart through the mechanism of intein splicing. The employment of protein ligases is rapidly expanding, revolutionizing therapeutics and biological tools, indeed demonstrating that nature is the greatest gift of all.

Author Contributions: Y.R. drafted the manuscript and prepared the figures. H.L. assisted in editing the manuscript and preparing the figures. Q.Z. conceptualized the review, edited the manuscript, and supervised the project. All authors have read and agreed to the published version of the manuscript.

Funding: The research in Zheng's lab was financially supported by the NIH (R35 GM150676) and startup funds from Purdue University for Q.Z.

Conflicts of Interest: The authors declare no conflicts of interest.

References

1. Listov, D.; Goverde, C.A.; Correia, B.E.; Fleishman, S.J. Opportunities and Challenges in Design and Optimization of Protein Function. *Nat. Rev. Mol. Cell Biol.* **2024**, *25*, 639–653. [CrossRef] [PubMed]
2. Xiao, W.; Jiang, W.; Chen, Z.; Huang, Y.; Mao, J.; Zheng, W.; Hu, Y.; Shi, J. Advance in Peptide-Based Drug Development: Delivery Platforms, Therapeutics and Vaccines. *Signal Transduct. Target. Ther.* **2025**, *10*, 74. [CrossRef] [PubMed]
3. Ebrahimi, S.B.; Samanta, D. Engineering Protein-Based Therapeutics through Structural and Chemical Design. *Nat. Commun.* **2023**, *14*, 2411. [CrossRef] [PubMed]
4. Johnson, I.S. Human Insulin from Recombinant DNA Technology. *Science* **1983**, *219*, 632–637. [CrossRef]
5. Schellenberger, V.; Wang, C.; Geething, N.C.; Spink, B.J.; Campbell, A.; To, W.; Scholle, M.D.; Yin, Y.; Yao, Y.; Bogin, O.; et al. A Recombinant Polypeptide Extends the in Vivo Half-Life of Peptides and Proteins in a Tunable Manner. *Nat. Biotechnol.* **2009**, *27*, 1186–1190. [CrossRef]
6. Tokmakov, A.A.; Kurotani, A.; Takagi, T.; Toyama, M.; Shirouzu, M.; Fukami, Y.; Yokoyama, S. Multiple Post-Translational Modifications Affect Heterologous Protein Synthesis. *J. Biol. Chem.* **2012**, *287*, 27106–27116. [CrossRef]
7. Berrade, L.; Camarero, J.A. Expressed Protein Ligation: A Resourceful Tool to Study Protein Structure and Function. *Cell. Mol. Life Sci.* **2009**, *66*, 3909–3922. [CrossRef]
8. Kulkarni, S.S.; Sayers, J.; Premdjee, B.; Payne, R.J. Rapid and Efficient Protein Synthesis through Expansion of the Native Chemical Ligation Concept. *Nat. Rev. Chem.* **2018**, *2*, 0122. [CrossRef]
9. Bondalapati, S.; Jbara, M.; Brik, A. Expanding the Chemical Toolbox for the Synthesis of Large and Uniquely Modified Proteins. *Nat. Chem.* **2016**, *8*, 407–418. [CrossRef]
10. Jacobitz, A.W.; Kattke, M.D.; Wereszczynski, J.; Clubb, R.T. Sortase Transpeptidases: Structural Biology and Catalytic Mechanism. *Adv. Protein Chem. Struct. Biol.* **2017**, *109*, 223–264. [CrossRef]
11. Spirig, T.; Weiner, E.M.; Clubb, R.T. Sortase Enzymes in Gram-Positive Bacteria. *Mol. Microbiol.* **2011**, *82*, 1044–1059. [CrossRef] [PubMed]
12. Foster, T.J. Surface Proteins of Staphylococcus Aureus. *Microbiol. Spectr.* **2019**, *7*, 1–22. [CrossRef] [PubMed]
13. Amacher, J.F.; Antos, J.M. Sortases: Structure, Mechanism, and Implications for Protein Engineering. *Trends Biochem. Sci.* **2024**, *49*, 596–610. [CrossRef] [PubMed]
14. Mazmanian, S.K.; Liu, G.; Ton-That, H.; Schneewind, O. Staphylococcus Aureus Sortase, an Enzyme That Anchors Surface Proteins to the Cell Wall. *Science* **1999**, *285*, 760–763. [CrossRef]
15. Bradshaw, W.J.; Davies, A.H.; Chambers, C.J.; Roberts, A.K.; Shone, C.C.; Acharya, K.R. Molecular Features of the Sortase Enzyme Family. *FEBS J.* **2015**, *282*, 2097–2114. [CrossRef]
16. Antos, J.M.; Truttmann, M.C.; Ploegh, H.L. Recent Advances in Sortase-Catalyzed Ligation Methodology. *Curr. Opin. Struct. Biol.* **2016**, *38*, 111–118. [CrossRef]
17. Pihl, R.; Zheng, Q.; David, Y. Nature-Inspired Protein Ligation and Its Applications. *Nat. Rev. Chem.* **2023**, *7*, 234–255. [CrossRef]
18. Chen, I.; Dorr, B.M.; Liu, D.R. A General Strategy for the Evolution of Bond-Forming Enzymes Using Yeast Display. *Proc. Natl. Acad. Sci. USA* **2011**, *108*, 11399–11404. [CrossRef]
19. Hirakawa, H.; Ishikawa, S.; Nagamune, T. Ca^{2+}-Independent Sortase-A Exhibits High Selective Protein Ligation Activity in the Cytoplasm of *Escherichia coli*. *Biotechnol. J.* **2015**, *10*, 1487–1492. [CrossRef]
20. Wuethrich, I.; Peeters, J.G.C.; Blom, A.E.M.; Theile, C.S.; Li, Z.; Spooner, E.; Ploegh, H.L.; Guimaraes, C.P. Site-Specific Chemoenzymatic Labeling of Aerolysin Enables the Identification of New Aerolysin Receptors. *PLoS ONE* **2014**, *9*, e109883. [CrossRef]
21. Piotukh, K.; Geltinger, B.; Heinrich, N.; Gerth, F.; Beyermann, M.; Freund, C.; Schwarzer, D. Directed Evolution of Sortase A Mutants with Altered Substrate Selectivity Profiles. *J. Am. Chem. Soc.* **2011**, *133*, 17536–17539. [CrossRef] [PubMed]
22. Whedon, S.D.; Lee, K.; Wang, Z.A.; Zahn, E.; Lu, C.; Yapa Abeywardana, M.; Fairall, L.; Nam, E.; DuBois-Coyne, S.; De Ioannes, P.; et al. Circular Engineered Sortase for Interrogating Histone H3 in Chromatin. *J. Am. Chem. Soc.* **2024**, *146*, 33914–33927. [CrossRef] [PubMed]
23. Yang, Q.; Gao, Y.; Liu, X.; Xiao, Y.; Wu, M. A General Method to Edit Histone H3 Modifications on Chromatin Via Sortase-Mediated Metathesis. *Angew. Chem. Int. Ed.* **2022**, *61*, e202209945. [CrossRef] [PubMed]

24. Dorr, B.M.; Ham, H.O.; An, C.; Chaikof, E.L.; Liu, D.R. Reprogramming the Specificity of Sortase Enzymes. *Proc. Natl. Acad. Sci. USA* **2014**, *111*, 13343–13348. [CrossRef]
25. Zhulenkovs, D.; Jaudzems, K.; Zajakina, A.; Leonchiks, A. Enzymatic Activity of Circular Sortase A under Denaturing Conditions: An Advanced Tool for Protein Ligation. *Biochem. Eng. J.* **2014**, *82*, 200–209. [CrossRef]
26. Li, Z.; Tong, Z.; Gong, Q.; Ai, H.; Peng, S.; Chen, C.; Chu, G.-C.; Li, J.-B. The Expedient, CAET-Assisted Synthesis of Dual-Monoubiquitinated Histone H3 Enables Evaluation of Its Interaction with DNMT1. *Chem. Sci.* **2023**, *14*, 5681–5688. [CrossRef]
27. Li, W.; Cao, P.; Xu, P.; Sun, F.; Wang, C.; Zhang, J.; Dong, S.; Wilson, J.R.; Xu, D.; Fan, H.; et al. Rapid Reconstitution of Ubiquitinated Nucleosome Using a Non-Denatured Histone Octamer Ubiquitylation Approach. *Cell Biosci.* **2024**, *14*, 81. [CrossRef]
28. Wang, Z.A.; Whedon, S.D.; Wu, M.; Wang, S.; Brown, E.A.; Anmangandla, A.; Regan, L.; Lee, K.; Du, J.; Hong, J.Y.; et al. H2B Deacylation Selectivity: Exploring Chromatin's Dark Matter with an Engineered Sortase. *J. Am. Chem. Soc.* **2022**, *144*, 3360–3364. [CrossRef]
29. Xiao, Y.; Zou, K.; Yang, J.; Wu, M. Deciphering Histone H4 Lysine Acetylation and Methylation via Sortase-Mediated Semisynthesis. *Cell Rep. Phys. Sci.* **2023**, *4*, 101638. [CrossRef]
30. Zou, Z.; Ji, Y.; Schwaneberg, U. Empowering Site-Specific Bioconjugations In Vitro and In Vivo: Advances in Sortase Engineering and Sortase-Mediated Ligation. *Angew. Chem. Int. Ed.* **2024**, *63*, e202310910. [CrossRef]
31. Witte, M.D.; Cragnolini, J.J.; Dougan, S.K.; Yoder, N.C.; Popp, M.W.; Ploegh, H.L. Preparation of Unnatural N-to-N and C-to-C Protein Fusions. *Proc. Natl. Acad. Sci. USA* **2012**, *109*, 11993–11998. [CrossRef] [PubMed]
32. Kolb, H.C.; Finn, M.G.; Sharpless, K.B. Click Chemistry: Diverse Chemical Function from a Few Good Reactions. *Angew. Chem. Int. Ed.* **2001**, *40*, 2004–2021. [CrossRef]
33. Saxon, E.; Bertozzi, C.R. Cell Surface Engineering by a Modified Staudinger Reaction. *Science* **2000**, *287*, 2007–2010. [CrossRef] [PubMed]
34. He, J.; Liang, C.; Yu, X.-H.; Ma, X.; Qu, Y.; Zhuang, W.-R.; Li, W.; Nie, W.; Ren, Y.; Lei, Y.; et al. Chemistry-Enabled Intercellular Enzymatic Labeling for Monitoring the Immune Effects of Cytotoxic T Lymphocytes In Vivo. *Anal. Chem.* **2024**, *96*, 13996–14003. [CrossRef]
35. Ito, H.; Seishima, M. Regulation of the Induction and Function of Cytotoxic T Lymphocytes by Natural Killer T Cell. *J. Biomed. Biotechnol.* **2010**, *2010*, 641757. [CrossRef]
36. Nelson, J.W.; Chamessian, A.G.; McEnaney, P.J.; Murelli, R.P.; Kazmierczak, B.I.; Spiegel, D.A. Correction to A Biosynthetic Strategy for Re-Engineering the Staphylococcus Aureus Cell Wall with Non-Native Small Molecules. *ACS Chem. Biol.* **2011**, *6*, 971. [CrossRef]
37. Jiang, F.; Cai, C.; Wang, X.; Han, S. A Dual Biomarker-Targeting Probe Enables Signal-on Surface Labeling of *Staphylococcus aureus*. *Bioorg. Med. Chem. Lett.* **2023**, *93*, 129428. [CrossRef]
38. Self, W.H.; Wunderink, R.G.; Williams, D.J.; Zhu, Y.; Anderson, E.J.; Balk, R.A.; Fakhran, S.S.; Chappell, J.D.; Casimir, G.; Courtney, D.M.; et al. *Staphylococcus aureus* Community-Acquired Pneumonia: Prevalence, Clinical Characteristics, and Outcomes. *Clin. Infect. Dis.* **2016**, *63*, 300–309. [CrossRef]
39. Silversides, J.A.; Lappin, E.; Ferguson, A.J. Staphylococcal Toxic Shock Syndrome: Mechanisms and Management. *Curr. Infect. Dis. Rep.* **2010**, *12*, 392–400. [CrossRef]
40. Nguyen, G.K.T.; Kam, A.; Loo, S.; Jansson, A.E.; Pan, L.X.; Tam, J.P. Butelase 1: A Versatile Ligase for Peptide and Protein Macrocyclization. *J. Am. Chem. Soc.* **2015**, *137*, 15398–15401. [CrossRef]
41. Nguyen, G.K.T.; Wang, S.; Qiu, Y.; Hemu, X.; Lian, Y.; Tam, J.P. Butelase 1 Is an Asx-Specific Ligase Enabling Peptide Macrocyclization and Synthesis. *Nat. Chem. Biol.* **2014**, *10*, 732–738. [CrossRef] [PubMed]
42. Hemu, X.; Zhang, X.; Bi, X.; Liu, C.-F.; Tam, J.P. Butelase 1-Mediated Ligation of Peptides and Proteins. In *Enzyme-Mediated Ligation Methods*; Nuijens, T., Schmidt, M., Eds.; Springer: New York, NY, USA, 2019; pp. 83–109. [CrossRef]
43. Morgan, H.E.; Turnbull, W.B.; Webb, M.E. Challenges in the Use of Sortase and Other Peptide Ligases for Site-Specific Protein Modification. *Chem. Soc. Rev.* **2022**, *51*, 4121–4145. [CrossRef] [PubMed]
44. Nguyen, G.K.T.; Cao, Y.; Wang, W.; Liu, C.F.; Tam, J.P. Site-Specific N-Terminal Labeling of Peptides and Proteins Using Butelase 1 and Thiodepsipeptide. *Angew. Chem. Int. Ed.* **2015**, *54*, 15694–15698. [CrossRef] [PubMed]
45. Hemu, X.; Zhang, X.; Nguyen, G.K.T.; To, J.; Serra, A.; Loo, S.; Sze, S.K.; Liu, C.-F.; Tam, J.P. Characterization and Application of Natural and Recombinant Butelase-1 to Improve Industrial Enzymes by End-to-End Circularization. *RSC Adv.* **2021**, *11*, 23105–23112. [CrossRef]
46. Zhao, J.; Fan, R.; Jia, F.; Huang, Y.; Huang, Z.; Hou, Y.; Hu, S.-Q. Enzymatic Properties of Recombinant Ligase Butelase-1 and Its Application in Cyclizing Food-Derived Angiotensin I-Converting Enzyme Inhibitory Peptides. *J. Agric. Food Chem.* **2021**, *69*, 5976–5985. [CrossRef]
47. Nguyen, G.K.T.; Qiu, Y.; Cao, Y.; Hemu, X.; Liu, C.-F.; Tam, J.P. Butelase-Mediated Cyclization and Ligation of Peptides and Proteins. *Nat. Protoc.* **2016**, *11*, 1977–1988. [CrossRef]

48. Zhao, J.; Song, W.; Huang, Z.; Yuan, X.; Huang, Y.; Hou, Y.; Liu, K.; Jin, P.; Hu, S.-Q. "Top-down" Overexpression Optimization of Butelase-1 in *Escherichia coli* and Its Application in Anti-Tumor Peptides. *Int. J. Biol. Macromol.* **2024**, *276*, 133933. [CrossRef]
49. Zhao, J.; Ge, G.; Huang, Y.; Hou, Y.; Hu, S.-Q. Study on Activation Mechanism and Cleavage Sites of Recombinant Butelase-1 Zymogen Derived from *Clitoria ternatea*. *Biochimie* **2022**, *199*, 12–22. [CrossRef]
50. Singh, A.K.; Antonenko, A.; Kocyła, A.; Krężel, A. An Efficient and Easily Obtainable Butelase Variant for Chemoenzymatic Ligation and Modification of Peptides and Proteins. *Microb. Cell Fact.* **2024**, *23*, 325. [CrossRef]
51. Cui, Y.; Han, D.; Bai, X.; Shi, W. Development and Applications of Enzymatic Peptide and Protein Ligation. *J. Pept. Sci.* **2025**, *31*, e3657. [CrossRef]
52. Cao, Y.; Nguyen, G.K.T.; Chuah, S.; Tam, J.P.; Liu, C.-F. Butelase-Mediated Ligation as an Efficient Bioconjugation Method for the Synthesis of Peptide Dendrimers. *Bioconjug. Chem.* **2016**, *27*, 2592–2596. [CrossRef] [PubMed]
53. Galvez, A.F.; Chen, N.; Macasieb, J.; de Lumen, B.O. Chemopreventive Property of a Soybean Peptide (Lunasin) That Binds to Deacetylated Histones and Inhibits Acetylation1. *Cancer Res.* **2001**, *61*, 7473–7478. [PubMed]
54. Zhao, J.; Ge, G.; Huang, Y.; Hou, Y.; Hu, S.-Q. Butelase 1-Mediated Enzymatic Cyclization of Antimicrobial Peptides: Improvements on Stability and Bioactivity. *J. Agric. Food Chem.* **2022**, *70*, 15869–15878. [CrossRef] [PubMed]
55. Harris, K.S.; Durek, T.; Kaas, Q.; Poth, A.G.; Gilding, E.K.; Conlan, B.F.; Saska, I.; Daly, N.L.; van der Weerden, N.L.; Craik, D.J.; et al. Efficient Backbone Cyclization of Linear Peptides by a Recombinant Asparaginyl Endopeptidase. *Nat. Commun.* **2015**, *6*, 10199. [CrossRef]
56. Tang, T.M.S.; Cardella, D.; Lander, A.J.; Li, X.; Escudero, J.S.; Tsai, Y.-H.; Luk, L.Y.P. Use of an Asparaginyl Endopeptidase for Chemo-Enzymatic Peptide and Protein Labeling. *Chem. Sci.* **2020**, *11*, 5881–5888. [CrossRef]
57. Yang, R.; Wong, Y.H.; Nguyen, G.K.T.; Tam, J.P.; Lescar, J.; Wu, B. Engineering a Catalytically Efficient Recombinant Protein Ligase. *J. Am. Chem. Soc.* **2017**, *139*, 5351–5358. [CrossRef]
58. Rehm, F.B.H.; Tyler, T.J.; Xie, J.; Yap, K.; Durek, T.; Craik, D.J. Asparaginyl Ligases: New Enzymes for the Protein Engineer's Toolbox. *ChemBioChem* **2021**, *22*, 2079–2086. [CrossRef]
59. Rehm, F.B.H.; Harmand, T.J.; Yap, K.; Durek, T.; Craik, D.J.; Ploegh, H.L. Site-Specific Sequential Protein Labeling Catalyzed by a Single Recombinant Ligase. *J. Am. Chem. Soc.* **2019**, *141*, 17388–17393. [CrossRef]
60. Rehm, F.B.H.; Tyler, T.J.; de Veer, S.J.; Craik, D.J.; Durek, T. Enzymatic C-to-C Protein Ligation. *Angew. Chem. Int. Ed.* **2022**, *61*, e202116672. [CrossRef]
61. Rehm, F.B.H.; Tyler, T.J.; Yap, K.; Durek, T.; Craik, D.J. Improved Asparaginyl-Ligase-Catalyzed Transpeptidation via Selective Nucleophile Quenching. *Angew. Chem. Int. Ed.* **2021**, *60*, 4004–4008. [CrossRef]
62. Tang, J.; Hao, M.; Liu, J.; Chen, Y.; Wufuer, G.; Zhu, J.; Zhang, X.; Zheng, T.; Fang, M.; Zhang, S.; et al. Author Correction: Design of a Recombinant Asparaginyl Ligase for Site-Specific Modification Using Efficient Recognition and Nucleophile Motifs. *Commun. Chem.* **2024**, *7*, 121. [CrossRef] [PubMed]
63. Simon Tang, T.M.; Mason, J.M. Intracellular Application of an Asparaginyl Endopeptidase for Producing Recombinant Head-to-Tail Cyclic Proteins. *JACS Au* **2023**, *3*, 3290–3296. [CrossRef] [PubMed]
64. Wan, X.-C.; Zhang, Y.-N.; Zhang, H.; Chen, Y.; Cui, Z.-H.; Zhu, W.-J.; Fang, G.-M. Asparaginyl Endopeptidase-Mediated Peptide Cyclization for Phage Display. *Org. Lett.* **2024**, *26*, 2601–2605. [CrossRef] [PubMed]
65. Fottner, M.; Heimgärtner, J.; Gantz, M.; Mühlhofer, R.; Nast-Kolb, T.; Lang, K. Site-Specific Protein Labeling and Generation of Defined Ubiquitin-Protein Conjugates Using an Asparaginyl Endopeptidase. *J. Am. Chem. Soc.* **2022**, *144*, 13118–13126. [CrossRef]
66. Wanka, V.; Fottner, M.; Cigler, M.; Lang, K. Genetic Code Expansion Approaches to Decipher the Ubiquitin Code. *Chem. Rev.* **2024**, *124*, 11544–11584. [CrossRef]
67. de Veer, S.J.; Craik, D.J.; Rehm, F.B.H. Highly Efficient Transpeptidase-Catalyzed Isopeptide Ligation. *J. Am. Chem. Soc.* **2025**, *147*, 557–565. [CrossRef]
68. Gundemir, S.; Colak, G.; Tucholski, J.; Johnson, G.V.W. Transglutaminase 2: A Molecular Swiss Army Knife. *Biochim. Biophys. Acta (BBA) Mol. Cell Res.* **2012**, *1823*, 406–419. [CrossRef]
69. Keillor, J.W.; Clouthier, C.M.; Apperley, K.Y.P.; Akbar, A.; Mulani, A. Acyl Transfer Mechanisms of Tissue Transglutaminase. *Bioorg. Chem.* **2014**, *57*, 186–197. [CrossRef]
70. Sarkar, N.K.; Clarke, D.D.; Waelsch, H. An Enzymically Catalyzed Incorporation of Amines into Proteins. *Biochim. Biophys. Acta* **1957**, *25*, 451–452. [CrossRef]
71. Pincus, J.H.; Waelsch, H. The Specificity of Transglutaminase: II. Structural Requirements of the Amine Substrate. *Arch. Biochem. Biophys.* **1968**, *126*, 44–52. [CrossRef]
72. Griffin, M.; Casadio, R.; Bergamini, C.M. Transglutaminases: Nature's Biological Glues. *Biochem. J.* **2002**, *368*, 377–396. [CrossRef] [PubMed]
73. Aaron, L.; Torsten, M. Microbial Transglutaminase: A New Potential Player in Celiac Disease. *Clin. Immunol.* **2019**, *199*, 37–43. [CrossRef] [PubMed]

74. Yang, P.; Wang, X.; Ye, J.; Rao, S.; Zhou, J.; Du, G.; Liu, S. Enhanced Thermostability and Catalytic Activity of Streptomyces Mobaraenesis Transglutaminase by Rationally Engineering Its Flexible Regions. *J. Agric. Food Chem.* **2023**, *71*, 6366–6375. [CrossRef] [PubMed]
75. Ye, J.; Yang, P.; Zhou, J.; Du, G.; Liu, S. Efficient Production of a Thermostable Mutant of Transglutaminase by Streptomyces Mobaraensis. *J. Agric. Food Chem.* **2024**, *72*, 4207–4216. [CrossRef]
76. Marx, C.K.; Hertel, T.C.; Pietzsch, M. Random Mutagenesis of a Recombinant Microbial Transglutaminase for the Generation of Thermostable and Heat-Sensitive Variants. *J. Biotechnol.* **2008**, *136*, 156–162. [CrossRef]
77. Buettner, K.; Hertel, T.C.; Pietzsch, M. Increased Thermostability of Microbial Transglutaminase by Combination of Several Hot Spots Evolved by Random and Saturation Mutagenesis. *Amino Acids* **2012**, *42*, 987–996. [CrossRef]
78. Yokoyama, K.; Utsumi, H.; Nakamura, T.; Ogaya, D.; Shimba, N.; Suzuki, E.; Taguchi, S. Screening for Improved Activity of a Transglutaminase from Streptomyces Mobaraensis Created by a Novel Rational Mutagenesis and Random Mutagenesis. *Appl. Microbiol. Biotechnol.* **2010**, *87*, 2087–2096. [CrossRef]
79. Wang, X.; Du, J.; Zhao, B.; Wang, H.; Rao, S.; Du, G.; Zhou, J.; Chen, J.; Liu, S. Significantly Improving the Thermostability and Catalytic Efficiency of Streptomyces Mobaraenesis Transglutaminase through Combined Rational Design. *J. Agric. Food Chem.* **2021**, *69*, 15268–15278. [CrossRef]
80. Malešević, M.; Migge, A.; Hertel, T.C.; Pietzsch, M. A Fluorescence-Based Array Screen for Transglutaminase Substrates. *ChemBioChem* **2015**, *16*, 1169–1174. [CrossRef]
81. Wang, X.; Xu, K.; Fu, H.; Chen, Q.; Zhao, B.; Zhao, X.; Zhou, J. Enhancing Substrate Specificity of Microbial Transglutaminase for Precise Nanobody Labeling. *Synth. Syst. Biotechnol.* **2025**, *10*, 185–193. [CrossRef]
82. Siegel, M.; Khosla, C. Transglutaminase 2 Inhibitors and Their Therapeutic Role in Disease States. *Pharmacol. Ther.* **2007**, *115*, 232–245. [CrossRef] [PubMed]
83. Sadiki, A.; Liu, S.; Vaidya, S.R.; Kercher, E.M.; Lang, R.T.; McIsaac, J.; Spring, B.Q.; Auclair, J.R.; Zhou, Z.S. Site-Specific Conjugation of Native Antibody: Transglutaminase-Mediated Modification of a Conserved Glutamine While Maintaining the Primary Sequence and Core Fc Glycan via Trimming with an Endoglycosidase. *Bioconjug. Chem.* **2024**, *35*, 465–471. [CrossRef] [PubMed]
84. Dickgiesser, S.; Rieker, M.; Mueller-Pompalla, D.; Schröter, C.; Tonillo, J.; Warszawski, S.; Raab-Westphal, S.; Kühn, S.; Knehans, T.; Könning, D.; et al. Site-Specific Conjugation of Native Antibodies Using Engineered Microbial Transglutaminases. *Bioconjug. Chem.* **2020**, *31*, 1070–1076. [CrossRef] [PubMed]
85. Spidel, J.L.; Vaessen, B.; Albone, E.F.; Cheng, X.; Verdi, A.; Kline, J.B. Site-Specific Conjugation to Native and Engineered Lysines in Human Immunoglobulins by Microbial Transglutaminase. *Bioconjug. Chem.* **2017**, *28*, 2471–2484. [CrossRef]
86. Lambert, J.M.; Morris, C.Q. Antibody–Drug Conjugates (ADCs) for Personalized Treatment of Solid Tumors: A Review. *Adv. Ther.* **2017**, *34*, 1015–1035. [CrossRef]
87. Bolzati, C.; Spolaore, B. Enzymatic Methods for the Site-Specific Radiolabeling of Targeting Proteins. *Molecules* **2021**, *26*, 3492. [CrossRef]
88. Jeger, S.; Zimmermann, K.; Blanc, A.; Grünberg, J.; Honer, M.; Hunziker, P.; Struthers, H.; Schibli, R. Site-Specific and Stoichiometric Modification of Antibodies by Bacterial Transglutaminase. *Angew. Chem. Int. Ed.* **2010**, *49*, 9995–9997. [CrossRef]
89. Dick, L.W., Jr.; Qiu, D.; Mahon, D.; Adamo, M.; Cheng, K.-C. C-Terminal Lysine Variants in Fully Human Monoclonal Antibodies: Investigation of Test Methods and Possible Causes. *Biotechnol. Bioeng.* **2008**, *100*, 1132–1143. [CrossRef]
90. Kuraishi, C.; Sakamoto, J.; Yamazaki, K.; Susa, Y.; Kuhara, C.; Soeda, T. Production of Restructured Meat Using Microbial Transglutaminase without Salt or Cooking. *J. Food Sci.* **1997**, *62*, 488–490. [CrossRef]
91. Feng, Y.; Liang, X.; Zhang, J.; Shi, P.; Cao, C.; Zhang, H.; Liu, Q.; Kong, B. Underlying Mechanisms and Combined Effects of Transglutaminase and κ-Carrageenan on the Quality Profiles and In Vitro Digestibility of Frankfurters. *Food Hydrocoll.* **2024**, *147*, 109344. [CrossRef]
92. Vasić, K.; Knez, Ž.; Leitgeb, M. Transglutaminase in Foods and Biotechnology. *Int. J. Mol. Sci.* **2023**, *24*, 12402. [CrossRef] [PubMed]
93. Zinina, O.; Merenkova, S.; Galimov, D.; Okuskhanova, E.; Rebezov, M.; Khayrullin, M.; Anichkina, O. Effects of Microbial Transglutaminase on Technological, Rheological, and Microstructural Indicators of Minced Meat with the Addition of Plant Raw Materials. *Int. J. Food Sci.* **2020**, *2020*, 8869401. [CrossRef] [PubMed]
94. Baugreet, S.; Kerry, J.P.; Brodkorb, A.; Gomez, C.; Auty, M.; Allen, P.; Hamill, R.M. Optimisation of Plant Protein and Transglutaminase Content in Novel Beef Restructured Steaks for Older Adults by Central Composite Design. *Meat Sci.* **2018**, *142*, 65–77. [CrossRef] [PubMed]
95. Chau, V.; Tobias, J.W.; Bachmair, A.; Marriott, D.; Ecker, D.J.; Gonda, D.K.; Varshavsky, A. A Multiubiquitin Chain Is Confined to Specific Lysine in a Targeted Short-Lived Protein. *Science* **1989**, *243*, 1576–1583. [CrossRef]
96. Berndsen, C.E.; Wolberger, C. New Insights into Ubiquitin E3 Ligase Mechanism. *Nat. Struct. Mol. Biol.* **2014**, *21*, 301–307. [CrossRef]

97. Ramachandran, S.; Ciulli, A. Building Ubiquitination Machineries: E3 Ligase Multi-Subunit Assembly and Substrate Targeting by PROTACs and Molecular Glues. *Curr. Opin. Struct. Biol.* **2021**, *67*, 110–119. [CrossRef]
98. Zheng, N.; Shabek, N. Ubiquitin Ligases: Structure, Function, and Regulation. *Annu. Rev. Biochem.* **2017**, *86*, 129–157. [CrossRef]
99. Toma-Fukai, S.; Shimizu, T. Structural Diversity of Ubiquitin E3 Ligase. *Molecules* **2021**, *26*, 6682. [CrossRef]
100. Jeong, Y.; Oh, A.-R.; Jung, Y.H.; Gi, H.; Kim, Y.U.; Kim, K. Targeting E3 Ubiquitin Ligases and Their Adaptors as a Therapeutic Strategy for Metabolic Diseases. *Exp. Mol. Med.* **2023**, *55*, 2097–2104. [CrossRef]
101. Wang, X.S.; Cotton, T.R.; Trevelyan, S.J.; Richardson, L.W.; Lee, W.T.; Silke, J.; Lechtenberg, B.C. The Unifying Catalytic Mechanism of the RING-between-RING E3 Ubiquitin Ligase Family. *Nat. Commun.* **2023**, *14*, 168. [CrossRef]
102. Lechtenberg, B.C.; Rajput, A.; Sanishvili, R.; Dobaczewska, M.K.; Ware, C.F.; Mace, P.D.; Riedl, S.J. Structure of a HOIP/E2~ubiquitin Complex Reveals RBR E3 Ligase Mechanism and Regulation. *Nature* **2016**, *529*, 546–550. [CrossRef] [PubMed]
103. Timms, R.T.; Mena, E.L.; Leng, Y.; Li, M.Z.; Tchasovnikarova, I.A.; Koren, I.; Elledge, S.J. Author Correction: Defining E3 Ligase–Substrate Relationships through Multiplex CRISPR Screening. *Nat. Cell Biol.* **2024**, *26*, 305. [CrossRef] [PubMed]
104. Ishida, T.; Ciulli, A. E3 Ligase Ligands for PROTACs: How They Were Found and How to Discover New Ones. *SLAS Discov.* **2021**, *26*, 484–502. [CrossRef] [PubMed]
105. Pettersson, M.; Crews, C.M. PROteolysis TArgeting Chimeras (PROTACs)—Past, Present and Future. *Drug Discov. Today Technol.* **2019**, *31*, 15–27. [CrossRef]
106. Moreau, K.; Coen, M.; Zhang, A.X.; Pachl, F.; Castaldi, M.P.; Dahl, G.; Boyd, H.; Scott, C.; Newham, P. Proteolysis-Targeting Chimeras in Drug Development: A Safety Perspective. *Br. J. Pharmacol.* **2020**, *177*, 1709–1718. [CrossRef]
107. Ye, P.; Chi, X.; Cha, J.-H.; Luo, S.; Yang, G.; Yan, X.; Yang, W.-H. Potential of E3 Ubiquitin Ligases in Cancer Immunity: Opportunities and Challenges. *Cells* **2021**, *10*, 3309. [CrossRef]
108. Zou, Y.; Ma, D.; Wang, Y. The PROTAC Technology in Drug Development. *Cell Biochem. Funct.* **2019**, *37*, 21–30. [CrossRef]
109. Li, R.; Liu, M.; Yang, Z.; Li, J.; Gao, Y.; Tan, R. Proteolysis-Targeting Chimeras (PROTACs) in Cancer Therapy: Present and Future. *Molecules* **2022**, *27*, 8828. [CrossRef]
110. Pasieka, A.; Diamanti, E.; Uliassi, E.; Laura Bolognesi, M. Click Chemistry and Targeted Degradation: A Winning Combination for Medicinal Chemists? *ChemMedChem* **2023**, *18*, e202300422. [CrossRef]
111. Yang, C.; Tripathi, R.; Wang, B. Click Chemistry in the Development of PROTACs. *RSC Chem. Biol.* **2024**, *5*, 189–197. [CrossRef]
112. Chang, M.; Gao, F.; Pontigon, D.; Gnawali, G.; Xu, H.; Wang, W. Bioorthogonal PROTAC Prodrugs Enabled by On-Target Activation. *J. Am. Chem. Soc.* **2023**, *145*, 14155–14163. [CrossRef] [PubMed]
113. Gao, J.; Hou, B.; Zhu, Q.; Yang, L.; Jiang, X.; Zou, Z.; Li, X.; Xu, T.; Zheng, M.; Chen, Y.-H.; et al. Author Correction: Engineered Bioorthogonal POLY-PROTAC Nanoparticles for Tumour-Specific Protein Degradation and Precise Cancer Therapy. *Nat. Commun.* **2022**, *13*, 4978. [CrossRef] [PubMed]
114. Teng, X.; Zhao, X.; Dai, Y.; Zhang, X.; Zhang, Q.; Wu, Y.; Hu, D.; Li, J. ClickRNA-PROTAC for Tumor-Selective Protein Degradation and Targeted Cancer Therapy. *J. Am. Chem. Soc.* **2024**, *146*, 27382–27391. [CrossRef] [PubMed]
115. Dragovich, P.S. Degrader-Antibody Conjugates. *Chem. Soc. Rev.* **2022**, *51*, 3886–3897. [CrossRef]
116. Guo, Y.; Li, X.; Xie, Y.; Wang, Y. What Influences the Activity of Degrader—Antibody Conjugates (DACs). *Eur. J. Med. Chem.* **2024**, *268*, 116216. [CrossRef]
117. Maneiro, M.; Forte, N.; Shchepinova, M.M.; Kounde, C.S.; Chudasama, V.; Baker, J.R.; Tate, E.W. Antibody–PROTAC Conjugates Enable HER2-Dependent Targeted Protein Degradation of BRD4. *ACS Chem. Biol.* **2020**, *15*, 1306–1312. [CrossRef]
118. Pillow, T.H.; Adhikari, P.; Blake, R.A.; Chen, J.; Del Rosario, G.; Deshmukh, G.; Figueroa, I.; Gascoigne, K.E.; Kamath, A.V.; Kaufman, S.; et al. Antibody Conjugation of a Chimeric BET Degrader Enables in Vivo Activity. *ChemMedChem* **2020**, *15*, 17–25. [CrossRef]
119. Cotton, A.D.; Nguyen, D.P.; Gramespacher, J.A.; Seiple, I.B.; Wells, J.A. Development of Antibody-Based PROTACs for the Degradation of the Cell-Surface Immune Checkpoint Protein PD-L1. *J. Am. Chem. Soc.* **2021**, *143*, 593–598. [CrossRef]
120. Gramespacher, J.A.; Cotton, A.D.; Burroughs, P.W.W.; Seiple, I.B.; Wells, J.A. Roadmap for Optimizing and Broadening Antibody-Based PROTACs for Degradation of Cell Surface Proteins. *ACS Chem. Biol.* **2022**, *17*, 1259–1268. [CrossRef]
121. Kane, P.M.; Yamashiro, C.T.; Wolczyk, D.F.; Neff, N.; Goebl, M.; Stevens, T.H. Protein Splicing Converts the Yeast TFP1 Gene Product to the 69-kdDSubunit of the Vacuolar H+-Adenosine Triphosphatase. *Science* **1990**, *250*, 651–657. [CrossRef]
122. Hirata, R.; Ohsumk, Y.; Nakano, A.; Kawasaki, H.; Suzuki, K.; Anraku, Y. Molecular Structure of a Gene, VMA1, Encoding the Catalytic Subunit of H(+)-Translocating Adenosine Triphosphatase from Vacuolar Membranes of Saccharomyces Cerevisiae. *J. Biol. Chem.* **1990**, *265*, 6726–6733. [CrossRef] [PubMed]
123. Perler, F.B.; Davis, E.O.; Dean, G.E.; Gimble, F.S.; Jack, W.E.; Neff, N.; Noren, C.J.; Thorner, J.; Belfort, M. Protein Splicing Elements: Inteins and Exteins--a Definition of Terms and Recommended Nomenclature. *Nucleic Acids Res.* **1994**, *22*, 1125. [CrossRef]
124. Wang, H.; Wang, L.; Zhong, B.; Dai, Z. Protein Splicing of Inteins: A Powerful Tool in Synthetic Biology. *Front. Bioeng. Biotechnol.* **2022**, *10*, 810180. [CrossRef] [PubMed]

125. Nanda, A.; Nasker, S.S.; Mehra, A.; Panda, S.; Nayak, S. Inteins in Science: Evolution to Application. *Microorganisms* **2020**, *8*, 2004. [CrossRef] [PubMed]
126. Wood, D.W.; Belfort, M.; Lennon, C.W. Inteins—Mechanism of Protein Splicing, Emerging Regulatory Roles, and Applications in Protein Engineering. *Front. Microbiol.* **2023**, *14*, 1305848. [CrossRef]
127. Friedel, K.; Popp, M.A.; Matern, J.C.J.; Gazdag, E.M.; Thiel, I.V.; Volkmann, G.; Blankenfeldt, W.; Mootz, H.D. A Functional Interplay between Intein and Extein Sequences in Protein Splicing Compensates for the Essential Block B Histidine. *Chem. Sci.* **2019**, *10*, 239–251. [CrossRef]
128. Volkmann, G.; Mootz, H.D. Recent Progress in Intein Research: From Mechanism to Directed Evolution and Applications. *Cell Mol. Life Sci.* **2012**, *70*, 1185–1206. [CrossRef]
129. Dawson, P.E.; Muir, T.W.; Clark-Lewis, I.; Kent, S.B.H. Synthesis of Proteins by Native Chemical Ligation. *Science* **1994**, *266*, 776–779. [CrossRef]
130. Pinto, F.; Thornton, E.L.; Wang, B. An Expanded Library of Orthogonal Split Inteins Enables Modular Multi-Peptide Assemblies. *Nat. Commun.* **2020**, *11*, 1529. [CrossRef]

MDPI AG
Grosspeteranlage 5
4052 Basel
Switzerland
Tel.: +41 61 683 77 34

BioChem Editorial Office
E-mail: biochem@mdpi.com
www.mdpi.com/journal/biochem

www.ingramcontent.com/pod-product-compliance
Lightning Source LLC
Chambersburg PA
CBHW041357210926
51850CB00013B/3107